# Molecular Biology of Long Non-coding RNAs

Ahmad M. Khalil · Jeff Coller
Editors

# Molecular Biology of Long Non-coding RNAs

Springer

*Editors*
Ahmad M. Khalil
Jeff Coller
School of Medicine
Case Western Reserve University
Cleveland, OH
USA

ISBN 978-1-4939-4528-3 ISBN 978-1-4614-8621-3 (eBook)
DOI 10.1007/978-1-4614-8621-3
Springer New York Heidelberg Dordrecht London



Printed on acid-free paper

Springer is part of Springer Science+Business Media (www.springer.com)

# Preface

The mammalian genome encodes both coding and non-coding transcripts that work synergistically to build and organize cellular structures, and regulate gene expression patterns, which ultimately determine cell identity and function. While coding transcripts serve mostly as templates for protein synthesis, non-coding RNA transcripts, which by definition lack significant protein-coding capacity, participate in a wide range of cellular functions. These functions include organization of protein synthesis (e.g., ribosomal RNAs and transfer RNAs), regulation of protein synthesis (e.g., micro RNAs), and regulation of gene expression at the transcriptional and post-transcriptional level (e.g., long non-coding RNAs).

This book will focus on the recently discovered and less understood class of long non-coding RNAs (lncRNAs). This class of non-coding RNAs has only been recently characterized on a genome-wide scale, and only a small fraction of total transcripts is functionally characterized to date. lncRNAs are generally defined as RNA polymerase II transcripts that are longer than 200 nucleotides but lack significant protein-coding capacity. lncRNAs are capped, spliced, and polyadenylated; however, a large fraction of lncRNAs are retained in the nucleus. Both experimental and bioinformatics analyses of the promoters of lncRNAs indicate that they are regulated by the same transcription factors as protein-coding genes. Also, many lncRNAs share a similar chromatin signature to protein-coding genes, suggesting that the transcription of lncRNAs follows the same rules as protein-coding genes.

Although only a small fraction of lncRNAs has been functionally characterized, the functions and mechanisms of lncRNAs appear to be diverse. Some lncRNAs (e.g., Xist and Tsix) are involved in the regulation of X chromosome inactivation (Xi) in mammalian females. For example, the lncRNA Xist (X inactive specific transcript) is required for the initiation and maintenance of Xi, which results in the inactivation of $\sim 80\ \%$ of protein-coding genes on the inactive X chromosome. While Xist regulate gene expression on the X chromosome, other lncRNAs also regulate gene expression but throughout the genome. For example, the lncRNA HOTAIR, which is transcribed from the HOX-C locus, regulate gene expression not of nearby genes, but of genes in the HOX-D cluster and other genes scattered throughout the genome. Xist and HOTAIR are two examples of a number of lncRNAs that have been studied to date that regulate gene expression by guiding and recruiting chromatin modifying complexes to the genome either *in cis* or *in trans*. Since hundreds of lncRNAs are found to be associated with chromatin-modifying complexes, it is likely that this is one of the major mechanisms of lncRNAs-mediated gene regulation. However, it is still not known how some lncRNAs exert their effects *in cis* while others *in trans*.

A few lncRNAs have been shown to exert their effects by associating with transcription factors and blocking their ability to bind specific genomic regions, and thus acting as decoys. Also, emerging evidence suggest that lncRNA can interact with microRNAs and act as "sponges" to block their ability to bind mRNAs. Finally, some lncRNAs are involved in the organization of cellular structures such as speckles and paraspeckles. The range of functions and mechanisms of lncRNAs is likely to be very diverse as discussed in the chapters of this book.

In this book, we have gathered a number of the world's experts on lncRNA to discuss new and exciting discoveries emerging from this new field. Topics range from the role of lncRNA in chromatin function, to possibilities of lncRNAs in disease. We also have discussions of lncRNAs outside mammalian organisms and highlight some of the new technologies that have come online to help study novel RNA transcripts. The lncRNA field is new and thus has the potential to be vast. We hope that the collected work, however, will provide the reader with an overview of what is known about lncRNAs and perhaps inspire new endeavors into this fascinating field.

Ahmad M. Khalil
Jeff Coller

# Contents

# Contributors

**Juan R. Alvarez-Dominguez** Department of Biology, Whitehead Institute for Biomedical Research, Cambridge, MA, USA

**Neal Bharadwaj** Department of Genetics, Stanford University, Stanford, CA, USA

**J. Mauro Calabrese** Department of Genetics, Carolina Center for Genome Sciences and Lineberger Comprehensive Cancer Center, The University of North Carolina, Chapel Hill, NC, USA

**Stormy Chamberlain** Department of Genetics and Developmental Biology, University of Connecticut Health Center, Farmington, CT, USA

**Daniel C. Factor** Department of Genetics and Genome Sciences, Case Western Reserve University School of Medicine, Cleveland, OH, USA

**Ellen Fortini** Centre for Medical Research, Western Australian Institute for Medical Research, The University of Western Australia, Crawley, WA, Australia

**Archa H. Fox** Centre for Medical Research, Western Australian Institute for Medical Research, The University of Western Australia, Crawley, WA, Australia

**Wenqian Hu** Whitehead Institute for Biomedical Research, Cambridge, MA, USA

**Fereshteh Jahaniani** Department of Genetics, Stanford University, Stanford, CA, USA

**Ruohan Li** Centre for Medical Research, Western Australian Institute for Medical Research, The University of Western Australia, Crawley, WA, Australia

**Nianwei Lin** Program for RNA Biology, Sanford-Burnham Medical Research Institute, La Jolla, CA, USA

**Harvey F. Lodish** Department of Biology, Whitehead Institute for Biomedical Research, Cambridge, MA, USA; Department of Bioengineering, Massachusetts Institute of Technology, Cambridge, MA, USA

**Terry Magnuson** Department of Genetics, Carolina Center for Genome Sciences and Lineberger Comprehensive Cancer Center, The University of North Carolina, Chapel Hill, NC, USA

**Kristen Martins-Taylor** Department of Genetics and Developmental Biology, University of Connecticut Health Center, Farmington, CT, USA

**Stephanie Nevins** Department of Genetics, Stanford University, Stanford, CA, USA

**Tariq M. Rana** Program for RNA Biology, Sanford-Burnham Medical Research Institute, La Jolla, CA, USA

**Varsha Rao** Department of Genetics, Stanford University, Stanford, CA, USA

**Jason Reuter** Department of Genetics, Stanford University, Stanford, CA, USA

**Michael Snyder** Department of Genetics, Stanford University, Stanford, CA, USA

**Damek Spacek** Department of Genetics, Stanford University, Stanford, CA, USA

**Paul J. Tesar** Department of Genetics and Genome Sciences, School of Medicine, Case Western Reserve University, Cleveland, OH, USA

**David Tollervey** Wellcome Trust Centre for Cell Biology, The University of Edinburgh, Edinburgh, Scotland, UK

**Alex Tuck** Wellcome Trust Centre for Cell Biology, The University of Edinburgh, Edinburgh, Scotland, UK

# Chromatin Regulation by Long Non-coding RNAs

**Daniel C. Factor, Paul J. Tesar and Ahmad M. Khalil**

## 1 The Hunt for Dark Matter in the Genome

The publication of the human genome in 2001 marked the beginning, rather than the conclusion, of a chapter in the understanding of human genetics (Venter et al. 2001). It had been understood for a time prior to this period that the central dogma of biology, that DNA functioned as a storage medium whereas, RNA existed solely as a means to template protein production, was an oversimplification. A wide variety of noncoding transcripts were known to exist and have vital roles prior to the widespread availability of sequencing data and technologies, but their diversity and number were underestimated. The initial publication of the human genome identified 26,588 high-confidence protein-coding transcripts, and the coding regions of these transcripts only accounted for 1.1 % of the genome (Venter et al. 2001). Attempts to determine the fraction of the genome that encodes RNA have suggested that as much as three-quarters of the mammalian genome is transcribed in at least one cell type (Bertone et al. 2004; Birney et al. 2007; Carninci et al. 2005; Djebali et al. 2012; Kapranov et al. 2010; Mercer et al. 2012; Okazaki et al. 2002; Ota et al. 2004; Rinn et al. 2003).

The extent of this transcribed "dark matter" unaccounted for by messenger RNAs (mRNAs) and known noncoding RNAs spurred numerous efforts to classify the remainder of the transcriptome. The dramatic drop in costs associated with microarray and sequencing technologies has resulted in the development of a crop of new methods aimed at more accurate prediction of transcript structure (Grabherr et al. 2011; Guttman et al. 2010; Howald et al. 2012; Trapnell et al. 2010) as well as precise localization of transcripts within the genome (Guttman et al. 2009; Jan et al. 2011; Khalil et al. 2009; Kodzius et al. 2006; Ng et al. 2005). A class of noncoding RNAs referred to as long noncoding RNAs or lncRNAs have proven to

D. C. Factor · P. J. Tesar · A. M. Khalil (✉)
Department of Genetics and Genome Sciences, Case Western Reserve University School of Medicine, Cleveland, OH, USA
e-mail: dr.ahmad.khalil@gmail.com

A. M. Khalil and J. Coller (eds.), *Molecular Biology of Long Non-coding RNAs*,
DOI: 10.1007/978-1-4614-8621-3_1, © Springer Science+Business Media New York 2013

be of particular interest. These transcripts are distinguished from other classes by their length and inability to produce protein. Similar to mRNAs, they are transcribed by RNA polymerase II, capped, spliced, and polyadenylated. lncRNA exons are evolutionarily conserved, but at a level less than that of protein-coding regions of the genome (Cabili et al. 2011; Derrien et al. 2012; Guttman et al. 2009; Khalil et al. 2009). Attempts to more carefully annotate the transcriptome have provided strong evidence for the existence of at least 9,277 lncRNA genes in human (Derrien et al. 2012). Some recent studies argue that lncRNAs are polycistronic messages encoding small peptides (Ingolia et al. 2011), but proteomic evidence has not supported this conclusion (Banfai et al. 2012).

Given the diversity and quantity of lncRNAs, it seems likely that their functions are as numerous as those of proteins; however, only a small fraction of these transcripts have been functionally characterized. Biologically, lncRNAs function in a wide variety of processes, including X-chromosome inactivation (XCI), genomic imprinting, development, and metastasis (Clark and Mattick 2011; Gupta et al. 2010; Moran et al. 2012; Ponting et al. 2009; Qureshi et al. 2010; Wang and Chang 2011). A lack of understanding of the features that allow lncRNAs to fulfill these roles has resulted in their being divided into classes by their relationship to other genomic features. It is unclear whether these distinctions are biologically relevant or simply convenient. Natural antisense transcripts (NATs) are transcribed from regions overlapping protein-coding genes in an antisense direction. Intronic lncRNAs are expressed from the introns of protein-coding genes. A recent study has suggested a third class of lncRNAs associated with protein-coding transcripts is transcribed in an antisense orientation originating from the protein-coding transcript's promoter region, rather than overlapping the gene body (Sigova et al. 2013). Finally, intervening lncRNAs (lincRNAs) are expressed from regions distal to known protein-coding genes (Guttman et al. 2009; Khalil et al. 2009). A subset of lincRNAs has been suggested to associate with distal enhancer elements (De Santa et al. 2010; Kim et al. 2010; Sigova et al. 2013).

One emerging theme from the functional studies that have been carried out is that many lncRNAs play roles in altering and maintaining the packaging of DNA in chromatin (Chu et al. 2011; Gupta et al. 2010; Khalil et al. 2009; Rinn et al. 2007; Tsai et al. 2010). In this chapter, we provide a brief overview of the importance of chromatin state, examine a variety of biological processes in which lncRNA regulation of chromatin state is important, discuss what is known of the mechanisms by which lncRNAs regulate chromatin state, and suggest questions that are likely to be fruitful in future investigation of this topic.

## 2 Chromatin Modifications Play Key Roles in Development and Cell Identity

Chromatin consists of a core repeating unit, the nucleosome, typically composed of two copies of each of the core histone proteins (H2A, H2B, H3, and H4) wrapped by 147 base pairs of DNA (Luger et al. 1997). Interactions between the nucleosomes, the underlying DNA, and a variety of other components are altered by targeted physical disruption or enzymatic modifications, which results in changes in the accessibility of the DNA sequence. Consequently, these changes in chromatin structure have dramatic effects on gene expression patterns and are vital in establishing cell identity (Bernstein et al. 2007). Chromatin state is remarkably cell-type specific, with some features displaying more divergence than protein-coding gene expression (Boyle et al. 2008; Heintzman et al. 2009; Song et al. 2011). This chromatin state is set up and maintained by a variety of proteins and complexes. Notable examples include the polycomb repressive complex 2 (PRC2) that catalyzes trimethylation of histone H3 lysine 27 (H3K27me3), a mark associated with transcriptional silencing (Rada-Iglesias et al. 2011; Schwartz and Pirrotta 2007; Zentner et al. 2011); the trithorax complexes (mixed lineage leukemia or MLL in human) that catalyze histone H3 lysine 4 trimethylation (H3K4me3), associated with transcriptional start sites (Heintzman et al. 2009, 2007; Wang et al. 2008); and the switch/sucrose non-fermentable (SWI/SNF) complex that catalyzes ATP-dependent histone remodeling.

Paradoxically, the complexes and proteins known to play roles in establishing and maintaining chromatin state are generally ubiquitously expressed. In many cases, the core components of these chromatin modifiers lack sequence specificity. For example, while the polycomb proteins bind to DNA sequences termed polycomb response elements (PREs) in *Drosophila*, no such element has been identified in mammals (Schwartz and Pirrotta 2007). However, lncRNAs are expressed in a more cell-type specific pattern than protein-coding genes, suggesting that lncRNAs may play a role in establishing or maintaining cell identity (Cabili et al. 2011; Guttman et al. 2009; Khalil et al. 2009). Many lncRNAs associate with chromatin-modifying enzymes and have roles in their proper distribution throughout the genome (Guil et al. 2012; Khalil et al. 2009; Zhao et al. 2010). It is possible that lncRNAs have taken on the role of providing spatial and temporal targeting of these complexes. In the next sections we discuss several examples of lncRNAs regulating gene expression and developmental processes through chromatin state (Fig. 1).

## 3 Xist and the Polycomb Complex in XCI

lncRNAs are not passive or transient components of chromatin, in fact they play a key role in one of the most dramatic chromatin compactions in development. In the early embryo of most female mammals, somatic cells undergo a process of random

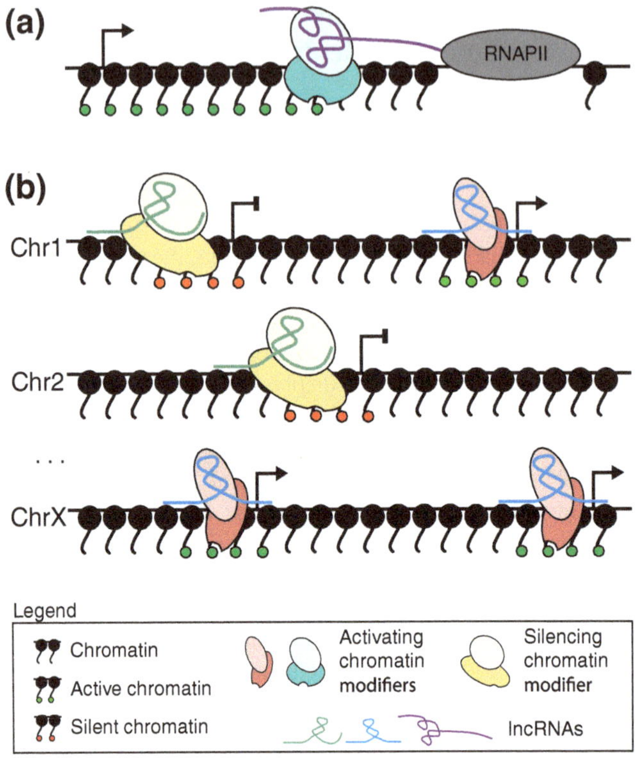


**Fig. 1** lncRNAs recruit chromatin-modifying complexes to specific genomic loci *in cis* and *in trans*. **a** The tethering model of *cis*-regulation of a region by a lncRNA. Here a lncRNA recruits an activating chromatin-modifying complex co-transcriptionally to reinforce its own expression. **b** lncRNAs regulate gene expression across the genome by recruiting chromatin-modifying complexes to specific loci and modulating chromatin state at those loci

XCI by which one randomly selected copy of the X-chromosome is silenced and compacted into the heterochromatic Barr body (Barr and Bertram 1949). This process is also referred to as dosage compensation, because silencing one of the two copies of the X-chromosome found in females equalizes gene expression levels to that of the single X-chromosome in males. While the process is incompletely understood, it is of particular interest in the lncRNA field because of the multilayered inter-regulation of a group of lncRNAs expressed from the X inactivation center (XIC) (Heard 2004). One of these lncRNAs, the X inactive specific transcript (*Xist*), was one of the first identified mammalian lncRNAs, and is thus one of the most studied (Brockdorff et al. 1992; Brown et al. 1992). *Xist* is required for both the initiation as well as the maintenance of XCI in vivo (Lee and Bartolomei 2013; Yildirim et al. 2013). A small repeat region within *Xist*, called RepA, is required for *Xist* mediated repression as it recruits the polycomb repressive complex 2 (PRC2) to the inactive X-chromosome (Zhao et al. 2008). By contrast, the lncRNA *Tsix* is transcribed antisense to *Xist* and plays a major role in maintaining an active state of the second X-chromosome (Lee et al. 1999; Sado et al. 2005).

Early in development, RepA and *Tsix* are transcribed from both XICs, and *Tsix* prevents RepA from recruiting PRC2 to the X-chromosomes (Zhao et al. 2008). *Tsix* also recruits the de novo DNA methyltransferase 3A (Dnmt3a) to the *Xist* promoter to prevent its transcription (Sado et al. 2005; Sun et al. 2006). Prior to gastrulation, RepA and PRC2 cooperate to initiate XCI at one of the XICs (Zhao et al. 2008). At this point, the inactive X-chromosome begins to express the full-length *Xist* transcript, which coats the inactive X, but not the active X-chromosome (Clemson et al. 1996). Following XCI, chromatin modifications associated with heterochromatin, including DNA methylation at CpG islands, H3K9me3 and H3K27me3, are found across the majority of the inactive X-chromosome (Lee and Bartolomei 2013). While great strides have been made in understanding the roles of lncRNAs in the process of XCI, many questions remain. *Tsix* is thought to continue to repress expression of the *Xist* transcript from the active X-chromosome, but it is unclear what prevents *Xist* from nucleating heterochromatin on the opposite allele, or in other regions of the genome *in trans*. Additionally, while *Xist* is thought to be the only gene expressed exclusively from the inactive X-chromosome, as many as 20 % of genes on the inactive X-chromosome escape silencing, and the mechanism behind this protection remains largely unknown (Carrel and Willard 2005; Khalil and Driscoll 2007).

## 4 lncRNAs and G9a in Genomic Imprinting

The process of imprinting is another key event in development that involves lncRNA control of allele-specific gene expression patterns through chromatin modifications. Similarly to XCI, genomic imprinting involves allele-specific regulation of gene expression; however, while the chromosome targeted in XCI is selected randomly, imprinting targets a specific allele based on the parent of origin. Additionally, many imprinted regions include an imprinting control region. This region plays a role analogous to that of the XIC in XCI, in that one or more lncRNAs are expressed from the imprinting control region to regulate expression of the nearby imprinted genes. In several cases, lncRNAs expressed from the imprinting control region interact with G9a, a histone H3 lysine 9 methyltransferase associated with gene silencing (Noma et al. 2001).

In mice the lncRNA *Air* is expressed exclusively from the paternal allele of an imprinted region that includes the *Igf2r*, *Slc22a2,* and *Slc22a3* genes (Nagano et al. 2008). All three genes are normally silenced at the paternal allele in the placenta; however, loss of *Air* expression results in re-expression of these genes (Sleutels et al. 2002). Loss of G9a is also associated with re-expression of *Slc22a3* from the silenced paternal allele (Nagano et al. 2008). G9a and *Air* interact, and both associate with the *Slc22a3* promoter; however, loss of *Air* causes a loss of G9a at the promoter, suggesting that the lncRNA *Air* recruits the G9a methyltransferase to the promoter (Nagano et al. 2008). Similarly, the lncRNA *Kcnq1ot1* recruits G9a to the imprinted Kcnq1 locus, along with the PRC2 complex and the DNA methyltransferase Dnmt1 (Pandey et al. 2008; Mohammad et al. 2010), suggesting that lncRNAs may control chromatin state at imprinted genes through a set of overlapping and independent mechanisms.

## 5 lncRNAs and Chromatin-Modifying Enzymes in Regulation of Hox Gene Expression

The Hox genes are a group of transcription factors first identified in *Drosophila melanogaster* that have a vital role in developmental patterning. These genes occur in 4 clusters in vertebrates, comprising a total of 39 genes (Mallo et al. 2010). Genomic distribution of the Hox genes from 3′ to 5′ within the clusters is related to their expression patterns along the anterior/posterior and proximal/distal axes of the organism, resulting in a complex system of epigenetic regulation, as chromatin becomes activated or silenced over successive cell divisions (Chang 2009; Kmita and Duboule 2003; Lemons and McGinnis 2006). Adding to the complexity, a large number of lncRNAs are also expressed from the Hox clusters (Bernstein et al. 2005; Carninci et al. 2005; Rinn et al. 2007; Sessa et al. 2007).

In 2007, Rinn et al. identified a lncRNA, Hox antisense intergenic RNA or *HOTAIR*, enriched in fibroblasts isolated from posterior and distal tissues (Rinn et al. 2007). They demonstrated that *HOTAIR* suppresses expression of a variety of protein-coding and noncoding loci within the HOXD cluster and interacts with PRC2 components Suz12 and EZH2 (Rinn et al. 2007). Further, loss of *HOTAIR* results in a decreases in PRC2 occupancy and the (H3K27me3) at the HOXD cluster. Importantly, the HOXB cluster did not lose silencing or PRC2 or H3K27me3 occupancy upon HOTAIR knockdown, indicating specificity of the *HOTAIR* silencing effect (Rinn et al. 2007). Later studies extended the function of *HOTAIR*, demonstrating that it is also capable of specifically silencing a set of loci genome-wide through interactions with both the PRC2 and LSD1/CoREST/REST complexes (Gupta et al. 2010; Khalil et al. 2009; Rinn et al. 2007; Tsai et al. 2010). Thus, by bridging and coordinating the recruitment of complementary histone modifying complexes, one lncRNA directs both removal of active chromatin marks and addition of silencing chromatin marks to a specific set of loci genome-wide, essentially fulfilling the role of a sequence specific transcription factor in a more information dense media (RNA, rather than protein). Interestingly, while the murine *Hotair* gene is 90 % identical to the human, it does not regulate expression of the Hoxd cluster, supporting the idea that the evolutionary flexibility of lncRNAs may contribute to the evolution of body patterning in closely related organisms (Schorderet and Duboule 2011).

On the other hand, Wang et al. identified a lncRNA that acts as a positive regulator of HOX gene expression in 2011 (Wang et al. 2011). This lncRNA, *HOTTIP*, for HOXA transcript at the distal tip, is also expressed in posterior and distal regions of the organism (Wang et al. 2011). In contrast to *HOTAIR*'s repressive action on a separate HOX cluster, *HOTTIP* activates neighboring genes in the 5′ region of the HOXA cluster and is required for proper distal limb development (Wang et al. 2011). Loss of *HOTTIP* causes a decrease in the occupancy of MLL complex components MLL1 and WDR5, as well as the activating H3K4me3 modification the complex deposits across this region (Wang et al. 2011). As with *HOTAIR*, this effect is specific to one HOX cluster, despite the similarity between the four (Wang et al. 2011). Interestingly, like many lncRNAs, *HOTTIP* is expressed at a very low level relative to protein-coding genes, averaging only 0.3 copies per cell according to single molecule FISH, raising questions about what kind of mechanism could reconcile the RNA's low expression with the dramatic regulatory effects observed (Wang et al. 2011).

## 6 lncRNAs Target Chromatin Modifiers to the Genome *in cis* and *in trans*

As we have discussed, lncRNAs play a role in a wide variety of biological processes through regulation of gene expression at the chromatin level. The significant effort exerted to identify modes of actions of lncRNAs has provided mechanistic insight into the functions of individual lncRNAs, while providing a glimpse of the potential roles of lncRNAs as a class (Mercer et al. 2009; Wang and Chang 2011).

There are many examples of natural antisense transcripts (NATs) that regulate gene expression *in cis* (Feng et al. 2006; Nagano et al. 2008; Pandey et al. 2008; Sleutels et al. 2002; Yap et al. 2010; Zhao et al. 2008). In most cases, NATs regulate the mRNA transcripts they overlap (He et al. 2008; Werner et al. 2009). Although it is not clear how NATs mediate their function as a class, many of these transcripts are bound to chromatin-modifying complexes and are potentially guiding these proteins to chromatin. Additionally, short RNAs are produced from many PRC2 target genes, interact with PRC2, and repress expression *in cis* (Kanhere et al. 2010).

As noted previously, some of the most studied roles of lncRNAs are in imprinting and XCI, both of which require regulating gene expression *in cis* in order to distinguish alleles. lncRNAs are particularly well suited to this role when compared to protein transcriptional regulators. Protein-coding genes are incapable of retaining information about their allele of origin, because they must be transported outside of the nucleus in order to be translated before the encoded protein can function. lncRNAs could function co-transcriptionally to regulate their allele of origin, which would explain the puzzling fact that lncRNAs can have dramatic knockdown phenotypes when they are expressed at levels lower than one copy per cell (Wang et al. 2011). The *HOTTIP* locus is normally brought into close proximity of its targets via chromatin looping, and exogenously expressed *HOTTIP* is incapable of transcriptionally activating those targets unless it is artificially tethered to them (Wang et al. 2011). Further, short half-lives, such as that of *Tsix,* are beneficial when diffusion of full-length transcripts has the potential to disrupt allelic expression patterns (Sun et al. 2006).

Alternatively, lncRNAs have also been shown to regulate gene expression *in trans*. In one study, six lncRNAs were knocked down, each resulting in significant differential expression of between 103 and 352 genes, none of which were within the nearest ten genes to either side of the lncRNA targeted, suggesting a *trans*-regulatory mechanism (Khalil et al. 2009). In a second study of 147 lncRNAs, 137 knockdowns caused significant differential expression, with a range between 20 and 936 genes disrupted per lncRNA (Guttman et al. 2011). In this study, only 8 lncRNAs affected a gene within 300 kb of its own locus. The lncRNA *HOTAIR* targets chromatin modifications to loci throughout the genome *in trans* (Gupta et al. 2010; Khalil et al. 2009; Rinn et al. 2007; Tsai et al. 2010). While less is understood about the mechanisms behind *trans* regulation by lncRNAs, there is evidence that a group of lncRNAs referred to as promoter RNAs interact with ribosomal DNA promoters through formation of DNA:DNA:RNA triplexes, suggesting that lncRNAs may possess inherent sequence specificity (Schmitz et al. 2010). A technique recently developed by Chu and colleagues to identify regions of the genome that are occupied by lncRNAs, referred to as chromatin isolation by RNA purification (ChIRP), provides evidence for a direct interaction between lncRNAs and chromatin (Chu et al. 2011). ChIRP demonstrated that *HOTAIR* and several other lncRNAs interact with specific DNA sequences, and that in the absence of its protein co-factor PRC2 *HOTAIR* is still capable of interacting with chromatin (Chu et al. 2011).

## 7 Summary and Implications

Elucidating how cell identity is established in mammalian systems is of great interest to the scientific community since it provides the means of reprogramming stem cells and fibroblasts into specific cell types for potential clinical use (Takahashi and Yamanaka 2006). The discovery of thousands of lncRNAs and their highly tissue-specific expression patterns suggest that lncRNAs may play a critical role in establishing and or maintaining cell identity. Although most lncRNAs are expressed at low levels, they can exert significant effects on cell identity by modulating chromatin structure by providing the targeting specificity for chromatin-modifying complexes to specific gene loci (Gupta et al. 2010; Huarte et al. 2010; Khalil et al. 2009; Khalil and Rinn 2011; Koziol and Rinn 2010; Rinn et al. 2007). Although the detailed mechanisms of lncRNA-mediated chromatin regulation are yet to be fully elucidated, recent studies suggest that lncRNAs can function both *in cis* as well as *in trans*, and thus exert their effects over large chromatin domains (Bertani et al. 2011; Chu et al. 2011; Guttman et al. 2009, 2011; Huarte et al. 2010; Khalil et al. 2009; Loewer et al. 2010; Rinn et al. 2007; Tsai et al. 2010; Wang et al. 2011).

Future studies are needed to determine how lncRNAs recognize specific protein complexes. For example, some lncRNAs interact specifically with PRC2 but not other complexes such as CoREST and SMCX (Khalil et al. 2009); however, it is not clear how such lncRNAs recognize PRC2. There is speculation in the field that lncRNAs recognize their protein partners via secondary structures, but since the prediction of secondary structures of lncRNAs remain at infancy, it is difficult to test this hypothesis. Also, another major question is how lncRNAs recognize specific genomic regions. Although there is some evidence for direct lncRNA-DNA interactions (Schmitz et al. 2010), further studies are needed to establish if this is the case for other lncRNAs.

## References

Banfai, B., Jia, H., Khatun, J., Wood, E., Risk, B., Gundling, W. E, Jr, et al. (2012). Long noncoding RNAs are rarely translated in two human cell lines. *Genome Research, 22*, 1646–1657.

Barr, M. L., & Bertram, E. G. (1949). A morphological distinction between neurones of the male and female, and the behaviour of the nucleolar satellite during accelerated nucleoprotein synthesis. *Nature, 163*, 676.

Bernstein, B. E., Kamal, M., Lindblad-Toh, K., Bekiranov, S., Bailey, D. K., Huebert, D. J., et al. (2005). Genomic maps and comparative analysis of histone modifications in human and mouse. *Cell, 120*, 169–181.

Bernstein, B. E., Meissner, A., & Lander, E. S. (2007). The mammalian epigenome. *Cell, 128*, 669–681.

Bertani, S., Sauer, S., Bolotin, E., & Sauer, F. (2011). The noncoding RNA Mistral activates Hoxa6 and Hoxa7 expression and stem cell differentiation by recruiting MLL1 to chromatin. *Molecular Cell, 43*, 1040–1046.

Bertone, P., Stolc, V., Royce, T. E., Rozowsky, J. S., Urban, A. E., Zhu, X., et al. (2004). Global identification of human transcribed sequences with genome tiling arrays. *Science, 306*, 2242–2246.

Birney, E., Stamatoyannopoulos, J. A., Dutta, A., Guigo, R., Gingeras, T. R., Margulies, E. H., et al. (2007). Identification and analysis of functional elements in 1 % of the human genome by the ENCODE pilot project. *Nature, 447*, 799–816.

Boyle, A. P., Davis, S., Shulha, H. P., Meltzer, P., Margulies, E. H., Weng, Z., et al. (2008). High-resolution mapping and characterization of open chromatin across the genome. *Cell, 132*, 311–322.

Brockdorff, N., Ashworth, A., Kay, G. F., McCabe, V. M., Norris, D. P., Cooper, P. J., Swift, S., & Rastan, S. (1992). The product of the mouse Xist gene is a 15 kb inactive X-specific transcript containing no conserved ORF and located in the nucleus. *Cell 71*, 515–526.

Brown, C. J., Hendrich, B. D., Rupert, J. L., Lafreniere, R. G., Xing, Y., Lawrence, J., et al. (1992). The human XIST gene: Analysis of a 17 kb inactive X-specific RNA that contains conserved repeats and is highly localized within the nucleus. *Cell, 71*, 527–542.

Cabili, M. N., Trapnell, C., Goff, L., Koziol, M., Tazon-Vega, B., Regev, A., et al. (2011). Integrative annotation of human large intergenic noncoding RNAs reveals global properties and specific subclasses. *Genes & Development, 25*, 1915–1927.

Carninci, P., Kasukawa, T., Katayama, S., Gough, J., Frith, M. C., Maeda, N., et al. (2005). The transcriptional landscape of the mammalian genome. *Science, 309*, 1559–1563.

Carrel, L., & Willard, H. F. (2005). X-inactivation profile reveals extensive variability in X-linked gene expression in females. *Nature, 434*, 400–404.

Chang, H. Y. (2009). Anatomic demarcation of cells: Genes to patterns. *Science, 326*, 1206–1207.

Chu, C., Qu, K., Zhong, F. L., Artandi, S. E., & Chang, H. Y. (2011). Genomic maps of long noncoding RNA occupancy reveal principles of RNA-chromatin interactions. *Molecular Cell, 44*, 667–678.

Clark, M. B., & Mattick, J. S. (2011). Long noncoding RNAs in cell biology. *Seminars in Cell & Developmental Biology, 22*, 366–376.

Clemson, C. M., McNeil, J. A., Willard, H. F., & Lawrence, J. B. (1996). XIST RNA paints the inactive X chromosome at interphase: Evidence for a novel RNA involved in nuclear/ chromosome structure. *The Journal of cell biology, 132*, 259–275.

De Santa, F., Barozzi, I., Mietton, F., Ghisletti, S., Polletti, S., Tusi, B. K., et al. (2010). A large fraction of extragenic RNA pol II transcription sites overlap enhancers. *PLoS Biology, 8*, e1000384.

Derrien, T., Johnson, R., Bussotti, G., Tanzer, A., Djebali, S., Tilgner, H., et al. (2012). The GENCODE v7 catalog of human long noncoding RNAs: Analysis of their gene structure, evolution, and expression. *Genome Research, 22*, 1775–1789.

Djebali, S., Davis, C. A., Merkel, A., Dobin, A., Lassmann, T., Mortazavi, A., et al. (2012). Landscape of transcription in human cells. *Nature, 489*, 101–108.

Feng, J., Bi, C., Clark, B. S., Mady, R., Shah, P., & Kohtz, J. D. (2006). The Evf-2 noncoding RNA is transcribed from the Dlx-5/6 ultraconserved region and functions as a Dlx-2 transcriptional coactivator. *Genes & Development, 20*, 1470–1484.

Grabherr, M. G., Haas, B. J., Yassour, M., Levin, J. Z., Thompson, D. A., Amit, I., et al. (2011). Full-length transcriptome assembly from RNA-Seq data without a reference genome. *Nature Biotechnology, 29*, 644–652.

Guil, S., Soler, M., Portela, A., Carrere, J., Fonalleras, E., Gomez, A., et al. (2012). Intronic RNAs mediate EZH2 regulation of epigenetic targets. *Nature Structural & Molecular Biology, 19*, 664–670.

Gupta, R. A., Shah, N., Wang, K. C., Kim, J., Horlings, H. M., Wong, D. J., et al. (2010). Long non-coding RNA HOTAIR reprograms chromatin state to promote cancer metastasis. *Nature, 464*, 1071–1076.

Guttman, M., Amit, I., Garber, M., French, C., Lin, M. F., Feldser, D., et al. (2009). Chromatin signature reveals over a thousand highly conserved large non-coding RNAs in mammals. *Nature, 458*, 223–227.

Guttman, M., Donaghey, J., Carey, B. W., Garber, M., Grenier, J. K., Munson, G., et al. (2011). lincRNAs act in the circuitry controlling pluripotency and differentiation. *Nature, 477*, 295–300.

Guttman, M., Garber, M., Levin, J. Z., Donaghey, J., Robinson, J., Adiconis, X., et al. (2010). Ab initio reconstruction of cell type-specific transcriptomes in mouse reveals the conserved multi-exonic structure of lincRNAs. *Nature Biotechnology, 28*, 503–510.

He, Y., Vogelstein, B., Velculescu, V. E., Papadopoulos, N., & Kinzler, K. W. (2008). The antisense transcriptomes of human cells. *Science, 322*, 1855–1857.

Heard, E. (2004). Recent advances in X-chromosome inactivation. *Current Opinion in Cell Biology, 16*, 247–255.

Heintzman, N. D., Hon, G. C., Hawkins, R. D., Kheradpour, P., Stark, A., Harp, L. F., et al. (2009). Histone modifications at human enhancers reflect global cell-type-specific gene expression. *Nature, 459*, 108–112.

Heintzman, N. D., Stuart, R. K., Hon, G., Fu, Y., Ching, C. W., Hawkins, R. D., et al. (2007). Distinct and predictive chromatin signatures of transcriptional promoters and enhancers in the human genome. *Nature Genetics, 39*, 311–318.

Howald, C., Tanzer, A., Chrast, J., Kokocinski, F., Derrien, T., Walters, N., et al. (2012). Combining RT-PCR-seq and RNA-seq to catalog all genic elements encoded in the human genome. *Genome Research, 22*, 1698–1710.

Huarte, M., Guttman, M., Feldser, D., Garber, M., Koziol, M. J., Kenzelmann-Broz, D., et al. (2010). A large intergenic noncoding RNA induced by p53 mediates global gene repression in the p53 response. *Cell, 142*, 409–419.

Ingolia, N. T., Lareau, L. F., & Weissman, J. S. (2011). Ribosome profiling of mouse embryonic stem cells reveals the complexity and dynamics of mammalian proteomes. *Cell, 147*, 789–802.

Jan, C. H., Friedman, R. C., Ruby, J. G., & Bartel, D. P. (2011). Formation, regulation and evolution of Caenorhabditis elegans 3′UTRs. *Nature, 469*, 97–101.

Kanhere, A., Viiri, K., Araujo, C. C., Rasaiyaah, J., Bouwman, R. D., Whyte, W. A., et al. (2010). Short RNAs are transcribed from repressed polycomb target genes and interact with polycomb repressive complex-2. *Molecular Cell, 38*, 675–688.

Kapranov, P., St Laurent, G., Raz, T., Ozsolak, F., Reynolds, C. P., Sorensen, P. H. et al. (2010). The majority of total nuclear-encoded non-ribosomal RNA in a human cell is 'dark matter' un-annotated RNA. *BMC Biology 8*, 149.

Khalil, A. M., & Driscoll, D. J. (2007). Trimethylation of histone H3 lysine 4 is an epigenetic mark at regions escaping mammalian X inactivation. *Epigenetics: official journal of the DNA Methylation Society, 2*, 114–118.

Khalil, A. M., Guttman, M., Huarte, M., Garber, M., Raj, A., Rivea Morales, D., et al. (2009). Many human large intergenic noncoding RNAs associate with chromatin-modifying complexes and affect gene expression. *Proceedings of the National Academy of Sciences USA, 106*, 11667–11672.

Khalil, A. M., & Rinn, J. L. (2011). RNA-protein interactions in human health and disease. *Seminars in Cell & Developmental Biology, 22*, 359–365.

Kim, T. K., Hemberg, M., Gray, J. M., Costa, A. M., Bear, D. M., Wu, J., et al. (2010). Widespread transcription at neuronal activity-regulated enhancers. *Nature, 465*, 182–187.

Kmita, M., & Duboule, D. (2003). Organizing axes in time and space; 25 years of colinear tinkering. *Science, 301*, 331–333.

Kodzius, R., Kojima, M., Nishiyori, H., Nakamura, M., Fukuda, S., Tagami, M., et al. (2006). CAGE: Cap analysis of gene expression. *Nature Methods, 3*, 211–222.

Koziol, M. J., & Rinn, J. L. (2010). RNA traffic control of chromatin complexes. *Current Opinion in Genetics & Development, 20*, 142–148.

Lee, J. T., & Bartolomei, M. S. (2013). X-Inactivation, imprinting, and long noncoding RNAs in health and disease. *Cell, 152*, 1308–1323.

Lee, J. T., Davidow, L. S., & Warshawsky, D. (1999). Tsix, a gene antisense to Xist at the X-inactivation centre. *Nature Genetics, 21*, 400–404.

Lemons, D., & McGinnis, W. (2006). Genomic evolution of Hox gene clusters. *Science, 313*, 1918–1922.

Loewer, S., Cabili, M. N., Guttman, M., Loh, Y. H., Thomas, K., Park, I. H., et al. (2010). Large intergenic non-coding RNA-RoR modulates reprogramming of human induced pluripotent stem cells. *Nature Genetics, 42*, 1113–1117.

Luger, K., Mader, A. W., Richmond, R. K., Sargent, D. F., & Richmond, T. J. (1997). Crystal structure of the nucleosome core particle at 2.8 A resolution. *Nature, 389*, 251–260.

Mallo, M., Wellik, D. M., & Deschamps, J. (2010). Hox genes and regional patterning of the vertebrate body plan. *Dev Biol, 344*, 7–15.

Mercer, T. R., Dinger, M. E., & Mattick, J. S. (2009). Long non-coding RNAs: Insights into functions. *Nature Reviews Genetics, 10*, 155–159.

Mercer, T. R., Gerhardt, D. J., Dinger, M. E., Crawford, J., Trapnell, C., Jeddeloh, J. A., et al. (2012). Targeted RNA sequencing reveals the deep complexity of the human transcriptome. *Nature Biotechnology, 30*, 99–104.

Mohammad, F., Mondal, T., Guseva, N., Pandey, G. K., & Kanduri, C. (2010). Kcnq1ot1 noncoding RNA mediates transcriptional gene silencing by interacting with Dnmt1. *Development, 137*, 2493–2499.

Moran, V. A., Perera, R. J., & Khalil, A. M. (2012). Emerging functional and mechanistic paradigms of mammalian long non-coding RNAs. *Nucleic Acids Research 40*, 6391–6400.

Nagano, T., Mitchell, J. A., Sanz, L. A., Pauler, F. M., Ferguson-Smith, A. C., Feil, R., et al. (2008). The air noncoding RNA epigenetically silences transcription by targeting G9a to chromatin. *Science, 322*, 1717–1720.

Ng, P., Wei, C. L., Sung, W. K., Chiu, K. P., Lipovich, L., Ang, C. C., et al. (2005). Gene identification signature (GIS) analysis for transcriptome characterization and genome annotation. *Nature Methods, 2*, 105–111.

Noma, K., Allis, C. D., & Grewal, S. I. (2001). Transitions in distinct histone H3 methylation patterns at the heterochromatin domain boundaries. *Science, 293*, 1150–1155.

Okazaki, Y., Furuno, M., Kasukawa, T., Adachi, J., Bono, H., Kondo, S., et al. (2002). Analysis of the mouse transcriptome based on functional annotation of 60,770 full-length cDNAs. *Nature, 420*, 563–573.

Ota, T., Suzuki, Y., Nishikawa, T., Otsuki, T., Sugiyama, T., Irie, R., et al. (2004). Complete sequencing and characterization of 21,243 full-length human cDNAs. *Nature Genetics, 36*, 40–45.

Pandey, R. R., Mondal, T., Mohammad, F., Enroth, S., Redrup, L., Komorowski, J., et al. (2008). Kcnq1ot1 antisense noncoding RNA mediates lineage-specific transcriptional silencing through chromatin-level regulation. *Molecular Cell, 32*, 232–246.

Ponting, C. P., Oliver, P. L., & Reik, W. (2009). Evolution and functions of long noncoding RNAs. *Cell, 136*, 629–641.

Qureshi, I. A., Mattick, J. S., & Mehler, M. F. (2010). Long non-coding RNAs in nervous system function and disease. *Brain Research, 1338*, 20–35.

Rada-Iglesias, A., Bajpai, R., Swigut, T., Brugmann, S. A., Flynn, R. A., & Wysocka, J. (2011). A unique chromatin signature uncovers early developmental enhancers in humans. *Nature, 470*, 279–283.

Rinn, J. L., Euskirchen, G., Bertone, P., Martone, R., Luscombe, N. M., Hartman, S., et al. (2003). The transcriptional activity of human chromosome 22. *Genes & Development, 17*, 529–540.

Rinn, J. L., Kertesz, M., Wang, J. K., Squazzo, S. L., Xu, X., Brugmann, S. A., et al. (2007). Functional demarcation of active and silent chromatin domains in human HOX loci by noncoding RNAs. *Cell, 129*, 1311–1323.

Sado, T., Hoki, Y., & Sasaki, H. (2005). Tsix silences Xist through modification of chromatin structure. *Developmental Cell, 9*, 159–165.

Schmitz, K. M., Mayer, C., Postepska, A., & Grummt, I. (2010). Interaction of noncoding RNA with the rDNA promoter mediates recruitment of DNMT3b and silencing of rRNA genes. *Genes & Development, 24*, 2264–2269.

Schorderet, P., & Duboule, D. (2011). Structural and functional differences in the long non-coding RNA hotair in mouse and human. *PLoS Genetics, 7*, e1002071.

Schwartz, Y. B., & Pirrotta, V. (2007). Polycomb silencing mechanisms and the management of genomic programmes. *Nature Reviews Genetics, 8*, 9–22.

Sessa, L., Breiling, A., Lavorgna, G., Silvestri, L., Casari, G., & Orlando, V. (2007). Noncoding RNA synthesis and loss of Polycomb group repression accompanies the colinear activation of the human HOXA cluster. *RNA, 13*, 223–239.

Sigova, A. A., Mullen, A. C., Molinie, B., Gupta, S., Orlando, D. A., Guenther, M. G., et al. (2013). Divergent transcription of long noncoding RNA/mRNA gene pairs in embryonic stem cells. *Proceedings of the National Academy of Sciences USA, 110*, 2876–2881.

Sleutels, F., Zwart, R., & Barlow, D. P. (2002). The non-coding Air RNA is required for silencing autosomal imprinted genes. *Nature, 415*, 810–813.
Song, L., Zhang, Z., Grasfeder, L. L., Boyle, A. P., Giresi, P. G., Lee, B. K., et al. (2011). Open chromatin defined by DNaseI and FAIRE identifies regulatory elements that shape cell-type identity. *Genome Research, 21*, 1757–1767.
Sun, B. K., Deaton, A. M., & Lee, J. T. (2006). A transient heterochromatic state in Xist preempts X inactivation choice without RNA stabilization. *Molecular Cell, 21*, 617–628.
Takahashi, K., & Yamanaka, S. (2006). Induction of pluripotent stem cells from mouse embryonic and adult fibroblast cultures by defined factors. *Cell, 126*, 663–676.
Trapnell, C., Williams, B. A., Pertea, G., Mortazavi, A., Kwan, G., van Baren, M. J., et al. (2010). Transcript assembly and quantification by RNA-Seq reveals unannotated transcripts and isoform switching during cell differentiation. *Nature Biotechnology, 28*, 511–515.
Tsai, M. C., Manor, O., Wan, Y., Mosammaparast, N., Wang, J. K., Lan, F., et al. (2010). Long noncoding RNA as modular scaffold of histone modification complexes. *Science, 329*, 689–693.
Venter, J. C., Adams, M. D., Myers, E. W., Li, P. W., Mural, R. J., Sutton, G. G., et al. (2001). The sequence of the human genome. *Science, 291*, 1304–1351.
Wang, K. C., & Chang, H. Y. (2011). Molecular mechanisms of long noncoding RNAs. *Molecular Cell, 43*, 904–914.
Wang, K. C., Yang, Y. W., Liu, B., Sanyal, A., Corces-Zimmerman, R., Chen, Y., et al. (2011). A long noncoding RNA maintains active chromatin to coordinate homeotic gene expression. *Nature, 472*, 120–124.
Wang, Z., Zang, C., Rosenfeld, J. A., Schones, D. E., Barski, A., Cuddapah, S., et al. (2008). Combinatorial patterns of histone acetylations and methylations in the human genome. *Nature Genetics, 40*, 897–903.
Werner, A., Carlile, M., & Swan, D. (2009). What do natural antisense transcripts regulate? *RNA biology 6*, 43–48.
Yap, K. L., Li, S., Munoz-Cabello, A. M., Raguz, S., Zeng, L., Mujtaba, S., et al. (2010). Molecular interplay of the noncoding RNA ANRIL and methylated histone H3 lysine 27 by polycomb CBX7 in transcriptional silencing of INK4a. *Molecular Cell, 38*, 662–674.
Yildirim, E., Kirby, J. E., Brown, D. E., Mercier, F. E., Sadreyev, R. I., Scadden, D. T., et al. (2013). Xist RNA is a potent suppressor of hematologic cancer in mice. *Cell, 152*, 727–742.
Zentner, G. E., Tesar, P. J., & Scacheri, P. C. (2011). Epigenetic signatures distinguish multiple classes of enhancers with distinct cellular functions. *Genome Research, 21*, 1273–1283.
Zhao, J., Ohsumi, T. K., Kung, J. T., Ogawa, Y., Grau, D. J., Sarma, K., et al. (2010). Genome-wide identification of polycomb-associated RNAs by RIP-seq. *Molecular Cell, 40*, 939–953.
Zhao, J., Sun, B. K., Erwin, J. A., Song, J. J., & Lee, J. T. (2008). Polycomb proteins targeted by a short repeat RNA to the mouse X chromosome. *Science 322*, 750–756.

# Regulation of Eukaryotic Cell Differentiation by Long Non-coding RNAs

**Juan R. Alvarez-Dominguez, Wenqian Hu and Harvey F. Lodish**

## 1 Introduction

The transfer of information from DNA to proteins is mediated by both RNA and protein components. Historically, our understanding of how these components act stems from a landmark model proposed by Jacob and Monod over half a century ago (Jacob and Monod 1961). According to this model, *structural* genes are transcribed into mRNA that acts as a template for protein synthesis, and this process is controlled by the products of *regulator* genes. The biochemical identity of these regulatory products was unclear at the time, but evidence that these could be either RNA or protein was widely discussed then. In the 50 years that followed a dominant view of proteins as the main regulators emerged, propelled by their ease of detection and manipulation compared to RNA, which is less abundant and more unstable. However, recent improvements in our ability to sequence entire genomes and detect their RNA transcripts now indicate greater roles for RNA regulators than previously anticipated.

The sequencing of various eukaryotic genomes resulted in the surprising finding that the number of protein-coding genes does not appear to vary significantly across metazoans, despite significant differences in developmental complexity. In contrast, the proportion of noncoding DNA (including introns) does seem to increase with developmental complexity, after accounting for varying ploidy (Mattick 2004; Taft et al. 2007). This led some to hypothesize that increasing

J. R. Alvarez-Dominguez · W. Hu · H. F. Lodish (✉)
Whitehead Institute for Biomedical Research, Cambridge, MA 02142, USA
e-mail: lodish@wi.mit.edu

J. R. Alvarez-Dominguez · H. F. Lodish
Department of Biology, Massachusetts Institute of Technology,
Cambridge, MA 02142, USA

H. F. Lodish
Department of Bioengineering, Massachusetts Institute of Technology,
Cambridge, MA 02142, USA

A. M. Khalil and J. Coller (eds.), *Molecular Biology of Long Non-coding RNAs*,
DOI: 10.1007/978-1-4614-8621-3_2, © Springer Science+Business Media New York 2013

amounts of RNA regulators, originating from these noncoding DNA regions, could have played a major role in giving rise to the diversity of cell differentiation programs that underlie development in multicellular organisms (Amaral and Mattick 2008). This model required that these regions be transcribed, and that many of the resulting RNA molecules act as functional regulators.

Evidence that noncoding DNA regions are indeed transcribed became clear as the focus of the field shifted from sequencing genomes to cataloging their transcriptomes. We now know that for every eukaryote examined the majority of the genome is transcribed, albeit at widely varying levels (Kapranov et al. 2007; Jacquier 2009). Only a small portion of the RNA species detected can be recognized as protein-coding, however, or as previously characterized classes of ncRNA, raising the possibility that some of the newly identified transcribed regions may actually encode novel functional ncRNAs.

Many of the previously unknown ncRNAs are longer than 200 nt and are thus classified as lncRNAs. As a class, lncRNAs accumulate to significant levels and resemble mRNAs in several structural features, such as splicing of similarly sized exons, therefore presenting a clear opportunity for detailed characterization. Consequently, attention over the past few years has shifted toward functionally characterizing lncRNAs, both through dedicated single-gene studies and large-scale approaches.

The following sections discuss those approaches to the study of lncRNA regulators in the context of cell differentiation in yeast, plants, and animals, focusing on selected examples that illustrate recent advances in the field. In particular, we highlight several mammalian differentiation systems where lncRNAs are increasingly recognized as an important layer of regulation during development.

## 2 Discovery and Characterization of Functional Long Noncoding RNAs

The first lncRNA to be characterized as such was described in the context of mouse embryonic development (Brannan et al. 1990). H19 was identified as a product of RNA Polymerase II, enriched in the fetal liver and in cardiac and skeletal muscle and that becomes strongly repressed after birth. H19 was capped and polyadenylated but contained no large open reading frame (ORF) for translation. Rather, it contained only small sporadic ORFs that were not evolutionary conserved, could not perform template translation in vivo, and did not produce detectable polypeptides. Shortly after, many more examples of this novel type of RNA were characterized in diverse eukaryotes, including Xist in mouse and human (Brockdorff et al. 1992; Brown et al. 1992), meiRNA in yeast (Watanabe and Yamamoto 1994), and roX1 in flies (Meller and Wu et al. 1997).

Over the following decade, the development of constantly improving technologies for transcriptome analysis propelled new efforts to detect and characterize lncRNAs at a global scale (Bertone et al. 2004; Carninci et al. 2005; Li et al. 2006; Kapranov et al. 2007; Dinger et al. 2008; Mercer et al. 2008a). These efforts increased exponentially the number of transcripts classified as lncRNAs. However, doubts about their biological relevance also grew, fueling a new era of technological innovations and fundamentally novel approaches for the genome-wide discovery of *bona fide* lncRNAs, as well as for the characterization of their functions.

## *2.1 Detecting and Identifying lncRNAs*

The advent of entire genome sequences precipitated a number of collaborative efforts that set out to survey their full transcriptional output (Tjaden et al. 2002; Yamada et al. 2003; Bertone et al. 2004; Stolc et al. 2004; Carninci et al. 2005; Stolc et al. 2005; David et al. 2006; Li et al. 2006; Birney et al. 2007; Nagalakshmi et al. 2008; Wilhelm et al. 2008). These efforts drove the rapid adaptation of classic gene expression profiling techniques into large-scale approaches of ever-increasing throughput, as occurred for Cap analysis of Gene Expression (CAGE) (Shiraki et al. 2003), microarrays, (Selinger et al. 2000) and cDNA sequencing (Mortazavi et al. 2008). Regardless of the technical approach, the unequivocal outcome of surveying multiple eukaryotic transcriptomes was that only a small portion of the detected transcripts could be recognized as protein-coding or as previously characterized species of small ncRNAs (such as rRNA, tRNA, snoRNA, microRNA, or piRNA). This fueled much excitement over the potential biological functions of the newly discovered RNAs (Kapranov et al. 2007; Amaral et al. 2008; Berretta and Morillon 2009; Jacquier 2009; Mercer et al. 2009). Since the number of uncharacterized loci easily surpassed that of protein-coding genes, it was speculated that their increase in number along the eukaryotic phylogeny may explain large differences in developmental complexity among eukaryotes with otherwise comparable numbers of protein-coding genes and protein families (Mattick 2004; Prasanth and Spector 2007).

Preliminary clues about the potential functionality of uncharacterized RNAs first emerged for those that were well-expressed and longer than 200 nt (putative lncRNAs). First, analysis of their sequence conservation showed clear evidence of evolutionary constraints (Pheasant and Mattick 2007; Ponjavic et al. 2007). Second, expression profiling indicated that many lncRNAs exhibit regulated and cell-type specific expression patterns during development (Blackshaw et al. 2004; Stolc et al. 2004; Inagaki et al. 2005; Ravasi et al. 2006; Dinger et al. 2008). Third, individual lncRNA candidates were found to localize to specific subcellular structures (Brown et al. 1992; Mercer et al. 2008b; Nagano et al. 2008; Clemson et al. 2009; Redrup et al. 2009; Sasaki et al. 2009; Sunwoo et al. 2009). However, considering that both the expression and the conservation of putative lncRNAs appeared to be much less than those of known mRNAs, uncertainty about their origin and biological relevance persisted. One technical concern was that, given the propensity of reverse transcriptase for spurious second-strand production during first-strand cDNA synthesis, catalogs of putative lncRNAs could be plagued by spurious antisense transcripts that were experimental artifacts (Perocchi et al. 2007; Ozsolak and Milos 2011). Another important concern was that many of the newly identified RNAs were simply transcriptional noise (Huttenhofer et al. 2005; Ponjavic et al. 2007; Struhl 2007), nonfunctional byproducts of the transcription of neighboring loci (including enhancers) (Struhl 2007; Ebisuya et al. 2008; De Santa et al. 2010; Kim et al. 2010), or actually encoding small functional peptides (Galindo et al. 2007; Dinger et al. 2008; Ingolia et al. 2011). Clearly, additional evidence was needed to distinguish biologically relevant lncRNA candidates from technical or biological noise.

A strategy devised by Guttman and colleagues to address these issues was to focus only on intergenic regions showing evidence of stable expression, as assayed by a signature of chromatin marks correlated with stable Pol II transcription (Guttman et al. 2009). This signature consisted of a short stretch of H3K4me3, indicative of Pol II initiation, followed by a longer stretch of H3K36me3, marking the region of Pol II elongation. The strategy identified in four mouse cell types about 1500 intergenic lncRNA (lincRNA) loci that were 5 kB or greater in length and that did not overlap protein-coding genes, microRNAs, or siRNAs. Their products were polyadenylated and primarily multiexonic transcripts with little or no protein-coding potential and strong evidence of 5′ capping. This subset of mouse lncRNAs indeed showed higher expression and conservation than previous collections, and a number of them were putatively associated with various developmental processes through correlative expression analysis. Extending the approach yielded about 1800 human lincRNAs (Khalil et al. 2009).

There are important limitations to using the chromatin signature approach for *de novo* discovery of lncRNAs, however. Not all loci actively transcribed by Pol II are marked by this K4-K36 signature; a study in mouse found that ~25 % of lincRNA or mRNA transcripts identified by RNA-seq alone are not (Guttman et al. 2010), and in human the number appears to be greater (Cabili et al. 2011). Conversely, not all regions with a detectable K4-K36 domain correspond to gene bodies; some correspond to transcribed enhancers (De Santa et al. 2010; Cabili et al. 2011), and close examination of existing lncRNA catalogs indicate that ~10–15 % actually overlap enhancers (Cabili et al. 2011). Moreover, it is possible that some lncRNAs are transcribed by RNA polymerase III (see (White 2011) for discussion) and thus lack chromatin marks characteristic of Pol II transcription.

Subsequent studies are now employing a combination of strategies for the discovery of stably expressed, reliable lncRNAs (Guttman et al. 2010; Cabili et al. 2011; Ulitsky et al. 2011; Derrien et al. 2012; Nam and Bartel 2012). Detection and assembly of *de novo* lncRNA transcript models are most frequently conducted by RNA-seq alone. Evidence of full-length independent transcriptional units is then sought by augmenting these models with evidence of transcript boundaries from orthogonal approaches. For example, transcriptional start sites can be determined directly through CAGE analysis or inferred from H3K4 marks.

Similarly, the 3′ ends can be mapped by poly(A)-position profiling or inferred by computational detection of motifs for poly(A) addition. The availability of paired-end sequencing reads can also allow for assessment of previously unappreciated connectivity between lncRNA and protein-coding genes. Constantly improving combination strategies are thus being used to obtain increasingly reliable collections of lncRNA genes in various organisms, resulting in a rapidly growing number of lncRNAs with recognized functions (see Amaral et al. 2011) for a comprehensive database.

## *2.2 Excluding Functional Protein-Coding Capacity*

A distinctive feature of lncRNAs is that they do not possess functional protein-coding capacity, which means that they produce no protein products. To determine whether this is true, the golden standard is to assess if polypeptides are produced from any ORF of candidate transcripts (Banfai et al. 2012). However, due to technical difficulties, such as the detection of low-abundance putative target polypeptides, or the absence of corresponding antibodies, the coding capacity of a newly identified RNA transcript is usually determined indirectly by computational and biochemical approaches (see Dinger et al. 2008 for review).

Computationally, evaluating coding potential can be done at a global scale by examining candidate transcripts for presence and conservation of ORFs, by looking for homology to known protein domains, and by scrutinizing any putative ORFs for biases in codon usage or in frequency of codon substitution through evolution. The presence of ORFs in a transcript is a necessary but not sufficient qualification for coding capacity. A putative ORF may occur purely by chance in any stretch of sequence, with the probability of such chance event increasing with sequence length (Dinger et al. 2008). Alternatively, a putative ORF may be a vestige of former coding capacity (Ponting et al. 2009). Indeed, *bona fide* lncRNAs such as human XIST and H19 do contain ORFs as long as 172 and 256 amino acids, respectively, but these are not evolutionary conserved and fail to template polypeptide synthesis in vivo (Brannan et al. 1990; Brockdorff et al. 1992). In the case of Xist, it is believed to have originated in part from genes that formerly coded for proteins (Duret et al. 2006).

To distinguish functional from spurious ORFs using computational methods, candidates can be tested for hallmark features of functional coding sequences. For example, known protein-coding regions typically display organism-specific differences in the frequency of occurrence of synonymous codons. The absence of such codon usage bias from a putative ORF can thus be used to argue that the ORF is unlikely to be functional. Evolutionary analysis can also be used to evaluate functional coding potential (see Lin et al. (2008) for review). Coding regions are under purifying selection to retain synonymous over nonsynonymous codon substitutions to preserve their function. Noncoding regions, in contrast, experience no such selection and thus typically exhibit similar frequencies of synonymous and nonsynonymous substitutions. The absence of a codon substitution bias inferred from the multispecies alignment of a putative ORF sequence can thus be used as evidence against functional coding capacity (Lin et al. 2011). However, approaches based on evolutionary analysis may fail to identify newly evolved functional ORFs. To address this, methods that do not require cross-species comparisons should be considered (Dinger et al. 2008), together with a direct inspection of homology in protein domain databases (i.e., BLASTX).

Collectively, computational approaches are a powerful and cost-efficient way of testing the coding potential of large collections of candidate lncRNAs. Those candidates that pass computational tests, however, ultimately require experimental verification of their noncoding status. Such status implies that a candidate transcript is not associated with actively translocating ribosomes, and this can be tested by examining its presence on polysomes through polysome fractionation analysis (Warner et al. 1963). This approach employs sucrose density gradients and ultracentrifugation to fractionate cell lysates. RNA transcripts associated with ribosomes predominantly sediment with the greatest velocity, whereas nonribosomal-associated transcripts remain at the top of the gradient. Care should be taken when interpreting these outcomes, however. If a transcript remains at the top of the gradient, it can be either a noncoding transcript or a translationally repressed protein-coding one. Conversely, if a transcript sediments with a higher velocity through the gradient, it only implies that the transcript is associated with large particles, which can be ribosomes but also other large complexes. Specific disruption of translation, such as treatment with the translation elongation inhibitor puromycin, is required to discriminate between these two possibilities. An alternative approach is ribosome profiling followed by RNA sequencing, which can measure the density and occupancy of translocating ribosomes at high resolution (Ingolia et al. 2009; Ingolia et al. 2011). This technique is a very powerful one to detect specific association with ribosomes. However, it cannot assess whether or not the bound ribosomes are actively making polypeptides. Some functionally characterized lncRNAs do associate with ribosomes without producing polypeptides, including H19 and GAS5 (Li et al. 1998; Smith and Steitz 1998). Additional experiments are therefore required to address this issue.

The coding status of a transcript can also be inferred from its localization within cells, as determined by RNA in-situ fluorescence hybridization (RNA FISH) or by fractionation of cell homogenates into nuclear and cytoplasmic fractions. Transcripts predominantly resident in the nucleus such as Xist are strong candidates to be noncoding, because translation occurs in the cytoplasm. One caveat of these studies, however, is that they only reveal the steady-state localization of the transcript. If the transcript is rapidly shuttling or efficiently degraded only in one cellular compartment the information obtained from its steady-state localization may be misleading (see Grunwald et al. (2011) for review).

It is worth noting that even if a polypeptide is in fact produced from an RNA transcript, this alone does not rule out a function as an RNA regulator. Examples of transcripts with dual functions as mRNA and lincRNA regulator have indeed been described from bacteria to man (Chooniedass-Kothari et al. 2004; Kloc et al.

2005; Hube et al. 2006; Jenny et al. 2006; Wadler and Vanderpool 2007). Conversely, association of a functional lncRNA with ribosomes may lead to non-functional polypeptides if the lncRNA derives from an mRNA in the process of losing functional coding capacity, or if the lncRNA itself is in the process of gaining it (Dinger et al. 2008; Ulitsky et al. 2011). Importantly, evolutionary transitions between coding and noncoding functionality can be rather lineage-specific (Duret et al. 2006; Ulitsky et al. 2011), complicating the use of cross-species preservation of coding potential as evidence against noncoding function.

In cases where a transcript's function is actually known, its functional coding capacity can be directly tested by using frame-shift mutations to disrupt any putative ORFs and assessing whether the function of the transcript is compromised. If such function is independent of all putative ORFs, a strong claim can be made that functionally the transcript is indeed noncoding. Altogether, under ideal conditions, computational analyses augmented by dedicated experiments are needed to convincingly determine whether or not the increasing number of putative lncRNAs identified by large-scale studies function or not as RNA regulators.

## *2.3 Characterizing lncRNA Features*

As with protein-coding genes, lncRNAs comprise a variety of subclasses with diverse properties and functions. Preliminary efforts to define these subclasses have largely focused on the genomic positioning of lncRNA loci. Based on this criterion, lncRNAs can be classified as intergenic, antisense to protein-coding genes, or overlapping known noncoding elements (such as enhancers, introns of protein-coding genes, or known small ncRNA loci). Currently, most efforts to characterize lncRNAs have focused on the intergenic ones, which are easier than the other subgroups to unambiguously identify and perturb. This section thus focuses on lincRNAs as models for global lncRNA characterization.

Recent approaches for large-scale lincRNA discovery in mouse and human have laid the conceptual frameworks for annotating their structural, conservation, and expression features (Guttman et al. 2010; Cabili et al. 2011; Derrien et al. 2012). These have shown that structurally, lincRNAs have exons of comparable size to those of mRNAs but have fewer of them. These results in shorter transcript lengths and in fewer isoforms, which appear to be produced using canonical splice sites. These observations may be limited, however, by technical difficulties in retrieving reads spanning splice sites and in accurately defining full transcriptional units, due to the relatively short reads of current sequencing technologies or to assembly errors (Cabili et al. 2011; Ozsolak and Milos 2011). At their termini, lincRNAs show clear evidence of 5′ capping and 3′ polyadenylation, but to a lower extent than mRNAs (Guttman et al. 2009; Derrien et al. 2012).

Conservation analyses have revealed that, in general, lincRNAs show distinctive evidence of purifying selection in their primary sequence (Guttman et al. 2009; Khalil et al. 2009; Marques and Ponting 2009; Cabili et al. 2011; Ulitsky et al. 2011). However, while lincRNA sequence conservation across promoters is comparable with that of mRNAs, it is significantly lower across exons. This observation may be explained by the fact that lncRNA and protein-coding genes are subject to very different selective pressures. While protein-coding genes are under pressure to preserve the functional polypeptide information encoded in their exons, the pressure at lncRNA genes may be to preserve encoded secondary structure information, which can tolerate more sequence changes (Washietl et al. 2005; Maenner et al. 2010; He et al. 2011; Parker et al. 2011; Schorderet and Duboule 2011; Novikova et al. 2012), to preserve only short regulatory sequence elements (Duret et al. 2006; Marques and Ponting 2009), or simply to maintain the overall genomic position, length, and orientation in which they are transcribed (Ponting et al. 2009; Cabili et al. 2011; Ulitsky et al. 2011). Since this feature may complicate the identification of lncRNA orthologs across species, several approaches that integrate conservation of secondary structure or of synteny to the discovery and characterization of lncRNAs have been recently developed (Stanke et al. 2008; Gorodkin and Hofacker 2011). It is worth pointing out that several functionally characterized lincRNAs do show strong conservation of primary sequence from zebrafish to human (Guttman et al. 2009; Ponting et al. 2009; Sheik Mohamed et al. 2010; Ulitsky et al. 2011). Conservation alone, however, is neither necessary nor sufficient evidence for functionality. Indeed, many functional lncRNAs appear rapidly evolving among eukaryotes, and some are restricted to the primate lineage (Pollard et al. 2006; Amaral and Mattick 2008; Dinger et al. 2008; Marques and Ponting 2009; Derrien et al. 2012).

In terms of expression, lincRNAs seem on average expressed at lower levels but with higher tissue and cell type-specificity than mRNAs (Guttman et al. 2010; Cabili et al. 2011; Derrien et al. 2012). The latter observation might confound the former, however, in studies where diverse cell or tissue types are pooled together. That is, lncRNAs highly expressed in a minor cell type might not be detected. Alternatively, low levels of expression may simply result from cell-to-cell variability in synthesis or degradation of short-lived transcripts, or from cell cycle- or developmental stage-specific expression among unsynchronized cell populations. Measuring expression within single cells or conducting bulk assays in cells sorted by cell cycle and developmental markers may help resolve this caveat.

The seemingly exquisite spatial and temporal patterns of lncRNA expression in mammals suggest that some may function to help specify cell identity. Alternatively, such patterns may be a byproduct of the tissue-specific activity of neighboring genes, of enhancer elements or of entire chromosomal domains. Hence, experimental evidence in the form of targeted perturbations is needed to characterize the specific functions of lncRNAs during cell fate specification.

## 2.4 Probing lncRNA Functions During Cell Differentiation

Individual examples of lncRNAs that modulate developmental processes have been studied in detail over the past two decades. For instance, Xist plays a well-characterized essential role in X-chromosome inactivation in female mammals via epigenetic silencing (see Lee 2011 for review), and H19 regulates growth during embryogenesis via imprinting of the maternal Igf2 allele (see Gabory et al. 2010 for review). For the vast majority of lncRNAs recently identified by large-scale studies, however, their potential roles in development remain to be explored. Several interesting observations suggest that pursuing such studies may be worthwhile. First, the noncoding proportion of the transcriptome seems to increase with developmental complexity, suggesting that ncRNA regulators, including lncRNAs, may have contributed to the emergence of diverse gene expression programs underlying differentiation of specialized cells during organismal development (Mattick 2004; Prasanth and Spector 2007; Amaral and Mattick 2008; Mercer et al. 2009; Pauli et al. 2011). Second, given that lncRNAs as a class show greater tissue specificity than mRNAs (Cabili et al. 2011; Derrien et al. 2012), it seems conceivable that distinct collections of lncRNAs modulate the developmental programs of distinct tissues. Third, dysregulation of lncRNAs has been observed under many pathological conditions including cancer, heart disease, and Alzheimer's disease (Reviewed in Wapinski and Chang 2011), suggesting that abnormal expression of some of these transcripts may contribute to the development of pathophysiological cellular states.

Importantly, recent studies have shown that lncRNAs are capable of regulating gene expression via diverse mechanisms (Fig. 1). For example, lncRNAs can function as molecular scaffolds that recruit chromatin modifiers to target genes *in cis* or *in trans* and thereby modulate their expression (see Schmitt and Paro 2006; Koziol and Rinn 2010 for review). In addition, lncRNAs can also modulate post-transcriptional events such as mRNA splicing (Tripathi et al. 2010), translation (Beltran et al. 2008; Yoon et al. 2012), and degradation (Gong and Maquat 2011). Furthermore, some lncRNAs can impair the function of specific microRNAs and thus indirectly enhance the stability of the mRNAs normally downregulated by these miRNAs (Franco-Zorrilla et al. 2007a; Cesana et al. 2011; Karreth et al. 2011; Salmena et al. 2011). Detailed mechanistic examples of how lncRNAs regulate gene expression have been summarized in recent reviews (Wang and Chang 2011; Guttman and Rinn 2012; Rinn and Chang 2012). Such regulatory capacities thus render lncRNAs as likely players in the modulation of cell differentiation programs.

Over the past few years, loss-of-function and gain-of-function studies have revealed that many lncRNAs are indeed involved in cell differentiation processes throughout the eukaryotic lineage (Ponting et al. 2009; Wilusz et al. 2009; Wapinski and Chang 2011; Ietswaart et al. 2012). In multicellular eukaryotes these include, but are not limited to, self-renewal, apoptosis, and differentiation of pluripotent or lineage-restricted progenitors during embryogenesis or mature tissue homeostasis. In the next section, we discuss selected examples of lncRNAs implicated in the regulation of various cell differentiation processes, using as a guide the life cycle of multicellular organisms, from gametogenesis and through embryogenesis to adult tissue homeostasis. In particular, we focus on those examples illustrating recent advances in this fast-evolving field.

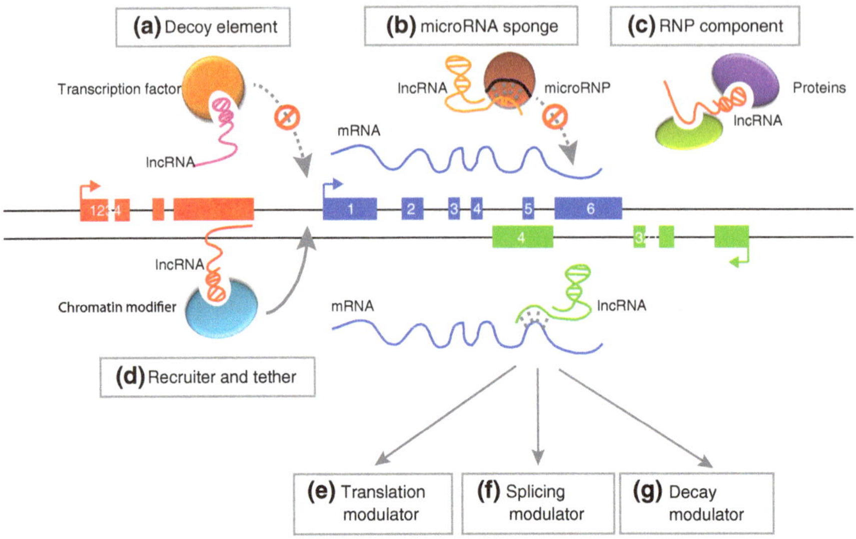


**Fig. 1** Mechanisms of lncRNA function. lncRNAs employ diverse mechanisms to regulate their targets. **a** Several lncRNAs act as decoy elements, titrating TFs away from their DNA targets by directly binding to them as target mimics. **b** Others work as decoys at the post-transcriptional level, titrating microRNA effector complexes away from their mRNA targets by containing target site mimics. lncRNAs whose microRNA target sites lack structural sequence features needed for transcript degradation have the net effect of 'sponging' their microRNA regulators. **c** Many lncRNAs bind specific combinations of proteins, such as chromatin modifiers or TFs, thus serving as scaffold elements within RNP. **d** Recruitment and tethering of chromatin modifying complexes to their DNA targets in *cis* has also emerged as a well-characterized function for a number of lncRNAs. Not depicted is recruitment in *trans*. A few lncRNAs appear to modulate direct post-transcriptional processing of their mRNA targets, including translation (**e**), splicing (**f**) and decay (**g**)

# 3 Examples of lncRNAs Implicated in Eukaryotic Cell Differentiation

## *3.1 Regulation of Gametogenesis by lncRNAs*

Differentiation of progenitor cells into gametes is essential for sexual reproduction in all eukaryotes. The initiation and execution of gametogenesis are normally triggered in response to specific developmental or environmental cues. Two key processes occur during gametogenesis: Meiosis, a specialized cell division which generates haploid cells from diploid precursors, and a developmental program by which the haploid precursors differentiate into mature gametes. Evidence from various organisms indicates that lncRNAs play key roles in the commitment to and execution of these key processes (Fig. 2 and Table 1).

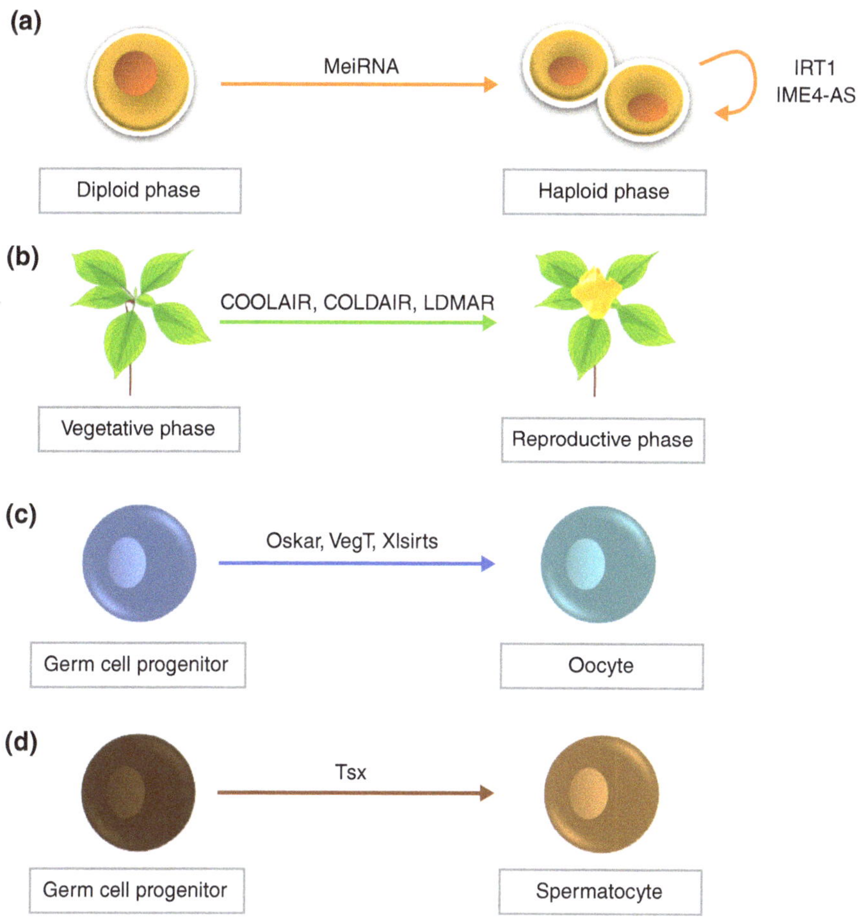


**Fig. 2** Regulation of gametogenesis by lncRNAs. Initiation and execution of gametogenesis are mediated by lncRNAs across eukaryotes. **a** Transitioning from diploid progenitors to haploid gametes in yeast requires MeiRNA during meiosis I. This transition is normally inhibited in haploid cells by the IRT1 and IME4-AS lncRNAs. **b** In *Arabidopsis thaliana*, flowering after the winter involves modulation by two lncRNAs, COOLAIR and COLDAIR. In rice, LDMAR is required for normal gametogenesis under long-day growth conditions. **c** Several lncRNAs contribute to oogenesis in *Xenopus laevis*, including Oskar and VegT, which also function as mRNAs, and the Xlsirts family of lncRNAs, which serve structural roles. **d** Normal spermatogenesis in mouse involves the Tsx lncRNA

The first example of an lncRNA regulator of meiosis emerged from studies of fission yeast two decades ago. Upon nutrient starvation, diploid *Schizosaccharomyces pombe* cells undergo meiosis and differentiate into stress-resistant gametes called spores. An RNA-binding protein, Mei2p, is essential for the meiotic phase by controlling pre-meiotic DNA synthesis and execution of the first meiotic division (Watanabe and Yamamoto 1994). A polyadenylated lncRNA, called

**Table 1** Examples of lncRNA regulators of Gametogenesis

| Name | Organism | Expression | Loss of function phenotype | Experimental manipulations | References |
|---|---|---|---|---|---|
| meiRNA | Fission yeast (Schizosaccharomyces pombe) | Diploid meiotic cells | Arrest prior to meiosis I; robust chromosome pairing impaired | Deletion; ectopic expression; directed mutagenesis; FISH; mislocalization | Watanabe and Yamamoto 1994; Yamashita et al. 1998; Sato et al. 2001; Shimada et al. 2003; Ding et al. 2012 |
| IRT1, IME4-AS | Budding yeast (Saccharomyces cerevisiae) | Haploid cells | Meiosis induced | Deletion; overexpression; directed mutagenesis; FISH | Hongay et al. 2006; Gelfand et al. 2011; van Werven, Neuert et al. 2012 |
| COOLAIR, COLDAIR | Arabidopsis thaliana | Reproductive phase during vernalization | Late flowering after vernalization | siRNA knockdown; directed mutagenesis; ectopic expression | Swiezewski et al. 2009; Liu et al. 2010; Heo and Sung 2011 |
| LDMAR | Rice (Oryza sativa) | Microspore mother cells; pollen cells | Programmed cell death in developing anthers; photoperiod-sensitive sterility | Overexpression; directed mutagenesis; FISH | Ding et al. 2012 |
| Oskar | Fruit fly (Drosophila melanogaster) | Developing oocytes | Sterility due to differentiation arrest | Directed mutagenesis; ecotpic expression; FISH | Jenny et al. 2006 |
| VegT, Xlsirts | Frog (Xenopus laevis) | Developing oocytes | Inpaired granule development due to collapse of cytokeratin cytoskeleton | Antisense oligo knockdown; FISH | Kloc et al. 2005; Kloc et al. 2007 |
| Tsx | Placental mammals | Meiotic germ cells, ES cells and brain | Mouse: smaller testes due to pachytene spermatocyte apoptosis in males; reduced feritlity in females | Mouse knockout; directed mutagenesis; FISH | Anguera et al. 2011 |

meiRNA, is required for Mei2p function during meiosis I. Deletion of meiRNA is inconsequential in haploid cells but causes diploid progenitors to arrest prior to meiosis I. meiRNA specifically binds to and enforces nuclear localization of Mei2p, which otherwise undergoes nucleocytoplasmic shuttling with longer cytoplasmic residence (Yamashita et al. 1998; Sato et al. 2001). Interestingly, meiRNA promotes nuclear retention of Mei2p via entrapment into a subnuclear granule structure anchored at the meiRNA locus itself (Shimada et al. 2003). The biological function of this RNP granule during meiosis remains unclear, but a recent study found that formation of the meiRNA granule favors chromosome pairing at homologous meiRNA loci during early meiotic prophase (Ding et al. 2012). Deletion of the meiRNA locus impairs robust homologous pairing and decreases the chromosome recombination frequency, while transposition to ectopic chromosomal sites favors pairing at these sites. The presence of the meiRNA locus is not required for proper chromosome segregation, however, suggesting that other yet uncharacterized chromosomal pairing sites may exist. Interestingly, such a role for meiRNA in chromosome pairing is seemingly independent of Mei2p recruitment. Thus, meiRNA is required for productive gametogenesis by promoting nuclear localization of the Mei2p meiotic regulator, while potentially participating directly or via another bound protein in homologous chromosome pairing.

Initiation of gametogenesis in diploid but non haploid cells (where it would be lethal) is a key cell fate decision. Remarkably, in the budding yeast *Saccharomyces cerevisiae* this decision depends on the activity of only two master transcription factors whose expression is controlled *in cis* via transcription of lncRNAs (Hongay et al. 2006; Gelfand et al. 2011; van Werven et al. 2012). Diploid yeasts normally initiate meiosis upon nutrient starvation by activating expression of the IME1 and IME4 transcription factors, which in turn enforces a meiotic differentiation program. Haploid cells, conversely, avoid a lethal meiosis by repressing IME1 and IME4 expression through transcription of the IRT1 and IME4-AS lncRNAs, respectively. The lncRNA IRT1 is located upstream of the IME1 promoter and the act of its transcription through the IME1 promoter represses the locus by establishing a repressive chromatin state. This effect is mediated by co-transcriptional recruitment of the Set2 histone methyltransferase and the Set3 histone deacetylase complex. The lncRNA IME4-AS, on the other hand, is located antisense to the IME4 locus and the act of its transcription through the IME4 locus represses it by preventing transcriptional elongation. Importantly, interfering with the expression of the IRT1 and IME4-AS lncRNAs is sufficient to induce lethal meiosis in haploid cells, whereas preventing their repression in diploid cells inhibited the capacity to carry out productive meiosis. Thus, regulation by lncRNAs is essential for controlling entry into gametogenesis.

Recent large-scale transcriptome annotation efforts are now identifying the full extent of gametogenesis-specific lncRNAs in budding and fission yeast (Watanabe et al. 2001; Miura et al. 2006; Wilhelm et al. 2008; Lardenois et al. 2011; Kim Guisbert et al. 2012). The loci of many of these lncRNAs are interleaved with those of gametogenesis-specific mRNAs in potentially interfering antisense or tandem orientations. Indeed, functional roles through transcriptional interference have been described for a few of them (Hongay et al. 2006; Gelfand et al. 2011; Chen et al. 2012; van Werven et al. 2012). The stability of all meiotic lncRNAs is tightly controlled by the activity of the nuclear exosome component Rrp6. After the onset of meiosis in budding yeast, the Rrp6 protein is degraded and this results in progressive accumulation of hundreds of meiosis-specific lncRNAs (Lardenois et al. 2011). Interestingly, Rrp6 is essential for pre-meiotic DNA synthesis, meiotic divisions, and subsequent spore formation. Thus, execution of the meiotic differentiation program involves dynamic regulation of the stability of protein and lncRNA components, and some of these are essential for its execution. The tight temporal control over the expression of many currently uncharacterized gametogenesis-specific lncRNAs suggests that they too may regulate this developmental process.

As with yeast, the onset of the reproductive phase in flowering plants is also modulated by lncRNAs. Transitioning to this phase (flowering) accompanies gametogenesis and is highly regulated by the integration of multiple cues, including light exposure and temperature. In *A. thaliana*, these cues converge on the activity of the FLOWERING LOCUS C (FLC), a master transcription factor in charge of repressing the flowering gene expression program. Prolonged exposure to winter cold promotes progressive silencing of FLC, which in turn aligns the onset of flowering with the favorable conditions of spring. This process is mediated by COOLAIR, an antisense lncRNA that encompasses the entire FLC locus (Swiezewski et al. 2009; Liu et al. 2010). Upregulation of COOLAIR silences sense transcription of FLC by promoting epigenetic silencing via the Polycomb Repressive Complex 2 (PRC2). Interestingly, this cold-inducible activity is sufficient to induce silencing of a heterologous reporter, and this effect seems to depend on the 3′ processing of COOLAIR. Another lncRNA, COLDAIR, also suppresses FLC (Heo and Sung 2011). COLDAIR is a ~1 kB capped but nonpolyadenylated lncRNA that is transcribed from the first intron of FLC. COLDAIR is expressed later than COOLAIR, but is required for robust cold-dependent epigenetic silencing of FLC. Moreover, unlike COOLAIR, COLDAIR appears to physically bind to the PRC2 complex and recruit it to the FLC locus. Hence, developing sexual reproductive capacity in both yeast and *Arabidopsis* involves modulation of key transcription factors by lncRNAs through their mediation of epigenetic modification. This paradigm is also manifested in rice, where a ~1.25 kB lncRNA called LDMAR is required for normal pollen development under long-day conditions (Ding et al. 2012). LDMAR is enriched in microspore mother cells and in pollen cells, suggesting a function in gametogenesis. Interestingly, loss of photoperiod-inducible LDMAR expression in a variant rice strain leads to programmed cell death in developing anthers and male sterility. Overexpression, conversely, leads to an increase in fertility under natural long days. The example of LDMAR thus links regulation of gametogenesis by an lncRNA to organismal fertility.

In the fruit fly *Drosophila melanogaster*, oogenesis is regulated by the oskar RNA, which has dual coding and noncoding functions (Jenny et al. 2006). Loss of oskar results in sterility due to early arrest of oocyte differentiation, a phenotype that can be rescued by expression of mutant oskar with a disrupted translation capacity. Expression of the oskar 3'UTR alone is in fact sufficient to rescue the deletion phenotype. Thus, as with LDMAR in rice, oskar acts as an RNA regulator of gametogenesis and its loss of function leads to sterility.

Other lncRNAs expressed in oocytes of the frog *X. laevis* serve structural roles (Kloc et al. 2005; Kloc et al. 2007). The Xlsirts family of repeat-containing lncRNAs and the bifunctional VegT mRNA are required for the integrity of the oocyte cytokeratin but not the actin cytoskeleton. Depletion of either type of lncRNA results in collapse of the cytokeratin network and impairs proper granule development. Such a role during germ cell development seems mediated by direct integration of lncRNAs into the oocyte cytokeratin structure itself.

The involvement of lncRNAs in gametogenesis in mammals is less well-characterized. However, regulation by small ncRNAs, including piRNAs and endo-siRNAs, is essential for germline specification and maintenance from worm to human (see Okamura and Lai 2008; Ghildiyal and Zamore 2009; Pauli et al. 2011 for review), suggesting that lncRNA regulators may also play key roles. In support of this, dozens of lncRNAs are specifically expressed in germline cells from worm to human (Inagaki et al. 2005; Ravasi et al. 2006; Cabili et al. 2011; Diez-Roux et al. 2011; Derrien et al. 2012; Nam and Bartel 2012).

In mice, lncRNAs are thought to be involved in germ cell development by virtue of their interference with mRNA gene expression. For example, the locus encoding the transcription factor Foxl2, critical for mammalian oogenesis, is antisense-overlapped by Foxl2OS, a ~4.5 kB lncRNA that is coordinately expressed with Foxl2 and is thus thought to act as a positive regulator (Cocquet et al. 2005). Similarly, the paternally expressed PEG1 locus, with roles in cellular growth regulation, is antisense-overlapped by PEG1-AS, a ~2.4 kB lncRNA found in testis and mature spermatozoa, where it is thought to modulate PEG1 activity (Li et al. 2002). Interestingly, PEG1 has been recently recognized as a selective suppressor of breast cancer metastasis and tumor reinitiation via its intronic microRNA-335 (Png et al. 2011). A recently described X-linked lncRNA, Tsx, appears to have specific functions in the germline (Anguera et al. 2011). Tsx is a spliced transcript conserved among mammals that escapes X inactivation and becomes enriched in male meiotic germ cells. Deletion of Tsx in male mice leads to apoptosis of spermatocytes during meiotic prophase I, resulting in smaller testes. Although much of the evidence for lncRNA regulation during mammalian gametogenesis remains correlative, the abundance and specificity of lncRNA expression in germ cells suggests that potential lncRNA modulators of gametogenesis remain to be characterized.

## 3.2 lncRNAs During Embryonic Stem Cell Maintenance and Differentiation

The fusion of sex gametes in metazoans begins the process of embryogenesis, whereby an embryo is produced from the fertilized egg. The early stages of this process give rise to pluripotent embryonic cells, which have the developmental plasticity of differentiating into all derivatives of the three primary germ layers (ectoderm, endoderm, and mesoderm). In culture, these pluripotent cells can generate embryonic stem (ES) cells that also have the unique capacity to produce all the cell types in an organism through division and differentiation. Maintaining pluripotency of ES cells requires delicate transcriptional regulation mediated by key transcription factors, such as Oct4, Sox2, and Nanog (see Young 2011 for review). In addition to these protein regulators, lncRNAs are also involved in modulating ES cell fate (Fig. 3 and Table 2).

In a recent study in mouse ES cells, Sheik Mohamed and coworkers focused on four highly conserved lncRNAs that are regulated by Oct4 and Nanog (Sheik Mohamed et al. 2010). Inhibition or misexpression of two of these, RNCR2 and AK14205, caused exit from the pluripotent state as evidenced by loss of pluripotency markers, upregulation of lineage-specific markers, cell proliferation, and morphology. These effects were accompanied by altered levels of Oct4 and Nanog themselves, suggesting that lncRNAs act in the regulatory networks that control ES cell pluripotency. This possibility was further examined at a larger scale by a study focusing on 147 putative lincRNAs identified in mouse ES cells by the K4-K36 chromatin signature (Guttman et al. 2009; Guttman et al. 2011). For about 90 % of the lincRNAs tested, knockdown using lentiviral-based shRNAs resulted in significant changes in the ES cell gene expression program. Importantly, 26 lincRNAs were specifically implicated in the maintenance of the pluripotent state, as assayed after knockdown by loss of pluripotency markers and cell morphology. Another 30 lincRNAs were also implicated in repressing specific differentiation programs, although their loss of function alone was not sufficient to cause differentiation. Importantly, expression of most of these lincRNAs is regulated by multiple combinations of ES cell-specific transcription factors, including Oct4, Sox2, Nanog, and Klf4. Furthermore, many of these lincRNAs appear to bind diverse combinations of chromatin regulatory proteins, potentially giving rise to

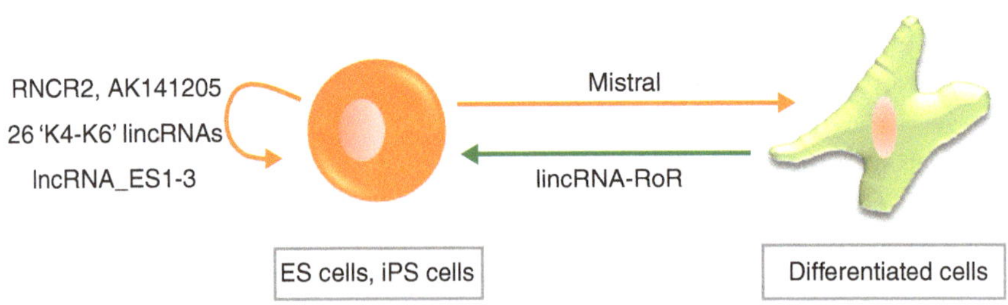


**Fig. 3** lncRNAs during embryonic stem cell maintenance and differentiation. Many lncRNAs are required for maintenance of ES pluripotency. These include 26 lincRNAs identified by a 'K4-K36' domain in mouse (see text) (Guttman et al. 2009; Guttman et al. 2011), three identified in human ES cells (lncRNA_ES1-3) and two that are highly conserved across mammals (RNRC2 and AK141205). ES cell differentiation, on the other hand, is associated with global changes including upregulation of lncRNAs such as Mistral. The process of dedifferentiation of specialized cells into iPS cells also employs lncRNAs such as linRNA-RoR

**Table 2** Examples of lncRNA regulators of embryonic stem cell maintenance and differentiation

| Name | Organism | Expression | Loss of function phenotype | Experimental manipulations | References |
|---|---|---|---|---|---|
| RNCR2, AK141205 | Placental mammals | ES cells | Mouse: changes in pluripotency marker expression; changes in lineage marker expression; loss of ES cell morphology; altered proliferation | siRNA knockdown; overexpression | Sheik Mohamed et al. 2010 |
| 26 'K4-K36' lincRNAs | Mouse (Mus musculus) | ES cells | Decreased expression of pluripotency markers; loss of ES cell morphology | shRNA knockdown | Guttman et al. 2009; Guttman et al. 2011 |
| lincRNA-RoR | Mouse (Mus musculus) | iPS cells | Impaired iPS cell generation | siRNA knockdown; overexpression | Loewer et al. 2010 |
| Mistral | Mouse (Mus musculus) | ES cells | Reduced expression of mesoderm lineage-associated genes | siRNA knockdown, FISH | Bertani et al. 2011 |
| lncRNA_ES1-3 | Human | ES cells | Downregulation of pluripotency markers; upregulation of lineage markers | siRNA knockdown | Ng et al. 2012 |

specific nuclear RNA–protein complexes. Such functions are conserved in human ES cells. A recent study focusing on differentiation of human ES cells into neurons identified three transcripts, lncRNA_ES1-3, that act in maintaining the pluripotent state (Ng et al. 2012). Knockdown of these lncRNAs by siRNA impairs pluripotency, as indicated by downregulation of pluripotency markers and upregulation of lineage markers. As with the 26 'K4-K36' lincRNAs, lncRNA_ES1-3 physically interact with chromatin modifiers of the Polycomb group. Surprisingly, they also appear to bind the pluripotency-associated transcription factor Sox2, suggesting that lncRNAs may also act as scaffolds for combinations of chromatin modifiers and transcription factors.

Collectively, these results implicate lncRNAs in the regulatory networks that maintain ES cell identity, potentially by assembling regulatory complexes of chromatin modifiers and/or transcription factors. The coding potential of all of these transcripts was only evaluated computationally, however, and so experimental evidence is still needed to verify that they function solely as noncoding RNA regulators.

Some lncRNAs are also involved in inducing ES cell pluripotency via reprogramming of somatic cells. Induced pluripotent stem (iPS) cells can be derived from terminally differentiated somatic cells by ectopic expression of key transcription factors such as Oct4, Nanog, Sox2, and c-Myc (see Stadtfeld and Hochedlinger 2010 for review). This cellular reprogramming is accompanied by extensive global remodeling of the epigenome (Hanna et al. 2010). Loewer et al. found that several lincRNAs contribute to this process of dedifferentiation (Loewer et al. 2010). Comparison of lincRNAs expressed in iPS cells versus those expressed in ES cells identified 10 that are specifically enriched in iPS cells. These lincRNAs also appear regulated by the pluripotency-associated master transcription factors Oct4 and Nanog, suggesting a functional role in the generation of iPS cells. In particular, inhibition of one such lincRNA, lincRNA-RoR, leads to a 2- to 8- fold decrease in iPS colony formation. This effect appears to be mediated by impaired growth and elevated apoptosis via p53. Conversely, over-expression results in a $\sim$2.5 fold increase in cellular reprogramming, a modest yet significant effect. These observations indicate that lncRNAs can modulate transcriptional programs associated with inducing or maintaining ES cell pluripotency, and that their impact on these processes can range from essential to subtle but detectable.

LncRNAs are also involved in modulating differentiation of ES cells. ES cell differentiation can be induced by treatment with retinoic acid (RA), which results in downregulation of pluripotency markers and activation of lineage-specific ones. These processes are mediated by epigenetic repressors belonging to the Polycomb group and by epigenetic activators belonging to the Trithorax group. A component of the latter, the H3K4 methyltransferase MLL1, interacts with lncRNA Mistral during activation of lineage-associated gene expression (Bertani et al. 2011). Mistral is an unspliced and polyadenylated 798 nt transcript upregulated during RA-induced ES cell differentiation. Knockdown of Mistral by siRNAs results in attenuated expression of broad-acting transcription factors that in turn activate genes associated with differentiation along the mesoderm lineage. This effect appears mediated by recruitment of the MLL1 epigenetic activator to the transcription factor loci via direct physical interaction with its methyltransferase domain. Hence, epigenetic modulation of gene expression via lncRNA cofactors seems to play a role during both ES cell pluripotency and differentiation.

## *3.3 lncRNAs as Regulators of Embryogenesis*

Differentiation of proliferating ES cells into early embryos requires precise temporal and spatial execution of multiple gene expression programs. The involvement of lncRNAs in modulating target gene expression predicts their involvement in commencing and executing these programs. Indeed, lncRNAs are essential to two of the earliest developmental programs during embryogenesis—dosage compensation and allelic imprinting (Fig. 4 and Table 3).

In order to equalize the dosage of X-linked genes between the sexes, early female mammalian embryos inactivate expression from one of the two copies of the X chromosome. This is achieved through epigenetic silencing of the entire chromosome mediated by a regulatory network of lncRNAs (see Lee 2011 for review). The best characterized of these is Xist, a polyadenylated transcript with multiple spliced isoforms that can reach $\sim$18–19 kB in mouse and human. Xist is exclusively expressed by the inactive X, from a region called the X inactivation

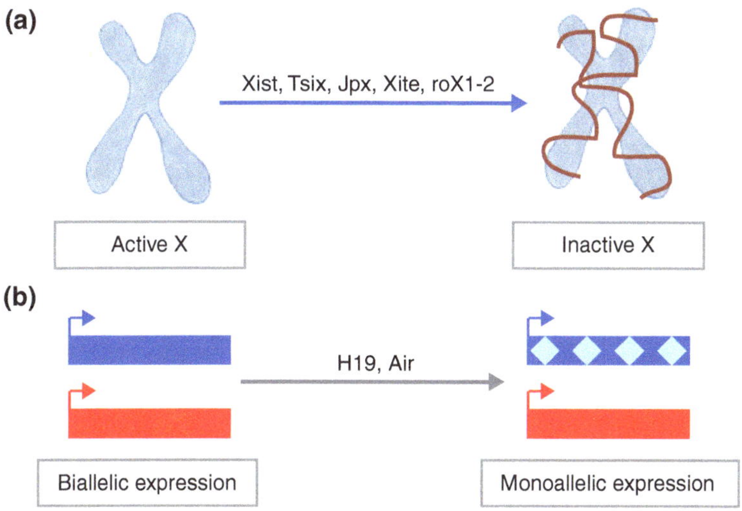


**Fig. 4** lncRNAs as regulators of embryogenesis. Several well-characterized lncRNAs act in developmental processes during early animal embryo development. **a** X epigenetic silencing is controlled by a network of lncRNAs that include Xist, Tsix, Jpx and Xite in mammals and roX1-2 in flies. **b** Imprinting of paternal alleles (blue) that influence growth during embryogenesis through the Igf2 network involve the lncRNAs H19 and Air

**Table 3** Examples of lncRNA regulators of embryogenesis

| Name | Organism | Expression | Loss of function phenotype | Experimental manipulations | References |
|---|---|---|---|---|---|
| Xist | Placental mammals | Embryonic and somatic cells | Mouse: epigenetic silencing of inactive X disrupted; embryonic lethal in females | Mouse knockout; directed mutagenesis; ectopic expression; siRNA knockdown; overexpression; FISH | Reviewed in Lee 2011 |
| Tsix | Placental mammals | Embryonic and somatic cells | Mouse: epigenetic repression of Xist disrupted; embryonic lethal in both sexes | Mouse knockout; directed mutagenesis; ectopic expression; shRNA knockdown; overexpression; FISH | Reviewed in Lee 2011 |
| Jpx | Placental mammals | Embryonic and somatic cells | Mouse: activation of Xist disrupted; embryonic lethal in females | Mouse knockout; directed mutagenesis; ectopic expression; shRNA knockdown; overexpression; FISH | Reviewed in Lee 2011 |
| Xite | Placental mammals | Embryonic and somatic cells | Mouse: downregulation of Tsix | Mouse knockout; directed mutagenesis; ectopic expression; overexpression; FISH | Reviewed in Lee 2011 |
| roX1-2 | Fruit fly (Drosophila melanogaster) | Male embryonic and somatic cells | Male embryonic lethality due to teduced expression of X-linked genes | Deletion; directed mutagenesis; ectopic expression; overexpression; FISH | Reviewed in Conrad and Akhtar 2011 |
| H19 | Placental mammals and marsupials | Fetal liver and fetal and adult cardiac and skeletal muscle; reactivated in cancer | Mouse: embryonic overgrowth due to disrupted imprining of maternal Igf2 | Mouse knockout; directed mutagenesis; ectopic expression; siRNA knockdown; overexpression; FISH | Reviewed in Gabory et al. 2010 |
| Air | Mouse (Mus musculus) and human | Fetal hear, lung and glial cells of developing brain | Mouse: embryonic growth defect due to disrupted imprinting of paternal Igf2r | Mouse knockout; directed mutagenesis; ectopic expression; FISH | Sleutels et al. 2002; Sleutels et al. 2003; Nagano et al. 2008 |

center (Xic), and is required for its silencing. After being transcribed Xist remains tethered to the Xic, an effect mediated by the YY1 RNA/DNA binding protein (Jeon and Lee 2011). Tethered Xist in turn recruits the chromatin repressor PRC2 through a structured RNA domain termed Repeat A (Zhao et al. 2008). PRC2 in turn facilitates the formation of heterochromatin via the histone modification H3K27me3, which spreads *in cis* throughout most of the X chromosome and thereby silences it. Thus, lncRNA Xist is essential for epigenetic silencing of the X chromosome during embryogenesis. Xist function is conserved in all placental mammals, despite showing limited conservation in primary sequence (Wutz 2011). Remarkably, marsupial mammals appear to have independently evolved the same function through an unrelated lncRNA (Grant et al. 2012).

Several other lncRNAs modulate X inactivation through their regulation of Xist expression (Lee 2011). For example, Tsix is transcribed antisense to the Xist locus from the active X, and its expression is anticorrelated with that of Xist. Transcription of Tsix leads to stable silencing of Xist *in cis* via recruitment of the DNA methyltransferase DNMT3A to the Xist promoter. Both Xist and Tsix are themselves regulated by the lncRNAs Jpx, and Xite, respectively. Jpx is required for Xist upregulation *in trans* at the inactive X, and its deletion is embryonically lethal in females. Xite, on the other hand, favors stable Tsix expression *in cis* at the active X. Collectively, these examples illustrate how a cascade of lncRNA interactions helps establish epigenetic states that in turn specify and maintain developmental fate.

As with mammals, flies also utilize lncRNA regulators for X chromosome dosage compensation during embryogenesis, although their compensation strategy is different. Instead of females silencing one of their X chromosomes, male flies must upregulate the majority of the genes on their single X chromosome. This is achieved by expression of two functionally redundant lncRNAs, roX1 and rox2, which direct the two-fold upregulation of most genes on the male X chromosome (see Conrad and Akhtar (2011) for review). These two lncRNAs directly associate with chromatin modifiers to form the dosage compensation complex (DCC). Binding of this complex to the X chromosome activates gene expression via acetylation of H4 histones. Double mutants in rox1 and rox2 experience reduced expression of X-linked genes, leading to male embryonic lethality. Both roX1 and roX2 mediate recruitment of the DCC to the X chromosome *in cis* or ectopically *in trans* by serving as structural components. Remarkably, additional lncRNAs expressed form the X chromosome can mediate recruitment of the DCC complex *in cis*. Thus, regulatory lncRNAs seem to mediate X-linked gene dosage throughout metazoans through epigenetic control.

Imprinting to ensure monoallelic expression is another developmental process mediated by lncRNAs during embryogenesis (see Barlow (2011) for review). This is typically achieved via epigenetic modification of promoter elements. For example, H19 controls embryonic imprinting of the maternal allele encoding the growth-regulator Igf2 (Gabory et al. 2010). H19 is a 2.3 kb lncRNA host to microRNA-675 that can be found in both nuclear and cytoplasmic compartments. In the nucleus, H19 is required for epigenetic silencing of Igf2, potentially via recruitment of the PRC2 repressor. In the cytoplasm, H19 has been proposed to downregulate Igf2 translation via sequestering mRNA binding-proteins that promote its translation. Lack of H19 causes embryonic overgrowth due to increased Igf2 dosage. In addition, H19 is reactivated in various cancers where it might influence tumor growth. Thus, H19 regulates growth during development and potentially disease by controlling Igf2 dosage. As with Xist, expression from the H19 locus is itself regulated by various other lncRNAs. Growth control is also regulated at the level of the Igf2 receptor, Igf2r, which is itself imprinted by another lncRNA, Air. Transcribed from the paternal allele in the second intron of the Igf2r locus, Air is essential for *cis*-imprinting of several genes on the paternal chromosome in a tissue-specific manner (Sleutels et al. 2002; Sleutels et al. 2003). This is mediated via recruitment of the histone methyltransferase G9a to target promoters (Nagano et al. 2008). Similarly, lncRNA Kcnq1ot1, a ~90 kb transcript expressed from the paternal chromosome, directs epigenetic silencing of multiple genes within the Kcnq1 domain (Mancini-Dinardo et al. 2006; Pandey et al. 2008). Kcnq1ot1 seems to recruit both G9a and PRC2 to repress expression of target loci *in cis,* analogous to the activity of Xist, H19, and Air. Collectively, these examples illustrate how epigenetic silencing mediated by lncRNAs plays an essential role during embryo growth and development.

### 3.4 Regulation of Hox Gene Expression by lncRNAs

Patterning of the body in developing animal embryos is regulated by the Hox family of genes (see Mallo et al. 2010 for review). These genes encode transcription factors that regulate a variety of developmental loci by binding to their regulatory elements via a protein domain known as the homeodomain. The genetic programs specified by these developmental loci in turn determine the body plan during embryogenesis. Precise temporal and spatial expression of Hox genes and accurate maintenance of their expression patterns are thus essential for animal development and cell fate determination. Consequently, Hox genes are subject to intensive regulation at both transcriptional and post-transcriptional levels (Pearson et al. 2005; Yekta et al. 2008). In addition to transcription factors and microRNAs, Hox gene clusters also encode hundreds of lncRNAs (Lipshitz et al. 1987; Rinn et al. 2007). Some of these lncRNAs play important roles in modulating the expression of Hox genes (Fig. 5 and Table 4).

Hox genes were first identified in the fruit fly *D. melanogaster* through mutations affecting segmental identities along the posterior–anterior body plan (Lewis 1978). Characterization of the function and regulation of the full range of fly Hox genes over the next decade led to the discovery of both the Polycomb group and Trithorax group of epigenetic regulators (see Ringrose and Paro 2004 for review). The Polycomb and Trithorax groups regulate Hox loci by maintaining their state of repressed or active transcription, respectively, through cell division cycles. They achieve this by establishing a repressed or active chromatin state throughout

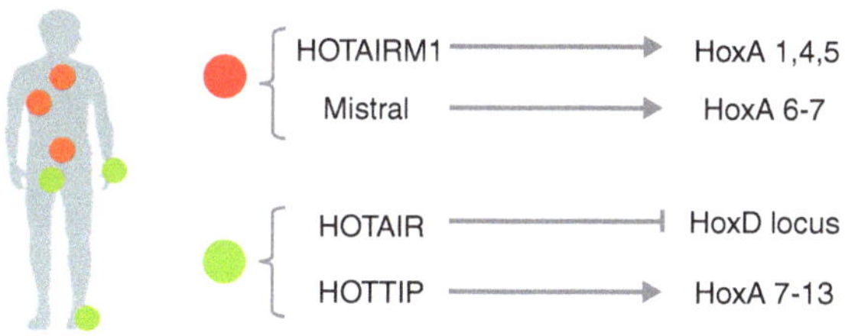


**Fig. 5** Regulation of Hox gene expression by lncRNAs. Differential expression of Hox genes across human body segments involves regulation by several lncRNAs. In cells with proximal and anterior positional identities (*red*), expression of HoxA genes 1–7 is promoted *in cis* by lncRNAs that include HOTAIRM1 and Mistral. In cells with distal and posterior positional identities (*green*), expression of HoxA genes 7–13 is promoted by HOTTIP *in cis*, while transcription of HoxD genes is repressed *in trans* by HOTAIR

cis-regulatory elements of their Hox loci targets that are a few hundred bases long and called Polycomb response elements (PREs). Close examination of PREs revealed that these elements are actually transcribed, and that the resulting lncRNAs exert regulatory functions (see Schmitt and Paro 2006 for review). Forcing transcription through silent PREs during embryogenesis switches their epigenetic state and leads to developmental abnormalities due to Hox gene mis-expression. The same phenotype is observed when transcription from active PREs is disrupted. Thus, production of lncRNAs mediates the epigenetic state at Hox loci PREs. In fact, the lncRNAs themselves appear to recruit Polycomb/Trithorax complexes to PREs, by remaining tethered to them and physically binding these complexes. These observations have led to a model whereby Polycomb/Trithorax regulators find their chromatin targets via direct interaction with the lncRNAs tethered to them (Hekimoglu and Ringrose 2009).

As with flies, regulation of Hox genes in mammals involves regulatory lncRNA components. There are 39 Hox genes in mammals, grouped into four chromosomal loci (HOXA to HOXD) that are expressed along the anterior–posterior axis of the body in a manner collinear with their genomic position from 3′ to 5′ of the cluster. Rinn et al. identified a 2.2 kb lncRNA called HOTAIR that can repress the HOXD locus *in trans* (Rinn et al. 2007; Tsai et al. 2010). HOTAIR is transcribed antisense to protein-coding genes at the HOXC cluster in cells with posterior and distal positional identities. Its knockdown results in upregulation of genes residing in the HOXD cluster, the strongest effect being a ~2-fold increase in HOXD10 expression. Such activation is accompanied by loss of epigenetic silencing as assayed by reduction in levels of H3K27me3. Repression of HOXD10 genes by HOTAIR is mediated by direct recruitment of PRC2 and of another chromatin modifying complex containing LSD1, a lysine demethylase which primarily targets H3K4. This role appears mediated by structural domains at the 5′ and 3′ ends of HOTAIR, consistent with greater evolutionary constraint on the inferred secondary structure than on the primary sequence among mammalian HOTAIR orthologs (He et al. 2011). Thus, the HOTAIR lncRNA acts to repress transcription

**Table 4** Examples of lncRNA regulators of Hox gene expression

| Name | Organism | Expression | Loss of function phenotype | Experimental manipulations | References |
|---|---|---|---|---|---|
| HOTAIR | Vertebrates | Posterior and distal tissues; upregulated in cancer | Mouse: derepression of HoxD locus; human: decrease in cancer invasiveness | Mouse knockout; directed mutagenesis; siRNA knockdown; overexpression; FISH | Rinn et al. 2007; Tsai et al. 2010; He et al. 2011; Gutschner and Diederichs 2012 |
| HOTTIP | Vertebrates | Posterior and distal tissues | Mouse: decreased expression of HoxA7-13; chicken: shortening of forelimb distal bone segments | siRNA and shRNA knockdown; ectopic expression; overexpression; FISH | Wang et al. 2011 |
| HOTAIRM1 | Placental mammals | Myeloid lineage; anterior and proximal tissues | Human: decreased activation of HoxA1,4,5 | siRNA and shRNA knockdown | Zhang et al. 2009 |
| Mistral | Mouse (Mus musculus) | ES cells; anterior and proximal tissues | Decreased expression of HoxA6-7 | siRNA knockdown, FISH | Bertani et al. 2011 |

of the HoxD locus via physical recruitment of chromatin modifiers *in trans*. Because HOTAIR recruits not only a Polycomb/Trithorax complex but also an unrelated chromatin modifier, this example laid the ground for an expanded model of lncRNAs as platforms for the assembly of functional combinations of chromatin modifiers in general (Koziol and Rinn 2010; Tsai et al. 2010). Recently HOTAIR has also been implicated in disease, as it is found overexpressed in a wide variety of cancers (Gutschner and Diederichs 2012). In breast and colorectal cancer, for example, HOTAIR appears to modulate tumor invasiveness by enhancing PRC2-mediated repression of genes that suppress metastasis (Gupta et al. 2010; Kogo et al. 2011). Therefore, HOTAIR plays a critical role during both development and cancer by helping specify gene expression programs.

In addition to repressing transcription, lncRNAs from mammalian Hox clusters can also facilitate transcriptional activation. Three lncRNAs from the HoxA cluster, HOTTIP, Mistral, and HOTAIRM1, have such capacity (Zhang et al. 2009; Bertani et al. 2011; Wang et al. 2011). HOTTIP resides in the 5′ tip of the HoxA locus. Although poorly expressed, this ~3.7 kb lncRNA can be specifically detected at distal/posterior sites in the embryo. The positive correlation between HOTTIP expression and that of its neighbors at the HoxA locus suggests that HOTTIP may modulate their activity. Consistent with this notion, inhibition of HOTTIP by siRNA results in 30–80 % reduction in the expression of the HoxA7-13 genes in a manner inversely proportional to their distance from HOTTIP. This reduction is associated with appearance of the repressive H3K27me3 chromatin mark and disappearance of the active H3K4me3 chromatin mark, accompanied by decreased occupancy of the WDR5/MLL1 complex, an epigenetic activator of the Trithorax group. Biochemical analysis revealed that WDR5 can specifically interact with HOTTIP and that this interaction causes target activation only when HOTTIP is physically positioned near them, as tested by tethering it to the promoter region of a report gene. Hence, HOTTIP helps maintain the active epigenetic state of the HoxA locus, and this effect depends on both direct association with the WDR5/MLL complex and immediate physical proximity. This is supported by detection of endogenous chromatin interactions between HOTTIP and target loci through chromosome conformation capture, and by the fact that its low copy number (<1 copy per cell measured by single-molecule RNA FISH) would limit significant activity *in trans*. To study HOTTIP function in vivo, Wang and colleagues injected retroviruses carrying shRNAs into the upper limb buds of early chicken embryos (Wang et al. 2011). Knockdown caused decreased expression of HoxA10-13, as expected, and this effect was most pronounced at the distal edge of developing limb buds, where the 5′ HoxA genes are most prominently expressed during normal conditions. Remarkably, by late embryonic stages this results in up to ~20 % reduction in distal limb bones, which exhibit notably abnormal morphology. Such dramatic phenotypes mirror those of mice lacking 5′ HoxA genes, indicating that Hox lncRNAs indeed contribute to organismal development by affecting Hox gene expression.

Cells at anterior and proximal locations express genes at the 3′ end of the HoxA locus rather than those at the 5′ tip. This is mediated by the other two lncRNA

**Fig. 6** lncRNAs during neural cell differentiation and brain development. Many lncRNAs are specifically enriched in the CNS and modulate differentiation of several cell types. **a** Development of excitatory neurons from neuron progenitors involves RMST and lncRNA_N1-3, whereas development of inhibitory interneurons is promoted by Evf2. **b** Differentiation of neural stem cells along the oligodendrocyte lineage can be promoted by lncRNA Nkx2.2AS. **c** Retinal cell development is modulated by several lncRNAs, including RNCR2, TUG1 and Vax2os1, which are required for proper formation of photoreceptor cells. **d** Two lncRNAs with conserved function from zebrafish to human (Cyrano and Megamind) are needed for proper brain and retina development ▶

activators, Mistral and HOTAIRM1. Mistral seems to recruit the WDR5/MLL1 complex to activate expression of its neighbors HoxA6 and HoxA7 (see Sect.3.2). As with HOTTIP, recruitment of the WDR5/MLL1 complex by Mistral can result in chromosome conformation changes that contribute to gene activation during cell differentiation. Transcription of the remaining 3′ HoxA genes, HoxA1-5, is influenced by HOTAIRM1 through an analogous mechanism. HOTAIRM1 was first characterized in the context of hematopoiesis, and so it is described in Sect. 3.7.

The examples reviewed here clearly indicate that, much like proteins and microRNAs, lncRNAs play important roles in repressing and activating target Hox genes. Consequently, lncRNA-mediated regulation can contribute to the precise temporal and spatial control of genes that specify the body plan in animals.

## *3.5 lncRNAs During Neural Cell Differentiation and Brain Development*

The development of neural tissues during animal embryogenesis involves a variety of cell differentiation processes executed under exquisite temporal and spatial control. Formation of the vertebrate central nervous system (CNS) alone involves the generation of millions of neurons with many distinct gene expression programs conferring distinct molecular and physiologic properties. There are two broad types of cells in this system: neurons and glia cells. These are generated from neural stem cells, which can be isolated from adult brain or derived from ES cells. As with the germ line, lncRNAs are strongly enriched in the CNS and play key roles during neural fate specification (Fig. 6 and Table 5).

The first clue concerning the importance of lncRNAs in neurogenesis came from the observation that in both fruit flies and mice hundreds of them are specifically expressed in the central nervous system (Inagaki et al. 2005; Mercer et al. 2008a). These include members of all known lncRNA subfamilies, such as intronic, antisense, and intergenic lncRNAs. In the mouse brain, detection by RNA FISH revealed that a great number of lncRNAs are expressed in specific cell types, neuroanatomical regions, and subcellular compartments. Such highly specific expression patterns suggested the possibility that some of these lncRNAs may

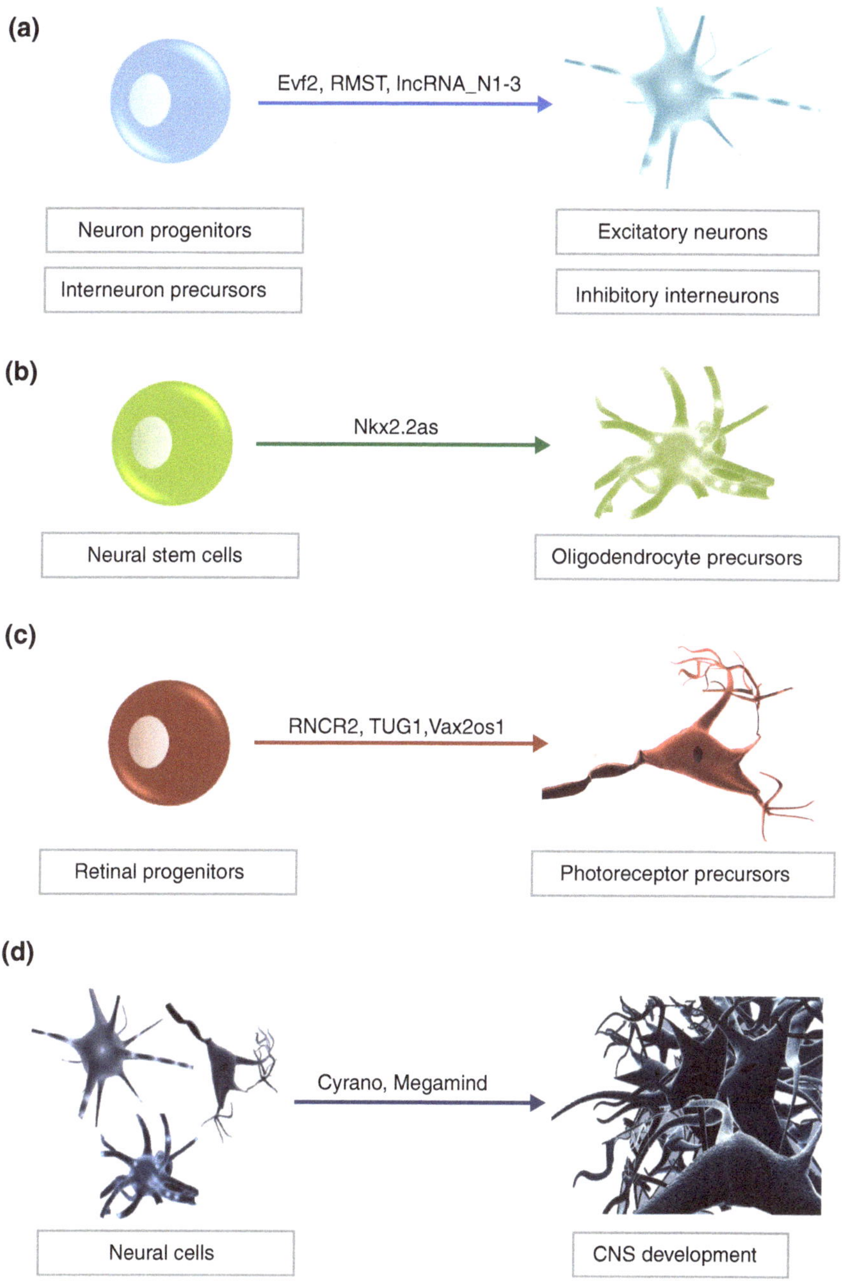
(a)
Evf2, RMST, lncRNA_N1-3
Neuron progenitors
Interneuron precursors
Excitatory neurons
Inhibitory interneurons
(b)
Nkx2.2as
Neural stem cells
Oligodendrocyte precursors
(c)
RNCR2, TUG1,Vax2os1
Retinal progenitors
Photoreceptor precursors
(d)
Cyrano, Megamind
Neural cells
CNS development

**Table 5** Examples of lncRNA regulators of neural cell differentiation and brain development

| Name | Organism | Expression | Loss of function phenotype | Experimental manipulations | References |
|---|---|---|---|---|---|
| Evf2 | Vertebrates | Embryonic ventral forebrain | Mouse: reduction of GABA-dependent neuronal circuitry due to lower number of GABAergic interneurons; synaptic inhibition | Mouse knockout; directed mutagenesis; ectopic expression; FISH | Feng et al. 2006; Bond et al. 2009 |
| Nkx2.2AS | Mouse (Mus musculus) | Neural stem cells | N/A | Overexpression; ectopic expression | Tochitani and Hayashizaki 2008 |
| RNCR2 | Vertebrates | Retinal progenitor cells | Mouse: increased differentiation into non-retinal lineages | shRNA knockdown; directed mutagenesis; ectopic expression; overexpression; FISH | Reviewed in Rapicavoli and Blackshaw 2009 |
| TUG1 | Placental mammals | Developing retina and brain | Mouse: disrupted photoreceptor formation due to impaired migration into outer nuclear layer and increased apoptosis | siRNA knockdown; FISH | Reviewed in Rapicavoli and Blackshaw 2009 |
| Vax2os1 | Placental mammals | Retinal progenitor cells | N/A | Directed mutagenesis; overexpression; FISH | Meola et al. 2012 |
| RMST; lncRNA_N1-3 | Human | Neuron progenitors | Impaired neuronal differentiation; downregulation of neuron markers; upregulation of glia markers | siRNA knockdown | Ng et al. 2012 |
| Cyrano; Megamind | Vertebrates | Central nervous system | Zebrafish: small head and eyes due to impaired neural tube and retinal development | Antisense oligo knockdown; directed mutagenesis; ectopic expression; FISH | Ulitsky et al. 2011 |

modulate the development and function of neural cell types. Consistent with this notion, transcriptome profiling during neurogenesis in mouse embryos confirmed that many lncRNAs are differentially expressed during neuronal-glial fate specification and during oligodendrocyte lineage maturation (Mercer et al. 2010). Functional studies have now characterized critical roles for several lncRNAs in modulating neural cell differentiation. For example, lncRNA Evf2 plays a well-characterized role during development of the hippocampus (Feng et al. 2006; Bond et al. 2009). Evf2 is a multiexonic and polyadenylated transcript expressed from the conserved intergenic region of the Dlx5 and Dlx6 loci. These loci encode homeodomain-containing transcription factors with critical roles in inhibitory interneuron differentiation and migration, as well as in limb patterning during development. Evf2 serves as a transcriptional activator of Dlx5, Dlx6, and Gad1 via direct recruitment of the Dlx2 and Mecp2 transcription factors to enhancer elements in the Dlx5/6 intergenic region. Ectopic expression of Evf2 increased expression of its targets, confirming a role as an RNA regulator with capacity to act *in trans*. Suppression of Evf2 in vivo by poly(A) site insertions reduced the number of GABAergic interneurons and compromised synaptic inhibition in the early postnatal hippocampus and dentate gyrus of mutant mice, a dramatic developmental phenotype. Although the numbers of GABAergic interneurons appeared to normalize in adult mutant mice, reduced synaptic inhibition persisted. Hence, Evf2 plays a critical role in the formation of GABA-dependent neuronal circuitry during early development of the hippocampus, by modulating the expression of key transcription factors that favor the GABAergic interneuron cell fate.

Analogously, the lncRNA Nkx2.2AS participates in neurogenesis by favoring differentiation of neural stem cells along the oligodendrocyte lineage (Tochitani and Hayashizaki 2008). Nkx2.2AS is a cytoplasmic transcript transcribed antisense to Nkx2.2, a master transcription factor of oligodendrocyte differentiation. Over-expression of Nkx2.2AS in cultured primary neural stem cells increased expression of Nkx2.2 by about 30 % and resulted in a modest increase in the formation of oligodendrocytes. Thus, Nkx2.2AS appears to favor the oligodendrocyte cell fate by enhancing Nkx2.2 expression, although no loss-of-function evidence is available as yet.

lncRNAs also play important roles during retinal cell development (see Rapicavoli and Blackshaw 2009 for review). For example, during embryogenesis the nuclear-retained lincRNA RNCR2 becomes specifically enriched in retinal progenitor cells. Knockdown by shRNAs resulted in differentiation of progenitor cells toward nonretinal cell lineages, such as amacrine cells, suggesting that RNCR2 is involved in retinal cell fate specification. The same effect was observed by mislocalization of RNCR2 to the cytoplasm, via fusion with an IRES-controlled GFP transgene, indicating that correct cellular localization of this lincRNA is important for its cellular function. RNCR2 seems to specifically interact with the SF1 splicing factor through conserved repeat sequences that resemble intron branch point motifs (Tsuiji et al. 2011). Binding of RNCR2 to SF1 in vitro can inhibit spliceosomal complex formation, suggesting that RNCR2 may exert its function by regulating splicing efficiency. Another well-characterized lncRNA regulator of retinal development is TUG1, a ~6.7 kb spliced and polyadenylated transcript that localizes to both nucleus and cytoplasm and is conserved throughout mammals. TUG1 is directly activated by Taurine, the master regulator of rod photoreceptor production. Downregulation of TUG1 by RNAi leads to disrupted photoreceptor formation due to impaired migration into the outer nuclear layer and increased apoptosis. Consistent with this phenotype, TUG1 is directly activated by p53 upon DNA damage and acts to repress a range of cell cycle genes via association with PRC2 (Guttman et al. 2009; Khalil et al. 2009). Analogously, Meola et al. (2012) recently reported that overexpression of the lncRNA Vax2os1 inhibited retinal progenitor cell proliferation. Vax2os1 is selectively expressed in the developing retina, and it appears to function through impairment of cell cycle progression and increased apoptosis.

Regulation by lncRNAs has also been studied during differentiation of human ES cells toward neuronal progenitor cells and ultimately neurons (Ng et al. 2012). About 35 lncRNAs were found to be upregulated during terminal neuron differentiation, suggesting potential roles in this process. Knockdown by siRNA of 4 of these, RMST and lncRNA_N1-3, resulted in significant changes of gene expression patterns and impairment of neuronal differentiation. Mechanistically, 3 of these lncRNAs appear to act in the regulation of chromatin state, as they are localized predominantly in the nucleus and bind the PRC2 complex. These examples indicate that as a group of gene expression regulators, lncRNAs play important yet diverse roles in neuronal differentiation both in culture and in vivo. The involvement of many lncRNAs in epigenetic control, and the fact that a large proportion of primate- and human-specific lncRNAs seem specifically enriched in the brain, predict that some might also be involved in maintaining proper neuronal function during complex physiological processes, such as long-term memory formation or behavioral patterns (Mercer et al. 2008b; Anguera et al. 2011; Lipovich et al. 2012).

Functional roles of lncRNAs during CNS development seem to be conserved from zebrafish to human. A recent study of hundreds of lincRNAs in the zebrafish *Danio rerio*, including 29 with detectable human orthologs, found 2 required for normal development of both brain and retina (Ulitsky et al. 2011). The first one, Cyrano, is a ~4.5 kb polyadenylated transcript conserved in mouse and humans that is expressed in brain, notochord, and subsequently spinal cord. Knockdown of Cyrano by antisense morpholinos caused small heads and eyes due in part to defects in neural tube opening and loss of retinal neuroD-positive cells. Remarkably, these defects could be rescued by ectopic expression of either zebrafish cyrano or its human or mouse orthologs. Cyrano harbors a 26 nt sequence highly conserved throughout vertebrates that mirrors a microRNA-7 binding site, suggesting that it might exert its function through microRNA regulation. The second lincRNA, Megamind, is a ~2.4 kb transcript antisense to an intron of birc6 that is specifically enriched in the brain. Knockdown of Megamind resulted in abnormal nervous system development such as smaller heads and eyes, enlarged brain ventricles (hydrocephalia), and loss of Neuro-D positive cells in the retina.

As with Cyrano, Megamind and its brain-specific expression are conserved in mouse and human, and its loss of function phenotype was rescued by either the zebrafish transcript or its human or mouse orthologs. Hence, lncRNA sequences, expression patterns and function during neural development appear conserved from zebrafish to human.

## 3.6 lncRNAs During Muscle Differentiation

Muscle cell differentiation is a well-characterized developmental program executed during both embryogenesis and adult tissue homeostasis. Many key transcription factors and microRNAs controlling the expression of genes involved in muscle growth, morphogenesis, and differentiation are well-characterized in both in vitro tissue culture and in vivo mouse models (see Braun and Gautel 2011 for review). In addition to protein and microRNA components, lncRNAs are also active regulators during muscle development (Fig. 7 and Table 6).

A number of lncRNAs are differentially expressed during differentiation of myoblasts into myotubes (Sunwoo, Dinger et al. 2009). Two of these, transcribed from the Neat1 locus, are strongly upregulated during myogenesis and play essential structural roles (Hutchinson et al. 2007; Clemson et al. 2009; Sasaki et al. 2009; Sunwoo et al. 2009). The Neat1 lncRNAs are single-exon, alternatively terminated and polyadenylated transcripts that are conserved throughout placental mammals. They localize to nuclear paraspeckles, granular structures that contain specific protein and RNA components and typically form during cell differentiation processes, although their function is presently unclear. Knockdown of Neat1 lncRNAs by antisense oligos disrupted existing paraspeckles and abolished their formation *de novo*, indicating an essential role as paraspeckle structural components. Nascent Neat1 lncRNAs nucleate formation of paraspeckles at their transcription sites by directly recruiting paraspeckle proteins, thereby being required for paraspeckle formation (Mao et al. 2011). Consistent with this function, mice

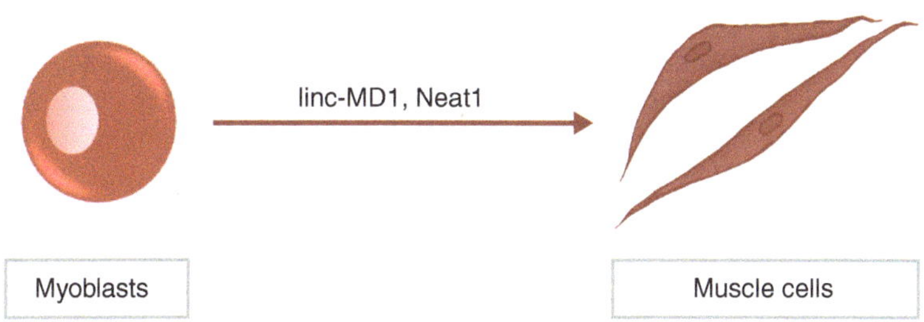


**Fig. 7** lncRNAs during muscle differentiation. Among lncRNAs differentially expressed during muscle development, two have been found to play essential roles. linc-MD1 is associated with activation of genes essential for muscle differentiation, whereas Neat1 is required for the formation of nuclear paraspeckles that accompanies myogenesis

**Table 6** Examples of lncRNA regulators of muscle differentiation

| Name | Organism | Expression | Loss of function phenotype | Experimental manipulations | References |
|---|---|---|---|---|---|
| linc-MD1 | Vertebrates | Myoblasts | Downregulation of genes asssociated with myogenesis | siRNA knockdown; directed mutagenesis; ectopic expression; overexpression | Cesana et al. 2011 |
| Neat1 | Placental mammals | Myoblasts | Disruption of nuclear paraspeckles | Atisense oligo knockdown; FISH | Sunwoo et al. 2009 |

deleted for Neat1 fail to form paraspeckles, but are viable and seem otherwise normal in terms of fertility, morphology, and proper differentiation of tissues that normally express Neat1 lncRNAs (Nakagawa et al. 2011). Thus, further research is needed to elucidate the function and biological relevance of paraspeckle assembly by Neat1 lncRNAs.

A recent report of cross-talk between lncRNA and microRNA function adds an interesting new layer of regulation to muscle cell fate determination. Cesana et al. characterized a muscle-specific lincRNA, linc-MD1, which inhibits two microRNAs important for muscle development, microRNA-133 and microRNA-135 (Cesana et al. 2011). linc-MD1 is an alternatively spliced, polyadenylated transcript that hosts microRNA-206 in one intron and microRNA-133b in one exon. Upon myoblast differentiation this small RNA-host lncRNA becomes activated by the master myogenic transcription factor MyoD. Unlike lincRNA regulators of epigenetic modification, however, linc-MD1 resides in the cytoplasm, suggesting that it may regulate cytoplasmic events. Inspection of its sequence revealed highly conserved binding sites for both microRNA-133 and microRNA-135. Functional studies indicated that linc-MD1 can "sponge" these two microRNAs during muscle differentiation and thus indirectly upregulate their mRNA targets, which include Mef2c and Maml1, which are required for normal muscle differentiation. Consistent with this function, inhibition of linc-MD1 compromises muscle differentiation, as assayed by reduced accumulation of myogenic markers. Overexpression, on the other hand, results in increased expression of these markers. Importantly, overexpression assays were conducted with a mutated linc-MD1 transcript from which microRNA-133b could not be released, indicating that microRNA-host lncRNA transcripts can have independent regulatory functions. The case of linc-MD1 illustrates a recently proposed model whereby endogenous RNAs can indirectly modulate each other by competing for the available pool of common microRNA regulators (Rubio-Somoza et al. 2011). Interestingly, linc-MD1 appears downregulated in Duchenne muscular dystrophy myoblasts, which are mutated in the dystrophin gene, and rescuing its levels via ectopic expression appears to partially restore normal myogenesis.

The opposite pattern is observed for another lncRNA, DBE-T, which is repressed in normal muscle cells but becomes active in facioscapulohumeral muscular dystrophy (FSHD) (Cabianca et al. 2012). FSHD is caused by a reduction in the copy number of the 3.3 kb repeat D4Z4. Under normal conditions, the D4Z4 repeat array is epigenetically silenced by the Polycomb group, resulting in a repressive chromatin state that leads to silencing of FSHD genes via long-range interactions. Under FSHD, loss of Polycomb-mediated silencing throughout the repeat array causes chromatin conformation changes that facilitate transcription of the upstream DBE-T locus. DBE-T lncRNA in turn recruits the Trithorax group protein Ash1L to the FSHD locus and coordinates de-repression of FSHD genes through long-range chromatin interactions. Thus, transcription of DBE-T mediates an epigenetic switch at the FSHD locus via direct recruitment of chromatin remodeling complexes. Interestingly, the FSHD locus shares several sequence features with Drosophila Polycomb/Trithorax response elements, which are also epigenetically switchable by virtue of lncRNA transcription (see Sect. 3.4). Therefore, roles for lncRNAs in driving chromatin architecture changes at Polycomb/Trithorax target loci that affect nearby gene expression are conserved from flies to human.

A recent study indicates that lncRNAs can also mediate mRNA decay pathways active during muscle differentiation. Using C2C12 myoblasts as an in vitro culture system, Gong et al. observed that two mRNA decay pathways, Staufen1-mediated mRNA decay (SMD) and nonsense-mediated mRNA decay (NMD), contribute to muscle differentiation by regulating the abundance of target mRNAs (Gong et al. 2009). Certain polyadenylated and cytoplasmic lncRNAs, termed ½-sbsRNAs, seem to trigger SMD by imperfect base-paring to the 3′ UTR of select target mRNAs through common Alu repeat elements. This lncRNA–mRNA interaction can recruit Staufen1, the key component of the SMD pathway, and thus lead to degradation of the targeted mRNA (Gong and Maquat 2011). ½-sbsRNAs are broadly expressed throughout human tissues, suggesting a ubiquitous role in mRNA decay. Hence, some cytoplasmic lncRNAs are able to modulate mRNA stability through the SMD pathway. The examples of linc-MD1 and ½-sbsRNAs provide evidence that in addition to regulating chromatin modification in the nucleus, some lncRNAs can also modulate microRNA activity and mRNA stability in the cytoplasm.

## *3.7 Modulation of Hematopoiesis by lncRNAs*

Hematopoiesis, the development process by which mature blood cells are generated from primary progenitors, is essential in all animals. In healthy humans, about two million erythrocytes must be generated every second to replace those lost by senescence, and overall numbers need to be maintained within a narrow physiological range. All of the hematopoietic effector cells (erythrocytes, myelocytes, and lymphocytes) are derived from hematopoietic stem cells within the fetal liver or the adult bone marrow through highly coordinated lineage specification and differentiation. Hematopoietic multipotent and lineage-determined progenitor cells

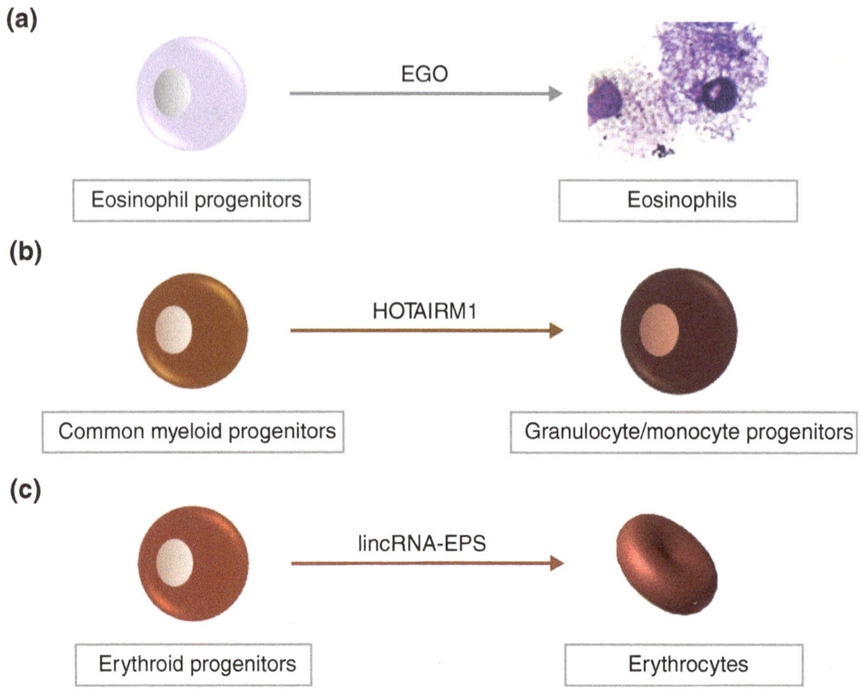


**Fig. 8** Modulation of hematopoiesis by lncRNAs. The differentiation of cells along several hematopoietic lineages is modulated by lncRNAs. **a** In the eosinophil lineage, lncRNA EGO is important for activating key regulators of eosinophil development. **b** Differentiation of common myeloid progenitors into the precursors of the granulocyte and monocyte lineages involves upregulation of HOTAIRM1. **c** In the erythroid lineage, lincRNA-EPS is essential for preventing apoptosis during red blood cell maturation

can be readily isolated using cell surface markers and have been extensively studied, making the hematopoietic system one of the best paradigms for studying cell lineage specification and differentiation in mammals (see Orkin and Zon 2008 for review). In addition to well-characterized transcription factors and microRNAs, recent evidence indicates that lncRNAs also modulate hematopoiesis, particularly during the development of cells of the myeloid lineages (Fig. 8 and Table 7).

An intronic lncRNA, EGO, was the first characterized hematopoietic lncRNA. EGO modulates the development of eosinophils (Wagner et al. 2007), one of the immune system components that plays a role in parasitic immunity and allergic diseases such as asthma. EGO is a conserved transcript derived from an intron of the ITPR1 gene. It is normally expressed in human CD34+ hematopoietic stem cells and becomes upregulated during their differentiation into eosinophils. Biochemical analysis indicates that the transcript is noncoding, as it does not associate with ribosomes. Knockdown of EGO by siRNAs in cultured CD34+ progenitors impaired the expression of genes important for eosinophil development, such as

**Table 7** Examples of lncRNA regulators of hematopoiesis

| Name | Organism | Expression | Loss of function phenotype | Experimental manipulations | References |
|---|---|---|---|---|---|
| EGO | Human | Eosinophils | Downregulation of major basic protein and eosinophil derived neurotoxin | siRNA knockdown | Wagner et al. 2007 |
| HOTAIRM1 | Placental mammals | Myeloid progenitors | Human: downregulation of genes associated with myelopoiesis | siRNA and shRNA knockdown | Zhang et al. 2009 |
| lincRNA-EPS | Placental mammals | Erythroblasts | Mouse: increased apoptosis; impaired enucleation | Directed mutagenesis; shRNA knockdown; ectopic expression | Hu et al. 2011 |

major basic protein and eosinophil-derived neurotoxin. These results suggest that EGO can modulate the differentiation of cells along the eosinophil lineage.

lncRNAs are also implicated in the regulation of myelopoiesis, the formation of granulocytes and monocytes. Zhang et al. (2009) identified a lincRNA (HOTAIRM1) in the HOXA cluster that is dramatically upregulated during retinoic acid—induced granulocytic differentiation of myeloid progenitor cells. Transcribed from the HOXA1/2 intergenic region, HOTAIRM1 is about 500 nt in length and does not associate with ribosomes. It exhibits coordinated expression with HoxA1 and HoxA2 along the body plan, suggesting that it might be involved in maintaining their active state. Knockdown of HOTAIRM1 inhibits RA-induced HoxA1 and HoxA4 activation during myeloid differentiation and specifically impairs the expression of several markers of differentiated myeloid cells, such as CD11b and CD18. This effect may be mediated through its interaction with various chromatin modifiers (Guttman et al. 2011). Hence, HOTAIRM1 modulates myelopoiesis potentially by regulating the epigenetic state of neighboring genes at the HoxA locus.

Recently, our group found one lincRNA that plays an essential role in the maturation of red blood cells (Hu et al. 2011). We performed transcriptome profiling on primary mouse erythroid cells at different developmental stages and found that hundreds of lncRNAs are differentially expressed during red blood cell differentiation. Among these we characterized one, lincRNA-EPS, which is specifically enriched in erythroid cells during terminal differentiation. lincRNA-EPS is a $\sim$2.5 kb capped and polyadenylated transcript that is alternatively spliced and resides in the nucleus. It becomes dramatically induced during the terminal differentiation of mouse erythroid cells from their progenitors both in vivo and in vitro. Knockdown by shRNAs resulted in elevated apoptosis and severely compromised differentiation. Conversely, ectopic expression protected erythroid progenitors from apoptosis triggered by erythropoietin starvation. These effects are mediated by a highly conserved region in the 3' terminal exon of lincRNA-EPS, which is sufficient for its antiapoptotic activity. Importantly, disruption of the putative short ORFs within the transcript does not alter function. Thus, this erythroid-specific lincRNA is required for red blood cell maturation by inhibiting apoptosis. Mechanistic studies suggest that LincRNA-EPS regulates apoptosis by repressing expression of a number of proapoptotic proteins, most prominently the caspase activating adaptor protein Pycard. Collectively, these examples illustrate that lncRNAs fulfill diverse regulatory functions that shape the development of hematopoietic cells of different lineages. Such functional capacities suggest that lncRNA dysregulation may be a factor contributing to blood disorders caused by developmental deficiencies.

## 3.8 lncRNAs and Maintenance of Adult Tissue Homeostasis

Besides regulating tissue development during embryogenesis, lncRNAs are also involved in the maintenance of mature tissues (Fig. 9 and Table 8). In a recent study, Kretz et al. (Kretz et al. 2012) demonstrated that lncRNA ANCR is required for suppressing differentiation of somatic progenitors in epidermal tissue, which typically are renewed in a weekly basis. Using high throughput transcriptome sequencing, they studied lncRNAs expressed during terminal differentiation of keratinocytes, adipocytes, and osteoblasts. Among more than 1000 dynamically expressed lncRNAs, they focused on one, ANCR, which shows dramatically reduced expression upon differentiation of all three cell types. ANCR is an intergenic 855 nt transcript that hosts both an intronic microRNA and an intronic snoRNA, which are present in the preprocessed but not in the mature transcript. Depletion in keratinocyte progenitors of the mature ANCR transcript by siRNAs resulted in upregulation of the epidermal differentiation program, including induction of early and late epidermal marker genes such as keratin 1 and filaggrin, in the absence of any differentiation stimuli. The same effects were observed upon knockdown of ANCR in regenerated, organotypic epidermal tissue that recapitulates normal epidermis structure and histology. Remarkably, loss of ANCR function within this recapitulated tissue also led to production of differentiation markers in the progenitor-rich epidermal basal layer compartment.

A similar role is fulfilled by another lncRNA, PINC, which is enriched in progenitor cells within the mammary gland (Ginger et al. 2001; Ginger et al. 2006; Shore et al. 2012). PINC is an alternatively spliced and polyadenylated transcript that can reside in the nucleus or in the cytoplasm depending on the stage of the cell cycle and that is expressed in luminal and alveolar progenitors within the mammary gland. Physiologically, PINC is strongly upregulated throughout pregnancy and becomes downregulated during late pregnancy and early lactation, when

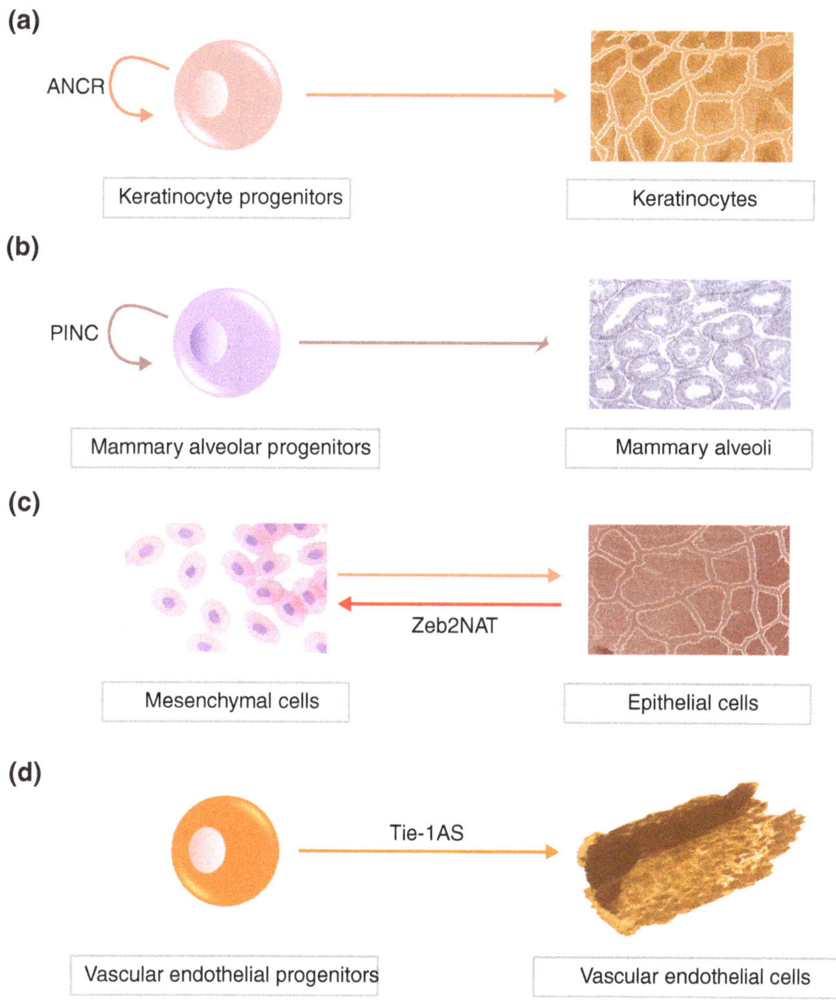


**Fig. 9** lncRNAs and maintenance of adult tissue homeostasis. lncRNAs are also involved in homeostasis of various tissues during adult life. **a** Within epidermis, differentiation of progenitors into keratinocytes is regularly suppressed by ANCR. **b** Within the mammary gland, development of alveoli is coordinated by PINC, which prevents premature differentiation during pregnancy. **c** Execution of EMT involves activation of Zeb2NAT as a key step to enforce global transcriptional changes. **d** Tie-1AS plays a role during vascular tube development by modulating the formation of the endothelial lining

alveolar cells derived from the mammary epithelium undergo terminal differentiation into milk-producing cells. Consistent with this pattern, PINC is activated in vivo by local stimulation of mammary gland tissue with estrogen and progesterone, and becomes downregulated in vitro when immortalized mammary epithelial cells are induced to differentiate by treatment with lactogenic hormones. In these cells, inhibition of PINC by siRNAs in the absence of differentiation stimuli

**Table 8** Examples of lncRNA regulators of adult tissue homeostasis

| Name | Organism | Expression | Loss of function phenotype | Experimental manipulations | References |
|---|---|---|---|---|---|
| ANCR | Human | Keratinocyte progenitors | Derepression of differentiation genes; differentiation within progenitor-specific epidermal layer | siRNA and shRNA knockdown | Kretz et al. 2012 |
| PINC | Placental mammals | Alveolar progenitors | Mouse: enhanced lactogenic differentiation | siRNA knockdown; overexpression; FISH | Ginger et al. 2001; Ginger et al. 2006; Shore et al. 2012 |
| Zeb2NAT | Mouse (Mus musculus) and human | Epithelial cells | N/A | Ectopic overexpression | Beltran et al. 2008 |
| Tie-1AS | Vertebrates | Vascular endothelial progenitors | Zebrafish: disrupts vascular tube integrity | Knockdown; ectopic exprssion | Li et al. 2010 |

affected their survival by limiting their cell cycle progression, whereas in the presence of such stimuli PINC knockdown seems to favor differentiation along the alveolar lineage. Overexpression of PINC, on the other hand, blocked alveolar differentiation. These effects seem mediated by repressing the expression of genes associated with alveologenesis via direct association with PRC2, likely through the coordinately expressed PRC2 subunit RbAp46. Consistent with such roles in alveolar differentiation during pregnancy and lactation, PINC is conserved throughout the mammalian lineage. Thus, as with ANCR, PINC acts to prevent adult lineage-determined progenitors from differentiating, likely via epigenetic repression of the differentiation program of their specific lineage.

Morphogenetic differentiation is another developmental process crucial for proper embryogenesis and adult tissue homeostasis. Regulation by lncRNAs has been documented in two important morphogenetic processes: the epithelial-to-mesenchymal transition (EMT) and the formation of the vascular endothelium. EMT is essential during embryogenesis for formation of mesoderm and the neural tube, and during epithelial cancer formation it is associated with increased proliferation and metastasis. During EMT, epithelial cells that adhere to one another in ordered layers via E-cadherin revert to a migratory and undifferentiated fate characteristic of mesenchymal cells. Beltran et al. found that an antisense lncRNA in the ZEB2 locus, Zeb2NAT, acts as a positive regulator of EMT (Beltran et al. 2008). Zeb2 is normally repressed in epithelial cells, and its activation along with that of Snail and Zeb1 can lead to EMT via downregulation of E-cadherin, which enforces global gene expression changes. Upregulation of Zeb2NAT by Snail1 can lead to Zeb2 activation via an unusual mechanism. The Zeb2NAT lncRNA appears to directly bind the Zeb2 pre-mRNA to prevent splicing of an intron containing an internal ribosome entry site. Retention of this site is in turn required for efficient translation of Zeb2 and thus for activation of the EMT differentiation program. Interestingly, Snail1 also represses E-cadherin by binding to its promoter, thus promoting EMT both directly and indirectly via Zeb2NAT-mediated translation of Zeb2.

In an analogous example, Li et al. (2010) described an antisense lncRNA, Tie-1AS, which seems to play a role during formation of the vascular endothelium, the inner lining of blood vessels. Tie-1AS is an evolutionary conserved, ~800 nt transcript expressed antisense to the Tie-1 gene, which encodes a cell surface tyrosine kinase receptor for angiopoietin ligands. Tie-1AS appears to regulate the mRNA levels of Tie-1 by formation of a Tie-1 and Tie1-AS RNA duplex. Transient transfection of Tie-AS disrupts vascular tube formation both in zebrafish in vivo and in human vascular endothelial progenitors in culture. Consistent with this phenotype, the ratio of Tie-1 mRNA versus Tie-1AS seems altered in pathological human vascular samples. This study suggests that modulation of Tie-1 levels by Tie-AS may be required for proper maintenance of vascular endothelial cells. However, loss-of-function experiments are needed to further clarify the physiological role of this antisense lncRNA. Hence, as with Zeb2NAT, direct interaction of antisense lncRNA Tie-1AS with its target mRNA serves to modulate somatic tissue morphogenesis during development and potentially during disease.

# 4 Future Perspectives and Outstanding Challenges

The rapid development and increasing affordability of techniques for large-scale transcriptome profiling over the past 10 years has yielded increasingly growing collections that already contain hundreds to thousands of lncRNA loci for every eukaryote examined (Bertone et al. 2004; Carninci et al. 2005; Birney et al. 2007; Kapranov et al. 2007; Guttman et al. 2009; Khalil et al. 2009; Cabili et al. 2011; Ulitsky et al. 2011; Derrien et al. 2012; Nam and Bartel 2012; Qu and Adelson 2012). Given the wide application of these technologies, it is likely that many more will be uncovered in the coming years. Characterization of the biological functions of these lncRNAs, however, has only been explored in detail for a small percentage of them. Ultimately, not all of the identified lncRNAs may be functional, and some of them may even be unproductive transcriptional noise (Struhl 2007; Ebisuya et al. 2008). Nonetheless, mounting evidence points toward an increasing number of lncRNAs with recognized biological functions in genome regulation under specific physiological and pathological contexts (Amaral and Mattick 2008; Dinger et al. 2008; Ponting et al. 2009; Wilusz et al. 2009; Orom and Shiekhattar 2011; Wapinski and Chang 2011; Rinn and Chang 2012). The selected examples from the previous sections demonstrated that lncRNAs are active players in the regulation of cell differentiation throughout the life cycle of eukaryotes, from the formation of unicellular gametes, through the development of specialized multicellular tissues, to the maintenance of these tissues in adult life. These observations provide tantalizing support to the hypothesis that increasing numbers of ncRNA regulators are responsible for the increase in developmental complexity from yeast to human. However, there still remain many unanswered questions about the origin, properties, mechanisms, and phenotypic consequence of lncRNAs implicated in development. How many RNA transcripts truly function as coding, non-coding or both? What properties should be used to group lncRNAs into functionally coherent families? How do lncRNAs achieve selective binding of protein, DNA or RNA partners in vivo? And ultimately, how important is lncRNA regulation for in vivo organismal development? Over the next sections, we present our own perspectives on some of these important questions.

## *4.1 Molecular Mechanisms of lncRNA Regulators of Cell Differentiation*

lncRNAs can modulate gene expression via diverse mechanisms (Fig. 1) (Wang and Chang 2011; Guttman and Rinn 2012; Moran et al. 2012). Of those lncRNAs currently implicated in cell differentiation processes, many seem to direct gene expression through recruitment of chromatin modifiers. This is consistent with multiple observations that chromatin modifiers, such as PRC2, can associate with a diversity of noncoding transcripts (Khalil et al. 2009; Zhao et al. 2010; Guttman et al. 2011; Ng et al. 2012). Interestingly, lncRNAs can function as scaffolds to recruit histone modification complexes (Koziol and Rinn 2010; Tsai et al. 2010), and lncRNAs in general exhibit richer tissue specificity than protein-coding genes (Cabili et al. 2011; Derrien et al. 2012). It thus seems tempting to speculate that one major, though not exclusive, function of lncRNAs during development is to promote, in a cell-type specific manner, assembly of specific combinations of ubiquitously expressed chromatin modifiers in target genomic regions, thereby controlling the epigenetic state with exquisite spatial and temporal precision. However, case by case analysis will be required to dissect in detail how specific binding to chromatin modifier partners is achieved in vivo, what sequence properties enable lncRNAs to then target these partners to specific areas in the genome, and what role does local chromatin conformation play in modulating these interactions.

An expanding toolbox of molecular approaches is rapidly becoming available to address these and other questions about lncRNA molecular mechanisms. Investigating these typically begins by first assessing subcellular localization. Cellular fractionation followed by RNA detection can be a cost-effective method to broadly distinguish nuclear-acting versus cytoplasmic lncRNAs. In addition, direct visualization of lncRNA by RNA-FISH can provide a high-resolution picture of localization to even smaller subcellular structures, such as the nucleolus, paraspeckles, or other granule RNA structures (Yamashita et al. 1998; Kloc et al. 2005; Nagano et al. 2008; Clemson et al. 2009; Sasaki et al. 2009; Sunwoo et al. 2009). In combination with other methods, such as DNA-FISH, immunofluorescence, or fluorescent protein tagging, RNA-FISH can also be used to detect lncRNAs in specific chromosomes or in regions of silent or active chromatin (Redrup et al. 2009; Reinius et al. 2010; Sexton et al. 2012), and can also be used to examine multimerization potential and colocalization with specific RNA or protein partners (Khalil et al. 2009; Chakraborty et al. 2012).

Importantly, *de novo* protein partners of lncRNAs can be identified via RNA-mediated pull-downs (Rinn et al. 2007; Huarte et al. 2010). Several powerful assays have also been recently developed to determine the genomic binding sites of nuclear-acting lncRNAs (Chu et al. 2011; Simon et al. 2011). These and other assays will greatly facilitate the exploration of lncRNA mechanisms within cell differentiation systems. Judging by the constant development and broad application of these assays, we predict that such exploration will greatly advance in the coming years.

## *4.2 Integrating lncRNAs to Known Regulatory Networks of Cell Differentiation*

Differentiation programs are exquisitely controlled at every stage by complex networks that respond to developmental and environmental signals. The examples discussed in this chapter argue that lncRNAs are likely to be integrated as key

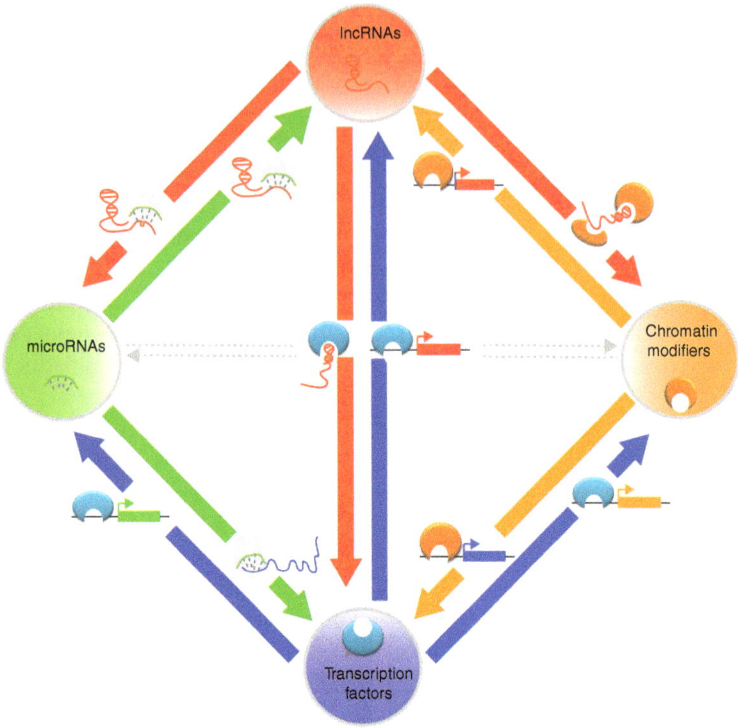


**Fig. 10** Integrating lncRNAs to known regulatory networks of cell differentiation. Integrating lncRNA functions with those of microRNAs, TFs and chromatin modifiers during cell differentiation will first require exploring their mutual regulatory relationships. Examples of some of these relationships are depicted. lncRNAs (*red* RNAs and *red arrows*). microRNAs (*green* RNAs and *green arrows*). TFs (*blue protein* and *blue arrows*). Chromatin modifiers (*orange proteins* and *orange arrows*). lncRNAs may regulate microRNAs or TFs as target site decoys, and they may also associate with chromatin modifying complexes as structural components in RNP complexes or as guides and tethers to their chromatin targets. microRNAs post-transcriptionally regulate RNA transcripts from TF, chromatin modifier or lncRNA loci by directly base-pairing to short stretches of sequence. TF control transcription of all the other regulators by directly binding their promoters. Similarly, chromatin modifiers enforce epigenetic states influencing expression from all the other network components. Not depicted are regulatory relationships between microRNAs and chromatin modifier components

components of these regulatory networks, on par with transcription factors, chromatin modifiers and microRNAs. Precisely how lncRNAs should be integrated can be answered by first exploring their regulatory relationships with other components (Fig. 10).

Expression of many lncRNAs modulating cell differentiation programs is indeed controlled by key transcription factors that serve as "master regulators" of those programs. This is the case for lncRNAs discussed in previous sections that are involved in ES cell pluripotency maintenance, rod photoreceptor differentiation, and muscle development. Interestingly, some lncRNA transcripts seem to physically bind transcription factors (Ng et al. 2012), suggesting that mutual modulation between lncRNAs and transcription factors is possible. Further progress in identifying the global binding sites of key transcription factors during cell differentiation, as well as the protein interactome of lncRNAs, will be of great help in reconstructing networks involving lncRNAs. Simply intersecting such datasets with transcriptome profiling along developmental pathways will be of great use in identifying candidate lncRNAs for functional studies.

Our present understanding of the relationship between lncRNAs and chromatin modifiers is governed by the constant observation of functionally productive physical associations between these factors. In fact, the prevalence of such functional partnerships throughout eukaryotes, as evidenced in the many examples presented here, has changed our understanding of how chromatin modifiers themselves operate. This is best illustrated in the case of Polycomb group proteins, which are now believed to recognize their target loci not through interactions with DNA but through interactions with RNA tethered to the DNA (Schmitt and Paro 2006; Hekimoglu and Ringrose 2009; Zhao et al. 2010). A growing body of evidence now suggests that this model might extend to several other classes of epigenetic modifiers (Koziol and Rinn 2010; Tsai et al. 2010; Spitale et al. 2011; Guttman and Rinn 2012). Thus, lncRNAs may be integrated into regulatory networks involving chromatin modifiers by serving as structural components, guides, and/or physical tethers. However, care should be placed in assuming such functions. Physical association by itself does not prove function, and detailed studies such as structure–function mapping are required for demonstrating functional relevance of lncRNA-chromatin modifier associations.

Several studies have also proposed that certain lncRNAs and microRNAs can regulate each other at the post-transcriptional level (Franco-Zorrilla et al. 2007b; Cesana et al. 2011; Karreth et al. 2011; Salmena et al. 2011; Ulitsky et al. 2011). Global identification of lncRNA targets of microRNAs remains in its earliest stages, however (Jeggari et al. 2012). Thus, identifying microRNAs and lncRNAs with complementary expression patterns during cell differentiation may generate candidate lncRNA-microRNA regulatory pairs to be tested for integration into regulatory networks. Such studies may not only serve to define such networks, but also to expand our understanding on how they contribute to development.

In comparing the role of lncRNAs with those of other factors involved in cell differentiation processes, it is important to note that, as with microRNAs, the biological effects of many lncRNAs tend to be rather mild, with 1.5–2-fold change in the expression of target loci upon lncRNA perturbation being a somewhat typical result. This may be in part due to limitations in achieving efficient knockdown of lncRNAs by current si/shRNA approaches. Alternatively, it may indicate that lncRNAs primarily act to fine tune target gene expression, much like microRNAs. In vivo knockout models of lncRNAs may thus be required to discriminate between these two possibilities, as discussed in the next section.

Compared to known transcription factors, chromatin modifiers, and microRNA regulators involved in cell differentiation, lncRNAs seem to employ a wider diversity of molecular mechanisms to modulate gene expression of their targets at the level of transcription, translation, and stability (Fig. 1). Therefore, it is not surprising that during cell differentiation lncRNAs may cooperate with, or sequester away, any of the other regulatory components to ensure precise gene expression at both the transcriptional and post-transcriptional levels.

## *4.3 In Vivo functions of lncRNAs*

Although we know that perturbation of many lncRNAs results in phenotypic changes during differentiation of in vitro cultured cells, our knowledge of the in vivo functions of lncRNAs remains limited. Several lncRNA-altered animals have been generated to bridge this gap in knowledge. Pioneering studies in non-mammalian vertebrate models have established essential developmental roles for conserved lncRNAs. For example, knockdown of lincRNAs Cyrano and Megamind severely impact CNS development in zebrafish, and such deficiencies can be rescued with their mouse and human orthologs (Ulitsky et al. 2011). Similarly, knockdown of HOTTIP in chicken embryos results in shortening and bending of distal bones (Wang et al. 2011). In vivo phenotypes in mouse knockout models, however, seem either more subtle or not immediately obvious. The strongest example so far is mice deleted for Evf2, which are delayed in forming GABAergic interneurons during early hippocampus development and thus exhibit compromised synaptic inhibition capacity (Bond et al. 2009). Similarly, strong effects are observed in male mice deleted for the X-linked Tsx, which show reduced fertility due to elevated apoptosis during spermatogenesis; they also display enhanced hippocampal short-term memory (Anguera et al. 2011). Examples of more moderate phenotypes have been found for H19 and Air, which regulate embryonic and early postnatal growth. Deleting H19, which mediates maternal imprinting of the growth regulator Igf2, results in embryonic weight increases of 10–20 % (Leighton et al. 1995; Ripoche et al. 1997; Wutz et al. 2001). Similarly, deleting Air, which is required for paternal imprinting of the Igf2 receptor Igf2, changes embryonic weight by about 20 % (Wutz et al. 2001). Conversely, deletion of Neat1 or Neat2, structural components of paraspeckles, causes no obvious phenotypes (Nakagawa et al. 2011; Eissmann et al. 2012; Nakagawa et al. 2012; Zhang et al. 2012). Both Neat1 and Neat2 are conserved throughout vertebrates and have been implicated in disease, but mice lacking them are viable, fertile, and exhibit normal histology in tissues normally enriched for Neat1 and/or Neat2. Similarly, a mouse deleted for the entire HoxC cluster, which contains HOTAIR, appears to show no obvious developmental defects (Schorderet and Duboule 2011).

The simplest explanation for these discrepancies is that as with many mRNAs and microRNAs, lncRNAs are functionally redundant and the loss of one can be readily compensated for by others with redundant function. Uncovering in vivo functions for some lncRNAs will thus require simultaneous mutation of previously identified or predicted redundant genes. Another explanation could be that, as with microRNAs, many lncRNAs function primarily to fine-tune gene expression, and so extreme conditions, such as physiological stress, are needed to bring about phenotypic consequences of their absence. These possibilities are not mutually exclusive, and they highlight the need for further investigation of lncRNAs under informative physiological conditions.

## References


Amaral, P. P., & M. B. Clark, et al. (2011). lncRNAdb: A reference database for long noncoding RNAs. *Nucleic Acids Research 39*(Database issue), D146–D151.

Amaral, P. P., Dinger, M. E., et al. (2008). The eukaryotic genome as an RNA machine. *Science, 319*(5871), 1787–1789.

Amaral, P. P., & Mattick, J. S. (2008). Noncoding RNA in development. *Mammalian Genome, 19*(7–8), 454–492.

Anguera, M. C., Ma, W., et al. (2011). Tsx produces a long noncoding RNA and has general functions in the germline, stem cells, and brain. *PLoS Genetics, 7*(9), e1002248.

Banfai, B., Jia, H., et al. (2012). Long noncoding RNAs are rarely translated in two human cell lines. *Genome Research, 22*(9), 1646–1657.

Barlow, D. P. (2011). Genomic imprinting: A mammalian epigenetic discovery model. *Annual Review of Genetics, 45*, 379–403.

Beltran, M., Puig, I., et al. (2008). A natural antisense transcript regulates Zeb2/Sip1 gene expression during Snail1-induced epithelial-mesenchymal transition. *Genes & Development, 22*(6), 756–769.

Berretta, J., & Morillon, A. (2009). Pervasive transcription constitutes a new level of eukaryotic genome regulation. *EMBO Reports, 10*(9), 973–982.

Bertani, S., Sauer, S., et al. (2011). The noncoding RNA Mistral activates Hoxa6 and Hoxa7 expression and stem cell differentiation by recruiting MLL1 to chromatin. *Molecular Cell, 43*(6), 1040–1046.

Bertone, P., Stolc, V., et al. (2004). Global identification of human transcribed sequences with genome tiling arrays. *Science, 306*(5705), 2242–2246.

Birney, E., Stamatoyannopoulos, J. A., et al. (2007). Identification and analysis of functional elements in 1% of the human genome by the ENCODE pilot project. *Nature, 447*(7146), 799–816.

Blackshaw, S., Harpavat, S., et al. (2004). Genomic analysis of mouse retinal development. *PLoS Biology, 2*(9), E247.

Bond, A. M., Vangompel, M. J., et al. (2009). Balanced gene regulation by an embryonic brain ncRNA is critical for adult hippocampal GABA circuitry. *Nature Neuroscience, 12*(8), 1020–1027.

Brannan, C. I., Dees, E. C., et al. (1990). The product of the H19 gene may function as an RNA. *Molecular and Cellular Biology, 10*(1), 28–36.

Braun, T., & Gautel, M. (2011). Transcriptional mechanisms regulating skeletal muscle differentiation, growth and homeostasis. *Nature Reviews Molecular Cell Biology, 12*(6), 349–361.

Brockdorff, N., Ashworth, A., et al. (1992). The product of the mouse Xist gene is a 15 kb inactive X-specific transcript containing no conserved ORF and located in the nucleus. *Cell, 71*(3), 515–526.

Brown, C. J., Hendrich, B. D., et al. (1992). The human XIST gene: analysis of a 17 kb inactive X-specific RNA that contains conserved repeats and is highly localized within the nucleus. *Cell, 71*(3), 527–542.

Cabianca, D. S., Casa, V., et al. (2012). A long ncRNA links copy number variation to a polycomb/trithorax epigenetic switch in FSHD muscular dystrophy. *Cell, 149*(4), 819–831.
Cabili, M. N., Trapnell, C., et al. (2011). Integrative annotation of human large intergenic noncoding RNAs reveals global properties and specific subclasses. *Genes & Development, 25*(18), 1915–1927.
Carninci, P., Kasukawa, T., et al. (2005). The transcriptional landscape of the mammalian genome. *Science, 309*(5740), 1559–1563.
Cesana, M., Cacchiarelli, D., et al. (2011). A long noncoding RNA controls muscle differentiation by functioning as a competing endogenous RNA. *Cell, 147*(2), 358–369.
Chakraborty, D., Kappei, D., et al. (2012). Combined RNAi and localization for functionally dissecting long noncoding RNAs. *Nature Methods, 9*(4), 360–362.
Chen, H. M., Rosebrock, A. P., et al. (2012). Repression of meiotic genes by antisense transcription and by Fkh2 transcription factor in Schizosaccharomyces pombe. *PLoS ONE, 7*(1), e29917.
Chooniedass-Kothari, S., Emberley, E., et al. (2004). The steroid receptor RNA activator is the first functional RNA encoding a protein. *FEBS Letters, 566*(1–3), 43–47.
Chu, C., Qu, K., et al. (2011). Genomic maps of long noncoding RNA occupancy reveal principles of RNA-chromatin interactions. *Molecular Cell, 44*(4), 667–678.
Clemson, C. M., Hutchinson, J. N., et al. (2009). An architectural role for a nuclear noncoding RNA: NEAT1 RNA is essential for the structure of paraspeckles. *Molecular Cell, 33*(6), 717–726.
Cocquet, J., Pannetier, M., et al. (2005). Sense and antisense Foxl2 transcripts in mouse. *Genomics, 85*(5), 531–541.
Conrad, T., & Akhtar, A. (2011). Dosage compensation in Drosophila melanogaster: epigenetic fine-tuning of chromosome-wide transcription. *Nature Reviews Genetics, 13*(2), 123–134.
David, L., Huber, W., et al. (2006). A high-resolution map of transcription in the yeast genome. *Proceedings of the National Academy of Science U S A, 103*(14), 5320–5325.
De Santa, F., Barozzi, I., et al. (2010). A large fraction of extragenic RNA pol II transcription sites overlap enhancers. *PLoS Biology, 8*(5), e1000384.
Derrien, T., Johnson, R., et al. (2012). The GENCODE v7 catalog of human long noncoding RNAs: Analysis of their gene structure, evolution, and expression. *Genome Research, 22*(9), 1775–1789.
Diez-Roux, G., Banfi, S., et al. (2011). A high-resolution anatomical atlas of the transcriptome in the mouse embryo. *PLoS Biology, 9*(1), e1000582.
Ding, D. Q., Okamasa, K., et al. (2012a). Meiosis-specific noncoding RNA mediates robust pairing of homologous chromosomes in meiosis. *Science, 336*(6082), 732–736.
Ding, J., Lu, Q., et al. (2012b). A long noncoding RNA regulates photoperiod-sensitive male sterility, an essential component of hybrid rice. *Proceedings of the National Academy of Science USA, 109*(7), 2654–2659.
Dinger, M. E., Amaral, P. P., et al. (2008a). Long noncoding RNAs in mouse embryonic stem cell pluripotency and differentiation. *Genome Research, 18*(9), 1433–1445.
Dinger, M. E., Pang, K. C., et al. (2008b). Differentiating protein-coding and noncoding RNA: Challenges and ambiguities. *PLoS Computational Biology, 4*(11), e1000176.
Duret, L., Chureau, C., et al. (2006). The Xist RNA gene evolved in eutherians by pseudogenization of a protein-coding gene. *Science, 312*(5780), 1653–1655.
Ebisuya, M., Yamamoto, T., et al. (2008). Ripples from neighbouring transcription. *Nature Cell Biology, 10*(9), 1106–1113.
Eissmann, M., & Gutschner, T. et al. (2012). Loss of the abundant nuclear non-coding RNA MALAT1 is compatible with life and development. *RNA Biology, 9*(8), 1076–1087.
Feng, J., Bi, C., et al. (2006). The Evf-2 noncoding RNA is transcribed from the Dlx-5/6 ultraconserved region and functions as a Dlx-2 transcriptional coactivator. *Genes & Development, 20*(11), 1470–1484.
Franco-Zorrilla, J. M., Valli, A., et al. (2007a). Target mimicry provides a new mechanism for regulation of microRNA activity. *Nature Genetics, 39*(8), 1033–1037.

Franco-Zorrilla, J. M., Valli, A., et al. (2007b). Target mimicry provides a new mechanism for regulation of microRNA activity. *Nature Genetics, 39*(8), 1033–1037.
Gabory, A., Jammes, H., et al. (2010). The H19 locus: role of an imprinted non-coding RNA in growth and development. *BioEssays, 32*(6), 473–480.
Galindo, M. I., Pueyo, J. I., et al. (2007). Peptides encoded by short ORFs control development and define a new eukaryotic gene family. *PLoS Biology, 5*(5), e106.
Gelfand, B., Mead, J., et al. (2011). Regulated antisense transcription controls expression of cell-type-specific genes in yeast. *Molecular and Cellular Biology, 31*(8), 1701–1709.
Ghildiyal, M., & Zamore, P. D. (2009). Small silencing RNAs: An expanding universe. *Nature Reviews Genetics, 10*(2), 94–108.
Ginger, M. R., Gonzalez-Rimbau, M. F., et al. (2001). Persistent changes in gene expression induced by estrogen and progesterone in the rat mammary gland. *Molecular Endocrinology, 15*(11), 1993–2009.
Ginger, M. R., Shore, A. N., et al. (2006). A noncoding RNA is a potential marker of cell fate during mammary gland development. *Proceedings of the National Academy of Science USA, 103*(15), 5781–5786.
Gong, C., Kim, Y. K., et al. (2009). SMD and NMD are competitive pathways that contribute to myogenesis: Effects on PAX3 and myogenin mRNAs. *Genes & Development, 23*(1), 54–66.
Gong, C., & Maquat, L. E. (2011). lncRNAs transactivate STAU1-mediated mRNA decay by duplexing with 3' UTRs via Alu elements. *Nature, 470*(7333), 284–288.
Gorodkin, J., & Hofacker, I. L. (2011). From structure prediction to genomic screens for novel non-coding RNAs. *PLoS Computational Biology, 7*(8), e1002100.
Grant, J., Mahadevaiah, S. K., et al. (2012). Rsx is a metatherian RNA with Xist-like properties in X-chromosome inactivation. *Nature, 487*(7406), 254–258.
Grunwald, D., Singer, R. H., et al. (2011). Nuclear export dynamics of RNA-protein complexes. *Nature, 475*(7356), 333–341.
Gupta, R. A., Shah, N., et al. (2010). Long non-coding RNA HOTAIR reprograms chromatin state to promote cancer metastasis. *Nature, 464*(7291), 1071–1076.
Gutschner, T., & Diederichs, S. (2012). The hallmarks of cancer: A long non-coding RNA point of view. *RNA Biology, 9*(6), 703–719.
Guttman, M., Amit, I., et al. (2009). Chromatin signature reveals over a thousand highly conserved large non-coding RNAs in mammals. *Nature, 458*(7235), 223–227.
Guttman, M., Donaghey, J., et al. (2011). lincRNAs act in the circuitry controlling pluripotency and differentiation. *Nature, 477*(7364), 295–300.
Guttman, M., Garber, M., et al. (2010). Ab initio reconstruction of cell type-specific transcriptomes in mouse reveals the conserved multi-exonic structure of lincRNAs. *Nature Biotechnology, 28*(5), 503–510.
Guttman, M., & Rinn, J. L. (2012). Modular regulatory principles of large non-coding RNAs. *Nature, 482*(7385), 339–346.
Hanna, J. H., Saha, K., et al. (2010). Pluripotency and cellular reprogramming: Facts, hypotheses, unresolved issues. *Cell, 143*(4), 508–525.
He, S., Liu, S., et al. (2011). The sequence, structure and evolutionary features of HOTAIR in mammals. *BMC Evolutionary Biology, 11*, 102.
Hekimoglu, B., & Ringrose, L. (2009). Non-coding RNAs in polycomb/trithorax regulation. *RNA Biology, 6*(2), 129–137.
Heo, J. B., & Sung, S. (2011). Vernalization-mediated epigenetic silencing by a long intronic noncoding RNA. *Science, 331*(6013), 76–79.
Hongay, C. F., Grisafi, P. L., et al. (2006). Antisense transcription controls cell fate in Saccharomyces cerevisiae. *Cell, 127*(4), 735–745.
Hu, W., Yuan, B., et al. (2011). Long noncoding RNA-mediated anti-apoptotic activity in murine erythroid terminal differentiation. *Genes & Development, 25*(24), 2573–2578.
Huarte, M., Guttman, M., et al. (2010). A large intergenic noncoding RNA induced by p53 mediates global gene repression in the p53 response. *Cell, 142*(3), 409–419.

Hube, F., Guo, J., et al. (2006). Alternative splicing of the first intron of the steroid receptor RNA activator (SRA) participates in the generation of coding and noncoding RNA isoforms in breast cancer cell lines. *DNA and Cell Biology, 25*(7), 418–428.

Hutchinson, J. N., Ensminger, A. W., et al. (2007). A screen for nuclear transcripts identifies two linked noncoding RNAs associated with SC35 splicing domains. *BMC Genomics, 8*, 39.

Huttenhofer, A., Schattner, P., et al. (2005). Non-coding RNAs: Hope or hype? *Trends in Genetics, 21*(5), 289–297.

Ietswaart, R., Wu, Z., et al. (2012). Flowering time control: Another window to the connection between antisense RNA and chromatin. *Trends in Genetics, 28*(9), 445–453.

Inagaki, S., Numata, K., et al. (2005). Identification and expression analysis of putative mRNA-like non-coding RNA in Drosophila. *Genes to Cells, 10*(12), 1163–1173.

Ingolia, N. T., Ghaemmaghami, S., et al. (2009). Genome-wide analysis in vivo of translation with nucleotide resolution using ribosome profiling. *Science, 324*(5924), 218–223.

Ingolia, N. T., Lareau, L. F., et al. (2011). Ribosome profiling of mouse embryonic stem cells reveals the complexity and dynamics of mammalian proteomes. *Cell, 147*(4), 789–802.

Jacob, F., & Monod, J. (1961). Genetic regulatory mechanisms in the synthesis of proteins. *Journal of Molecular Biology, 3*, 318–356.

Jacquier, A. (2009). The complex eukaryotic transcriptome: Unexpected pervasive transcription and novel small RNAs. *Nature Reviews Genetics, 10*(12), 833–844.

Jeggari, A., Marks, D. S., et al. (2012). miRcode: A map of putative microRNA target sites in the long non-coding transcriptome. *Bioinformatics, 28*(15), 2062–2063.

Jenny, A., Hachet, O., et al. (2006). A translation-independent role of oskar RNA in early Drosophila oogenesis. *Development, 133*(15), 2827–2833.

Jeon, Y., & Lee, J. T. (2011). YY1 tethers Xist RNA to the inactive X nucleation center. *Cell, 146*(1), 119–133.

Kapranov, P., Cheng, J., et al. (2007a). RNA maps reveal new RNA classes and a possible function for pervasive transcription. *Science, 316*(5830), 1484–1488.

Kapranov, P., Willingham, A. T., et al. (2007b). Genome-wide transcription and the implications for genomic organization. *Nature Reviews Genetics, 8*(6), 413–423.

Karreth, F. A., Tay, Y., et al. (2011). In vivo identification of tumor- suppressive PTEN ceRNAs in an oncogenic BRAF-induced mouse model of melanoma. *Cell, 147*(2), 382–395.

Khalil, A. M., Guttman, M., et al. (2009). Many human large intergenic noncoding RNAs associate with chromatin-modifying complexes and affect gene expression. *Proceedings of the National Academy of Science USA, 106*(28), 11667–11672.

Kim Guisbert, K. S., Zhang, Y., et al. (2012). Meiosis-induced alterations in transcript architecture and noncoding RNA expression in S. *cerevisiae*. *RNA, 18*(6), 1142–1153.

Kim, T. K., Hemberg, M., et al. (2010). Widespread transcription at neuronal activity-regulated enhancers. *Nature, 465*(7295), 182–187.

Kloc, M., Bilinski, S., et al. (2007). Organization of cytokeratin cytoskeleton and germ plasm in the vegetal cortex of Xenopus laevis oocytes depends on coding and non-coding RNAs: Three-dimensional and ultrastructural analysis. *Experimental Cell Research, 313*(8), 1639–1651.

Kloc, M., Wilk, K., et al. (2005). Potential structural role of non-coding and coding RNAs in the organization of the cytoskeleton at the vegetal cortex of Xenopus oocytes. *Development, 132*(15), 3445–3457.

Kogo, R., Shimamura, T., et al. (2011). Long noncoding RNA HOTAIR regulates polycomb-dependent chromatin modification and is associated with poor prognosis in colorectal cancers. *Cancer Research, 71*(20), 6320–6326.

Koziol, M. J., & Rinn, J. L. (2010). RNA traffic control of chromatin complexes. *Current Opinion in Genetics & Development, 20*(2), 142–148.

Kretz, M., Webster, D. E., et al. (2012). Suppression of progenitor differentiation requires the long noncoding RNA ANCR. *Genes & Development, 26*(4), 338–343.

Lardenois, A., Liu, Y., et al. (2011). Execution of the meiotic noncoding RNA expression program and the onset of gametogenesis in yeast require the conserved exosome subunit Rrp6. *Proceedings of the National Academy of Science USA, 108*(3), 1058–1063.

Lee, J. T. (2011). Gracefully ageing at 50, X-chromosome inactivation becomes a paradigm for RNA and chromatin control. *Nature Reviews Molecular Cell Biology, 12*(12), 815–826.

Leighton, P. A., Ingram, R. S., et al. (1995). Disruption of imprinting caused by deletion of the H19 gene region in mice. *Nature, 375*(6526), 34–39.

Lewis, E. B. (1978). A gene complex controlling segmentation in Drosophila. *Nature, 276*(5688), 565–570.

Li, K., Blum, Y., et al. (2010). A noncoding antisense RNA in tie-1 locus regulates tie-1 function in vivo. *Blood, 115*(1), 133–139.

Li, L., Wang, X., et al. (2006). Genome-wide transcription analyses in rice using tiling microarrays. *Nature Genetics, 38*(1), 124–129.

Li, T., Vu, T. H., et al. (2002). An imprinted PEG1/MEST antisense expressed predominantly in human testis and in mature spermatozoa. *Journal of Biological Chemistry, 277*(16), 13518–13527.

Li, Y. M., Franklin, G., et al. (1998). The H19 transcript is associated with polysomes and may regulate IGF2 expression in trans. *Journal of Biological Chemistry, 273*(43), 28247–28252.

Lin, M. F., Deoras, A. N., et al. (2008). Performance and scalability of discriminative metrics for comparative gene identification in 12 Drosophila genomes. *PLoS Computational Biology, 4*(4), e1000067.

Lin, M. F., Jungreis, I., et al. (2011). PhyloCSF: A comparative genomics method to distinguish protein coding and non-coding regions. *Bioinformatics, 27*(13), i275–i282.

Lipovich, L., & Dachet, F. et al. (2012). Activity-dependent human brain coding/non-coding gene regulatory networks. *Genetics, 192*(3):1133–1148.

Lipshitz, H. D., Peattie, D. A., et al. (1987). Novel transcripts from the Ultrabithorax domain of the bithorax complex. *Genes & Development, 1*(3), 307–322.

Liu, F., Marquardt, S., et al. (2010). Targeted 3' processing of antisense transcripts triggers Arabidopsis FLC chromatin silencing. *Science, 327*(5961), 94–97.

Loewer, S., Cabili, M. N., et al. (2010). Large intergenic non-coding RNA-RoR modulates reprogramming of human induced pluripotent stem cells. *Nature Genetics, 42*(12), 1113–1117.

Maenner, S., Blaud, M., et al. (2010). 2-D structure of the A region of Xist RNA and its implication for PRC2 association. *PLoS Biology, 8*(1), e1000276.

Mallo, M., Wellik, D. M., et al. (2010). Hox genes and regional patterning of the vertebrate body plan. *Developmental Biology, 344*(1), 7–15.

Mancini-Dinardo, D., Steele, S. J., et al. (2006). Elongation of the Kcnq1ot1 transcript is required for genomic imprinting of neighboring genes. *Genes & Development, 20*(10), 1268–1282.

Mao, Y. S., Sunwoo, H., et al. (2011). Direct visualization of the co-transcriptional assembly of a nuclear body by noncoding RNAs. *Nature Cell Biology, 13*(1), 95–101.

Marques, A. C., & Ponting, C. P. (2009). Catalogues of mammalian long noncoding RNAs: Modest conservation and incompleteness. *Genome Biology, 10*(11), R124.

Mattick, J. S. (2004). RNA regulation: A new genetics? *Nature Reviews Genetics, 5*(4), 316–323.

Meller, V. H., Wu, K. H., et al. (1997). roX1 RNA paints the X chromosome of male Drosophila and is regulated by the dosage compensation system. *Cell, 88*(4), 445–457.

Meola, N., Pizzo, M., et al. (2012). The long noncoding RNA Vax2os1 controls the cell cycle progression of photoreceptor progenitors in the mouse retina. *RNA, 18*(1), 111–123.

Mercer, T. R., Dinger, M. E., et al. (2008a). Noncoding RNAs in long-term memory formation. *Neuroscientist, 14*(5), 434–445.

Mercer, T. R., Dinger, M. E., et al. (2009). Long non-coding RNAs: Insights into functions. *Nature Reviews Genetics, 10*(3), 155–159.

Mercer, T. R., Dinger, M. E., et al. (2008b). Specific expression of long noncoding RNAs in the mouse brain. *Proceedings of the National Academy of Sciences USA, 105*(2), 716–721.

Mercer, T. R., Qureshi, I. A., et al. (2010). Long noncoding RNAs in neuronal-glial fate specification and oligodendrocyte lineage maturation. *BMC Neuroscience, 11*, 14.
Miura, F., Kawaguchi, N., et al. (2006). A large-scale full-length cDNA analysis to explore the budding yeast transcriptome. *Proceedings of the National Academy of Sciences USA, 103*(47), 17846–17851.
Moran, V. A., & Perera, R. J. et al. (2012). Emerging functional and mechanistic paradigms of mammalian long non-coding RNAs. *Nucleic Acids Research, 40*(14):6391–6400.
Mortazavi, A., Williams, B. A., et al. (2008). Mapping and quantifying mammalian transcriptomes by RNA-Seq. *Nature Methods, 5*(7), 621–628.
Nagalakshmi, U., Wang, Z., et al. (2008). The transcriptional landscape of the yeast genome defined by RNA sequencing. *Science, 320*(5881), 1344–1349.
Nagano, T., Mitchell, J. A., et al. (2008). The Air noncoding RNA epigenetically silences transcription by targeting G9a to chromatin. *Science, 322*(5908), 1717–1720.
Nakagawa, S., & Ip, J. Y. et al. (2012). Malat1 is not an essential component of nuclear speckles in mice. *RNA, 18*(8):1487–1499.
Nakagawa, S., Naganuma, T., et al. (2011). Paraspeckles are subpopulation-specific nuclear bodies that are not essential in mice. *Journal of Cell Biology, 193*(1), 31–39.
Nam, J. W., & Bartel, D. (2012). Long non-coding RNAs in C. *elegans*. *Genome Research, 22*(12), 2529–2540.
Ng, S. Y., Johnson, R., et al. (2012). Human long non-coding RNAs promote pluripotency and neuronal differentiation by association with chromatin modifiers and transcription factors. *EMBO Journal, 31*(3), 522–533.
Novikova, I. V., Hennelly, S. P., et al. (2012). Structural architecture of the human long non-coding RNA, steroid receptor RNA activator. *Nucleic Acids Research, 40*(11), 5034–5051.
Okamura, K., & Lai, E. C. (2008). Endogenous small interfering RNAs in animals. *Nature Reviews Molecular Cell Biology, 9*(9), 673–678.
Orkin, S. H., & Zon, L. I. (2008). Hematopoiesis: An evolving paradigm for stem cell biology. *Cell, 132*(4), 631–644.
Orom, U. A., & Shiekhattar, R. (2011). Noncoding RNAs and enhancers: Complications of a long-distance relationship. *Trends in Genetics, 27*(10), 433–439.
Ozsolak, F., & Milos, P. M. (2011). RNA sequencing: Advances, challenges and opportunities. *Nature Reviews Genetics, 12*(2), 87–98.
Pandey, R. R., Mondal, T., et al. (2008). Kcnq1ot1 antisense noncoding RNA mediates lineage-specific transcriptional silencing through chromatin-level regulation. *Molecular Cell, 32*(2), 232–246.
Parker, B. J., Moltke, I., et al. (2011). New families of human regulatory RNA structures identified by comparative analysis of vertebrate genomes. *Genome Research, 21*(11), 1929–1943.
Pauli, A., Rinn, J. L., et al. (2011). Non-coding RNAs as regulators of embryogenesis. *Nature Reviews Genetics, 12*(2), 136–149.
Pearson, J. C., Lemons, D., et al. (2005). Modulating Hox gene functions during animal body patterning. *Nature Reviews Genetics, 6*(12), 893–904.
Perocchi, F., & Xu, Z. Y. et al. (2007). Antisense artifacts in transcriptome microarray experiments are resolved by actinomycin D. *Nucleic Acids Research 35*(19), e128.
Pheasant, M., & Mattick, J. S. (2007). Raising the estimate of functional human sequences. *Genome Research, 17*(9), 1245–1253.
Png, K. J., Yoshida, M., et al. (2011). MicroRNA-335 inhibits tumor reinitiation and is silenced through genetic and epigenetic mechanisms in human breast cancer. *Genes & Development, 25*(3), 226–231.
Pollard, K. S., Salama, S. R., et al. (2006). An RNA gene expressed during cortical development evolved rapidly in humans. *Nature, 443*(7108), 167–172.
Ponjavic, J., Ponting, C. P., et al. (2007). Functionality or transcriptional noise? Evidence for selection within long noncoding RNAs. *Genome Research, 17*(5), 556–565.

Ponting, C. P., Oliver, P. L., et al. (2009). Evolution and functions of long noncoding RNAs. *Cell, 136*(4), 629–641.

Prasanth, K. V., & Spector, D. L. (2007). Eukaryotic regulatory RNAs: An answer to the 'genome complexity' conundrum. *Genes & Development, 21*(1), 11–42.

Qu, Z. P., & Adelson, D. L. (2012). Bovine ncRNAs are abundant, primarily intergenic, conserved and associated with regulatory genes. *PLoS One 7*(8), e42638.

Rapicavoli, N. A., & Blackshaw, S. (2009). New meaning in the message: Noncoding RNAs and their role in retinal development. *Developmental Dynamics, 238*(9), 2103–2114.

Ravasi, T., Suzuki, H., et al. (2006). Experimental validation of the regulated expression of large numbers of non-coding RNAs from the mouse genome. *Genome Research, 16*(1), 11–19.

Redrup, L., Branco, M. R., et al. (2009). The long noncoding RNA Kcnq1ot1 organises a lineage-specific nuclear domain for epigenetic gene silencing. *Development, 136*(4), 525–530.

Reinius, B., Shi, C., et al. (2010). Female-biased expression of long non-coding RNAs in domains that escape X-inactivation in mouse. *BMC Genomics, 11*, 614.

Ringrose, L., & Paro, R. (2004). Epigenetic regulation of cellular memory by the Polycomb and Trithorax group proteins. *Annual Review of Genetics, 38*, 413–443.

Rinn, J. L., & Chang, H. Y. (2012). Genome regulation by long noncoding RNAs. *Annual Review of Biochemistry, 81*, 145–166.

Rinn, J. L., Kertesz, M., et al. (2007). Functional demarcation of active and silent chromatin domains in human HOX loci by noncoding RNAs. *Cell, 129*(7), 1311–1323.

Ripoche, M. A., Kress, C., et al. (1997). Deletion of the H19 transcription unit reveals the existence of a putative imprinting control element. *Genes & Development, 11*(12), 1596–1604.

Rubio-Somoza, I., Weigel, D., et al. (2011). ceRNAs: miRNA target mimic mimics. *Cell, 147*(7), 1431–1432.

Salmena, L., Poliseno, L., et al. (2011). A ceRNA hypothesis: the Rosetta Stone of a hidden RNA language? *Cell, 146*(3), 353–358.

Sasaki, Y. T., Ideue, T., et al. (2009). MENepsilon/beta noncoding RNAs are essential for structural integrity of nuclear paraspeckles. *Proceedings of the National Academy of Sciences USA, 106*(8), 2525–2530.

Sato, M., Shinozaki-Yabana, S., et al. (2001). The fission yeast meiotic regulator Mei2p undergoes nucleocytoplasmic shuttling. *FEBS Letters, 499*(3), 251–255.

Schmitt, S., & Paro R. (2006). RNA at the steering wheel. *Genome Biology 7*(5), 218.

Schorderet, P., & Duboule, D. (2011). Structural and functional differences in the long non-coding RNA hotair in mouse and human. *PLoS Genetics, 7*(5), e1002071.

Selinger, D. W., Cheung, K. J., et al. (2000). RNA expression analysis using a 30 base pair resolution Escherichia coli genome array. *Nature Biotechnology, 18*(12), 1262–1268.

Sexton, T., Yaffe, E., et al. (2012). Three-dimensional folding and functional organization principles of the Drosophila genome. *Cell, 148*(3), 458–472.

Sheik Mohamed, J., Gaughwin, P. M., et al. (2010). Conserved long noncoding RNAs transcriptionally regulated by Oct4 and Nanog modulate pluripotency in mouse embryonic stem cells. *RNA, 16*(2), 324–337.

Shimada, T., Yamashita, A., et al. (2003). The fission yeast meiotic regulator Mei2p forms a dot structure in the horse-tail nucleus in association with the sme2 locus on chromosome II. *Molecular Biology of the Cell, 14*(6), 2461–2469.

Shiraki, T., Kondo, S., et al. (2003). Cap analysis gene expression for high-throughput analysis of transcriptional starting point and identification of promoter usage. *Proceedings of the National Academy of Sciences USA, 100*(26), 15776–15781.

Shore, A. N., Kabotyanski, E. B., et al. (2012). Pregnancy-Induced Noncoding RNA (PINC) associates with polycomb repressive complex 2 and regulates mammary epithelial differentiation. *PLoS Genetics, 8*(7), e1002840.

Simon, M. D., Wang, C. I., et al. (2011). The genomic binding sites of a noncoding RNA. *Proceedings of the National Academy of Sciences USA, 108*(51), 20497–20502.

Sleutels, F., Tjon, G., et al. (2003). Imprinted silencing of Slc22a2 and Slc22a3 does not need transcriptional overlap between Igf2r and Air. *EMBO Journal, 22*(14), 3696–3704.
Sleutels, F., Zwart, R., et al. (2002). The non-coding Air RNA is required for silencing autosomal imprinted genes. *Nature, 415*(6873), 810–813.
Smith, C. M., & Steitz, J. A. (1998). Classification of gas5 as a multi-small-nucleolar-RNA (snoRNA) host gene and a member of the 5'-terminal oligopyrimidine gene family reveals common features of snoRNA host genes. *Molecular and Cellular Biology, 18*(12), 6897–6909.
Spitale, R. C., Tsai, M. C., et al. (2011). RNA templating the epigenome: Long noncoding RNAs as molecular scaffolds. *Epigenetics, 6*(5), 539–543.
Stadtfeld, M., & Hochedlinger, K. (2010). Induced pluripotency: history, mechanisms, and applications. *Genes & Development, 24*(20), 2239–2263.
Stanke, M., Diekhans, M., et al. (2008). Using native and syntenically mapped cDNA alignments to improve de novo gene finding. *Bioinformatics, 24*(5), 637–644.
Stolc, V., Gauhar, Z., et al. (2004). A gene expression map for the euchromatic genome of Drosophila melanogaster. *Science, 306*(5696), 655–660.
Stolc, V., Samanta, M. P., et al. (2005). Identification of transcribed sequences in Arabidopsis thaliana by using high-resolution genome tiling arrays. *Proceedings of the National Academy of Sciences USA, 102*(12), 4453–4458.
Struhl, K. (2007). Transcriptional noise and the fidelity of initiation by RNA polymerase II. *Nature Structural & Molecular Biology, 14*(2), 103–105.
Sunwoo, H., Dinger, M. E., et al. (2009). MEN epsilon/beta nuclear-retained non-coding RNAs are up-regulated upon muscle differentiation and are essential components of paraspeckles. *Genome Research, 19*(3), 347–359.
Swiezewski, S., Liu, F., et al. (2009). Cold-induced silencing by long antisense transcripts of an Arabidopsis Polycomb target. *Nature, 462*(7274), 799–802.
Taft, R. J., Pheasant, M., et al. (2007). The relationship between non-protein-coding DNA and eukaryotic complexity. *BioEssays, 29*(3), 288–299.
Tjaden, B., Saxena, R. M., et al. (2002). Transcriptome analysis of Escherichia coli using high-density oligonucleotide probe arrays. *Nucleic Acids Research, 30*(17), 3732–3738.
Tochitani, S., & Hayashizaki, Y. (2008). Nkx2.2 antisense RNA overexpression enhanced oligodendrocytic differentiation. *Biochemical and Biophysical Research Communications, 372*(4), 691–696.
Tripathi, V., Ellis, J. D., et al. (2010). The nuclear-retained noncoding RNA MALAT1 regulates alternative splicing by modulating SR splicing factor phosphorylation. *Molecular Cell, 39*(6), 925–938.
Tsai, M. C., Manor, O., et al. (2010). Long noncoding RNA as modular scaffold of histone modification complexes. *Science, 329*(5992), 689–693.
Tsuiji, H., Yoshimoto, R., et al. (2011). Competition between a noncoding exon and introns: Gomafu contains tandem UACUAAC repeats and associates with splicing factor-1. *Genes to Cells, 16*(5), 479–490.
Ulitsky, I., Shkumatava, A., et al. (2011). Conserved function of lincRNAs in vertebrate embryonic development despite rapid sequence evolution. *Cell, 147*(7), 1537–1550.
van Werven, F. J., Neuert, G., et al. (2012). Transcription of two long noncoding RNAs mediates mating-type control of gametogenesis in budding yeast. *Cell, 150*(6), 1170–1181.
Wadler, C. S., & Vanderpool, C. K. (2007). A dual function for a bacterial small RNA: SgrS performs base pairing-dependent regulation and encodes a functional polypeptide. *Proceedings of the National Academy of Sciences USA, 104*(51), 20454–20459.
Wagner, L. A., Christensen, C. J., et al. (2007). EGO, a novel, noncoding RNA gene, regulates eosinophil granule protein transcript expression. *Blood, 109*(12), 5191–5198.
Wang, K. C., & Chang, H. Y. (2011). Molecular mechanisms of long noncoding RNAs. *Molecular Cell, 43*(6), 904–914.
Wang, K. C., Yang, Y. W., et al. (2011). A long noncoding RNA maintains active chromatin to coordinate homeotic gene expression. *Nature, 472*(7341), 120–124.

Wapinski, O., & Chang, H. Y. (2011). Long noncoding RNAs and human disease. *Trends in Cell Biology, 21*(6), 354–361.
Warner, J. R., Knopf, P. M., et al. (1963). A multiple ribosomal structure in protein synthesis. *Proceedings of the National Academy of Sciences USA, 49*, 122–129.
Washietl, S., Hofacker, I. L., et al. (2005). Mapping of conserved RNA secondary structures predicts thousands of functional noncoding RNAs in the human genome. *Nature Biotechnology, 23*(11), 1383–1390.
Watanabe, T., Miyashita, K., et al. (2001). Comprehensive isolation of meiosis-specific genes identifies novel proteins and unusual non-coding transcripts in Schizosaccharomyces pombe. *Nucleic Acids Research, 29*(11), 2327–2337.
Watanabe, Y., & Yamamoto, M. (1994). S. pombe mei2 + encodes an RNA-binding protein essential for premeiotic DNA synthesis and meiosis I, which cooperates with a novel RNA species meiRNA. *Cell, 78*(3), 487–498.
White, R. J. (2011). Transcription by RNA polymerase III: more complex than we thought. *Nature Reviews Genetics, 12*(7), 459–463.
Wilhelm, B. T., Marguerat, S., et al. (2008). Dynamic repertoire of a eukaryotic transcriptome surveyed at single-nucleotide resolution. *Nature, 453*(7199), 1239–1243.
Wilusz, J. E., Sunwoo, H., et al. (2009). Long noncoding RNAs: functional surprises from the RNA world. *Genes & Development, 23*(13), 1494–1504.
Wutz, A. (2011). Gene silencing in X-chromosome inactivation: advances in understanding facultative heterochromatin formation. *Nature Reviews Genetics, 12*(8), 542–553.
Wutz, A., Theussl, H. C., et al. (2001). Non-imprinted Igf2r expression decreases growth and rescues the Tme mutation in mice. *Development, 128*(10), 1881–1887.
Yamada, K., Lim, J., et al. (2003). Empirical analysis of transcriptional activity in the Arabidopsis genome. *Science, 302*(5646), 842–846.
Yamashita, A., Watanabe, Y., et al. (1998). RNA-assisted nuclear transport of the meiotic regulator Mei2p in fission yeast. *Cell, 95*(1), 115–123.
Yekta, S., Tabin, C. J., et al. (2008). MicroRNAs in the Hox network: an apparent link to posterior prevalence. *Nature Reviews Genetics, 9*(10), 789–796.
Yoon, J. H., & Abdelmohsen, K. et al. (2012). LincRNA-p21 suppresses target mRNA translation. *Molecular Cell, 47* (4), 648–655.
Young, R. A. (2011). Control of the embryonic stem cell state. *Cell, 144*(6), 940–954.
Zhang, B., Arun, G., et al. (2012). The lncRNA Malat1 is dispensable for mouse development but its transcription plays a cis-regulatory role in the adult. *Cell Reports, 2*(1), 111–123.
Zhang, X., Lian, Z., et al. (2009). A myelopoiesis-associated regulatory intergenic noncoding RNA transcript within the human HOXA cluster. *Blood, 113*(11), 2526–2534.
Zhao, J., Ohsumi, T. K., et al. (2010). Genome-wide identification of polycomb-associated RNAs by RIP-seq. *Molecular Cell, 40*(6), 939–953.
Zhao, J., Sun, B. K., et al. (2008). Polycomb proteins targeted by a short repeat RNA to the mouse X chromosome. *Science, 322*(5902), 750–756.

# Roles of Long Non-coding RNAs in X-Chromosome Inactivation

**J. Mauro Calabrese and Terry Magnuson**

## 1 Introduction

Female mammals silence the majority of genes along one of their two X chromosomes in a process termed X-chromosome inactivation (XCI). XCI likely evolved in mammals as the X and Y chromosome, once homologous autosomal pairs, diverged in sequence, largely through degeneration of the Y. This degeneration left males with only one functional copy of most X-linked genes, necessitating the development of a compensation process that would equalize X-linked gene dosage between the sexes (Livernois et al. 2012).

XCI is critical for mammalian development. Severe defects in the process are developmentally lethal, while abnormalities in X-chromosome dosage, which occur in about 1 of 500 live births, can be pleiotropic disorders, associated with forms of intellectual disabilities, infertility, and autoimmunity (Powell 2005). The importance of regulating X-linked gene dosage is underscored by the chromosomal counting process inherent to XCI. Regardless of the total number of X chromosomes an individual has, XCI ensures that one X per diploid genome remains active, with the remainder subject to inactivation, in both males and females. For example, XCI tends to silence two X's in tetraploid female cells, and only one in tetraploid male/female cell fusions (Monkhorst et al. 2008). In both cases, the ratio of one active X per diploid genome is maintained. Similarly, in humans, XCI shuts down two X's in females with three (Triple X Syndrome), and one X in males with two (Klinefelter's Syndrome); the sole X in females with Turner's syndrome remains active. These chromosomal abnormalities are often accompanied by chronic health issues (Powell 2005), indicating imperfect regulation of X-linked dosage. However, the intrinsic capability of mammalian cells, male or female, to sense and at least partially deal with abnormalities in

J. M. Calabrese (✉) · T. Magnuson
Department of Genetics, Carolina Center for Genome Sciences, and Lineberger Comprehensive Cancer Center, University of North Carolina, Chapel Hill, NC 27599, USA
e-mail: jmcalabr@email.unc.edu

A. M. Khalil and J. Coller (eds.), *Molecular Biology of Long Non-coding RNAs*, 
DOI: 10.1007/978-1-4614-8621-3_3, 

X-chromosome dosage is remarkable and speaks to the physiological importance of XCI.

In addition to its role in development and human health, XCI has emerged as a paradigm for epigenetic silencing mediated by noncoding RNA (ncRNA), given the critical role of *Xist* and other ncRNAs in the process. Advances in DNA sequencing technologies have led to the identification of thousands of ncRNAs expressed by the mammalian genome, many of which are developmentally regulated and conserved (Dunham et al. 2012; Derrien et al. 2012; Cabili et al. 2011). Early studies have shown these RNAs have critical functions in a range of biological processes, including stem cell maintenance, regulation of the DNA damage response, and developmental specification (Guttman and Rinn 2012). XCI was one of the first identified gene regulatory processes in mammals with a conserved role for ncRNAs (Brockdorff et al. 1992; Brown et al. 1992). Therefore, as the importance of ncRNA-mediated gene regulation has become broadly apparent, XCI has remained a flagship model for understanding ncRNA function. In the pages below, we describe the major features of XCI, with particular focus on the diverse roles that ncRNAs play in the process.

## 2 XCI Overview

In the mouse, historically the field's most utilized experimental model, XCI occurs in two waves during early development. The first is termed imprinted XCI, due to the exclusive inactivation of the paternally inherited X chromosome (Takagi and Sasaki 1975). Imprinted XCI occurs rapidly after formation of the zygote, initiating at the 4-cell stage of development, and nearing completion for some paternal loci at the formation of the early blastocyst, around the 32-cell stage (Kalantry et al. 2009; Okamoto et al. 2005; Patrat et al. 2009; Williams et al. 2011). This stark parent-of-origin bias appears to be independent of the meiotic sex chromosome inactivation that occurs in the male germline (Okamoto et al. 2005), and instead is due to an imprint placed on the maternal X during oocyte maturation, which somehow blocks XCI from occurring on the chromosome (Tada et al. 2000). Cells of the extraembryonic lineage propagate a paternally derived inactive X (Xi) throughout their existence (Takagi and Sasaki 1975; West et al. 1977). In contrast, XCI is reversed in the inner cell mass (ICM) of the blastocyst, which gives rise to the embryo proper (Mak et al. 2004; Okamoto et al. 2004). Postimplantation, XCI re-occurs in the epiblast, nearing completion around embryonic gestational day (E) 6.5 (Rastan 1982). In this second wave, termed random XCI, the choice to inactivate a given X is largely random and independent from its parent-of-origin (McMahon et al. 1983). Random XCI is maintained in all cells save the germline (Sugimoto and Abe 2007), resulting in adult females who are mosaics of paternally and maternally derived Xi's.

Not all mammals share the biphasic inactivation strategy of the mouse. While rats and cows show imprinted XCI in their extraembryonic tissue (Xue et al. 2002;

Wake et al. 1976), suggesting a mouse-like biphasic inactivation strategy, other eutherian mammals examined to date—humans, horses, and mules—appear to undergo random XCI in all lineages (Moreira de Mello et al. 2010; Wang et al. 2012). In contrast, metatherians, such as the kangaroo and opossum, inactivate their paternally inherited X in all tissues (Sharman 1971; Grant et al. 2012).

## 3 Control of XCI via the X-Inactivation Center

Studies of balanced chromosomal translocations in the mouse mapped the location of a single X-linked region that invariably tracked with inactivation of adjoining X-linked DNA, and often led to partial silencing of the fused autosome (Lyon, M. F., Searle A. G., & International Committee on Standardized Genetic Nomenclature for Mice 1989). Because of the region's ability to inactivate neighboring DNA, it was proposed to contain the *cis*-mediated genetic signals required to

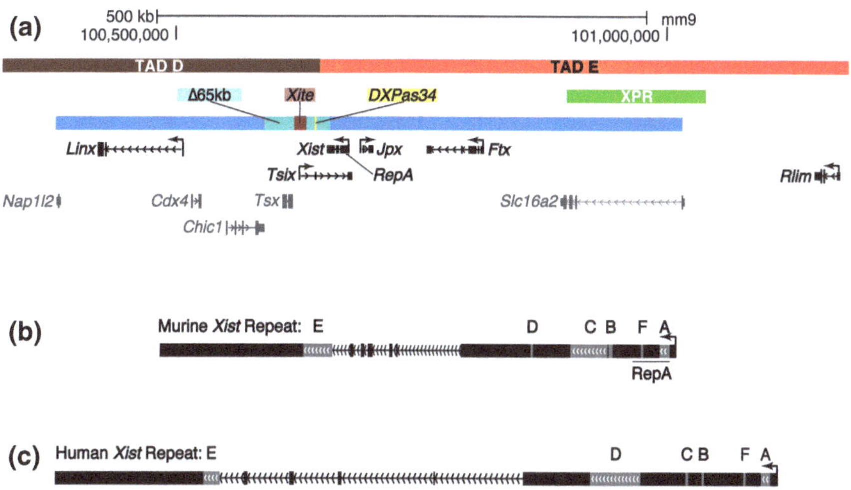


**Fig. 1** *Xist* and the X-inactivation center. **a** The protein coding genes, noncoding RNAs, and regulatory elements of the murine X-inactivation center, depicted to scale relative to UCSC genome build mm9. Genes and regulatory regions in black text denote those discussed in the text with documented or proposed roles in XCI. Genes in grey text have no known roles in XCI. Exons and introns are depicted as solid bars and hashed lines, respectively. Regulatory regions are depicted as colored bars above genes. Denoted TADs are those described in (Nora et al. 2012). The large blue bar spanning the majority of Fig. 1a denotes the genomic span of bacterial and yeast artificial chromosomes that recapitulate aspects of XCI when integrated as multicopy transgene arrays into mouse cell lines (Heard et al. 1999; Lee et al. 1996). **b, c** Mouse and human *Xist* genomic loci. Exons and introns are depicted as in (**a**). Exonic regions in grey mark the location of the six annotated *Xist* repeats, A through F, as described in (Brockdorff et al. 1992; Brown et al. 1992; Nesterova et al. 2001). The location of the RepA transcript within the murine *Xist* locus is underlined

initiate and maintain XCI, and was termed the X-inactivation center (Xic) (Fig. 1a; (Rastan and Brown 1990)). Subsequent analysis of structurally rearranged chromosomes in humans identified a single homologous Xic, as well (Brown et al. 1991). Since then, a range of genetic and cell biological experiments have defined several features contained within the Xic that are critical for proper execution of XCI, including a surprising number of ncRNAs and regulatory elements that produce ncRNA species. At the top of this regulatory cascade is *Xist*, which stands for Xi-specific transcript. *Xist* is essential for XCI, coating the otherwise inactive chromosome from which it was expressed. Several other ncRNAs have been identified within the Xic, including *Tsix*, *Jpx*, *Ftx, Linx*, and *RepA*, most of which have documented roles in XCI. Also, at least two critical regulatory regions within the Xic, *DXPas34*, and *Xite*, have themselves been documented to produce RNA. Most recently, it was discovered that a large ncRNA, termed *Xact*, is expressed from the active X specifically in human pluripotent cells. Together with a complex interplay of transacting factors, many of which remain undefined, the ncRNAs and regulatory elements over the X establish a remarkably robust system of dosage compensation that is capable of delivering a single active X (Xa) per diploid genome, even in the presence of chromosomal abnormalities (Table 1).

## 4 *Xist*, A Long Noncoding RNA Required for XCI

One of the more striking cytological features of the Xi is the coating of the chromosome by the *Xist* ncRNA, which can be visualized under a fluorescent microscope via RNA fluorescence in situ hybridization (FISH). *Xist* was initially

**Table 1** Proposed and validated functions of noncoding RNAs and regulatory elements associated with XCI

| Region | Classification | Proposed/Validated Function | Seminal Reference(s) |
|---|---|---|---|
| *Xist* | NcRNA | Master regulator of XCI | (Brockdorff et al. 1992; Brown et al. 1992; Brown et al. 1991) |
| *Jpx* | NcRNA | *Xist* activator | (Tian et al. 2010) |
| *Ftx* | NcRNA | *Xist* activator | (Chureau et al. 2011) |
| *Tsix* | NcRNA | *Xist* repressor | (Lee et al. 1999) |
| *DXPas34* | Reg. Element | *Tsix* activator | (Courtier et al. 1995; Heard et al. 1993) |
| *Xite* | Reg. Element | *Tsix* activator | (Ogawa and Lee 2003) |
| *Linx* | NcRNA | *Tsix* regulator | (Nora et al. 2012) |
| *RepA* | NcRNA | *Xist* activator, PRC2 recruitment | (Zhao et al. 2008) |
| *LINEs* | DNA/RNA | *Xist* spreading, gene silencing | (Chow et al. 2010) |
| *XACT* | NcRNA | Xa maintenance | (Vallot et al. 2013) |

identified as a candidate gene to control XCI because of its exclusive expression from the Xi and its chromosomal localization within the region defined as the Xic (Brown et al. 1991). Subsequent work defined the major characteristics of the gene in both human and mouse: It is approximately 17 kb in length, can be detected as spliced and polyadenylated, and is exclusively nuclear and untranslated (Brockdorff et al. 1992; Brown et al. 1992). Multiple spliceforms exist, some of which appear to lack polyA tails (Brown et al 1991, 1992; Hong et al. 2000; Ma and Strauss 2005; Memili et al. 2001). Consistent with its classification as a ncRNA, *Xist* lacks conserved open reading frames, but does contain up to six regions of tandemly arrayed repetitive sequence that may be responsible for aspects of its function (Brockdorff et al. 1992; Brown et al. 1992; Nesterova et al. 2001). These regions are on the order of 100 bp to 2 kb in length, and several are clearly conserved between mouse and human (Fig. 1b, c; (Brockdorff et al. 1992; Brown et al. 1992; Nesterova et al. 2001)).

Notably, recent work has identified an Xi-specific transcript in metatherian mammals, termed *Rsx* (Grant et al. 2012). *Rsx* does not share sequence homology with *Xist*, yet, similar to *Xist*, the RNA is expressed from the Xi, appears to coat the chromosome in cis, lacks open reading frames, and is enriched for tandemly repeated sequence at its 5′ end (Grant et al. 2012). This apparent functional conservation without sequence similarity suggests that ncRNA-mediated regulation of dosage compensation arose at least twice during mammalian evolution, highlighting the general utility of this regulatory strategy for the large-scale management of gene expression programs.

Genetic ablation of *Xist* demonstrated its critical role in XCI. Mouse embryonic stem cells (ESCs), which serve as a useful in vitro model because they have yet to undergo XCI, show complete, nonrandom inactivation of a wild-type over a mutant *Xist* allele during differentiation, which induces XCI in these cells (Penny et al. 1996). Similarly, maternal inheritance of an *Xist* deletion results in nonrandom inactivation of the wild-type, paternally inherited X in the mouse embryo. Paternal inheritance of this same deletion results in lethality due to failure of XCI in the extraembryonic lineages, where the wild-type, maternally inherited X is resistant to silencing (Marahrens et al. 1997). These studies indicate that an X-chromosome without *Xist* cannot undergo stable XCI.

While *Xist* coats the Xi in virtually every cell that contains one, the ncRNA is only required during the initiation and early maintenance of the process, at least in the mouse. Using an inducible *Xist* transgene integrated into an autosomal locus, Wutz and Jaenisch were able to show that *Xist* is only capable of gene silencing in ESCs up to 48 h postinduction of differentiation with retinoic acid. Before this time point silencing was reversible and dependent on continued expression of *Xist*, whereas afterwards XCI was irreversible even if *Xist* expression was extinguished (Wutz and Jaenisch 2000). The in vivo correlate of this time frame is unclear, but it is likely between E9.5 and 12.5, as deletion of *Xist* in mouse embryonic fibroblasts (MEFs), which are frequently derived from these developmental time points, does not result in X-reactivation (Csankovszki et al. 1999).

Other than gene silencing, coating of the Xi by *Xist* is the first documented cytological event during initiation of XCI in the mouse, and is seen as early as the four-cell stage of development (Okamoto et al. 2005). *Xist* stabilization and coating of the Xi is also observed at the onset of random XCI (Panning et al. 1997; Sheardown et al. 1997). The closely coupled timing of *Xist* coating and XCI's initiation strongly suggest a role for *Xist* in the earliest stages of XCI, including the initiation of the process.

Rigorous tests examining *Xist's* role in initiating XCI in the mouse have thus far yielded conflicting results. To address the question, Kalantry and colleagues measured the kinetics of gene silencing during the earliest stages of imprinted XCI (Kalantry et al. 2009). They made the surprising observation that several X-linked genes exhibited indistinguishable patterns of silencing between wild-type mice and those carrying a paternally inherited *Xist* deletion at the 8- and 16-cell stage of development. At these early time points, silencing of certain genes was more affected by *Xist* loss than others, whereas all genes were affected at later time points. The results suggest imprinted XCI can initiate in the absence of *Xist* in the mouse. Moreover, they support an evolutionary model of XCI, which posits that inactivation evolved in a piece-meal fashion over the X chromosome (Lahn and Page 1999); Kalantry and colleagues found that genes whose silencing was most affected by *Xist* loss were those thought to be subject to dosage compensation for the longest amount of evolutionary time (Kalantry et al. 2009). In complete contrast, using a similar mutant allele and examining a similar set of X-linked genes, Namekawa and colleagues found that imprinted XCI did not initiate in the absence of *Xist*, suggesting the opposite conclusion reached by Kalantry and colleagues: *Xist* triggers the initiation of imprinted XCI (Namekawa et al. 2010).

Methodological differences have been proposed to explain the discrepancy between these two studies (Namekawa et al. 2010; Brockdorff 2011). The two works also used different *Xist* mutant alleles. Whereas the mutant allele used by Kalantry and colleagues removed *Xist* exons 1 through 3, the mutant allele used by Namekawa and colleagues removed *Xist* exons 1 through 6 (Kalantry et al. 2009; Namekawa et al. 2010). Nonetheless, both alleles appear to be complete for loss of *Xist* function, making this difference unlikely to account for the discrepancy between the studies.

A favored explanation is that differences between inbred mouse strains account for the differential detection of *Xist*-independent processes during the initiation of imprinted XCI. Genetic background differences often affect phenotypes of mutant mice, due to the presence of modifier alleles that associate with particular mouse strains; notable examples of this include mutational analyses of the *Apc* and *Egfr* genes (Montagutelli 2000). Whereas Kalantry and colleagues utilized F1 hybrids of *M. m. musculus* and *M. m. molossinus* mice (Kalantry et al. 2009), Namekawa and colleagues utilized F1 hybrids of *M. m. musculus* and *M. m. castaneous* mice (Namekawa et al. 2010). Therefore, differences in modifier alleles between the *M. m. molossinus* and *M. m. castaneous* subspecies could have been responsible for the differential detection of *Xist* sensitivity during the initiation of imprinted XCI. Under this assumption, the studies conducted by Kalantry and Namekawa indicate that imprinted XCI can initiate in the absence of *Xist* over certain X-linked genes, but that the strength of *Xist*-independent initiation varies with genetic background, such that it is not detectable in *M. m. castaneous/musculus* hybrids (Kalantry et al. 2009; Namekawa et al. 2010).

Whether similar *Xist*-independent processes are involved in the initiation of random XCI is unclear. While many of the cytological features of the Xi are the same in cells subject to imprinted and random XCI (coating in *Xist* and histone H3-lysine27-tri-methylation (H3K27me3), late DNA replication, methylation of CpG islands), a major difference exists in how the future Xi is chosen between the two types of XCI. In imprinted XCI the identity of the Xi is pre-determined; in random XCI it is not. Careful quantification of cell growth and death rates during induction of random XCI via ESC differentiation showed that cells heterozygous for a mutant *Xist* only ever chose the wild-type X for inactivation (Royce-Tolland et al. 2010). This and other studies suggest *Xist* is required to trigger the initiation of random XCI in the mouse (Royce-Tolland et al. 2010; Clerc and Avner 1998; Gribnau et al. 2005; Lee and Lu 1999; Newall et al. 2001). Nevertheless, whether random XCI can initiate in the complete absence of functional *Xist* is still an open question. If it could, it would be predicted to be highly unstable in *Xist*'s absence, given that cells heterozygous for *Xist* mutations never appear to inactivate the mutant X (Penny et al. 1996; Marahrens et al. 1997; Royce-Tolland et al. 2010; Gribnau et al. 2005).

## 5 Spread of *Xist* Over the Xi

*Xist* is an unusual RNA in that it appears to coat the gene-dense regions of the Xi from which it is expressed (Chadwick and Willard 2004; Duthie et al. 1999; Mak et al. 2002). Genetic tagging experiments performed in cell fusions have shown *Xist* is retained on its chromosome of origin, suggesting the RNA spreads over the Xi only in *cis*, and cannot dissociate to bind other X's (Jonkers et al. 2008). This banded pattern of association is stable during metaphase in mouse but not in human (Duthie et al. 1999; Clemson et al. 1996). Curiously, in female MEFs expressing transgenic *Xist* from an autosomal locus, endogenously produced RNA diffuses away from its Xi of synthesis and accumulates over the integrated autosomal transgene (Jeon and Lee 2011). This phenomenon depends on a short conserved region at *Xist's* 5′ end, Repeat F (Nesterova et al. 2001; Jeon and Lee 2011). Whether *Xist* ever leaves its chromosome of synthesis in more natural settings is unclear, but these experiments indicate that diffusion is possible in certain scenarios.

Exactly how *Xist* manages to coat the gene-dense regions of the Xi is unclear. The X chromosome is significantly, and specifically, enriched in LINE repetitive elements relative to the autosomes. In mouse and human, 35 % of X-linked DNA is LINE-derived, as compared to 20 % of autosomal DNA. Other repetitive elements do not display similar enrichment levels (Fujita et al. 2011). At a minimum, this enrichment indicates that the X chromosome provides a favorable genomic environment for LINE insertions, and further suggests insertion of these elements has been co-opted in some way to facilitate XCI. Toward the latter suggestion, LINEs were initially proposed to serve as direct conduits, or booster elements, for the spread of *Xist* over the Xi (Lyon 1998). Studies of *Xist* expression from various autosomal loci have shown that high LINE-density positively correlates with the ability of *Xist* to spread across autosomes, supporting a role for LINEs in *Xist* coating (Chow et al. 2010; Popova et al. 2006; Tang et al. 2010). These elements likely affect the propagation of *Xist* indirectly, however, as analysis of chromosome spreads indicates *Xist* is absent over the most LINE-dense regions of the Xi, associating instead with the gene-dense regions of the chromosome (Chadwick and Willard 2004; Duthie et al. 1999; Mak et al. 2002).

In addition to the role that LINE-dense regions may play in the spread of *Xist* over the Xi, mounting evidence supports an important role for the nuclear matrix in the process. Disruption of chromatin structure via DNaseI and salt extraction does not alter *Xist* localization in human cells, suggesting an indirect interaction between the RNA and the Xi, potentially via the nuclear matrix (Clemson et al. 1996). Consistent with the nuclear matrix playing a role in *Xist's* coating of the Xi, a targeted siRNA screen identified the nuclear matrix protein Hnrnpu/SAF-A as required for the process. Knockdown of Hnrnpu/SAF-A results in destabilization of a long isoform of *Xist*, diffusion of a shorter isoform throughout the nucleus, and defective induction of XCI (Hasegawa et al. 2010). Hnrnpu/SAF-A has both RNA and DNA association domains, and it is possible that the protein serves as a direct interface between *Xist* and regions of the Xi (Hasegawa et al. 2010). In support of this model, this protein has been shown to coat the Xi in both mouse and human cells (Pullirsch et al. 2010; Helbig and Fackelmayer 2003).

A different screening approach led to the identification of SATB1 as a critical factor in the initiation of *Xist*-mediated silencing (Agrelo et al. 2009). The protein is known to be involved in the formation of chromatin loops, binding special AT-rich DNA sequences at nuclear matrix attachment regions, again implicating the nuclear matrix in *Xist's* coating of the Xi (Alvarez et al. 2000; de Belle et al. 1998). SATB1 localizes to the area surrounding the Xi and *Xist*, rather than directly over the chromosome (Agrelo et al. 2009). Based on these properties, it has been proposed that SATB1 could anchor together the gene-poor, LINE-dense regions of the Xi, which may, in turn, condense the Xi's gene-dense regions, and facilitate the spread of *Xist* RNA over the chromosome (Tattermusch and Brockdorff 2011). Recent work has shown that the most LINE-dense regions of the Xi are located adjacent to the *Xist* coat and gene-dense regions of the chromosome, consistent with such a model (Calabrese et al. 2012).

The transcription factor YY1 has been found to tether *Xist* to its site of synthesis on the Xi (Jeon and Lee 2011). This tethering depends on YY1 binding sites in the genomic DNA, located just upstream of Repeat F in the *Xist* locus (Jeon and Lee 2011). How this local tether relates to the nuclear matrix, or the spread of *Xist* over the Xi, is unclear. Immunofluorescence analysis indicates YY1 does not form a microscopically visible coat over the Xi, suggesting it is not directly involved in the spread of *Xist* beyond the Repeat F locus (Jeon and Lee 2011). However, siRNA knockdown of YY1 precludes *Xist* coating in MEFs, suggesting a critical role for local docking of *Xist* in the spread of the RNA over the Xi (Jeon and Lee 2011).

Multiple regions of the *Xist* RNA itself appear to mediate its ability to coat the Xi. A landmark study, in which a series of inducible *Xist* transgenes harboring various segmental deletions were inserted into the X-linked Hprt locus, found that no single region of *Xist* was directly responsible for its spread over the Xi (Wutz et al. 2002). In an endogenous setting, however, the spread of *Xist* is sensitive to specific disruptions. Two groups, using different antisense technologies predicted to disrupt RNA secondary structure, found that targeting of *Xist's* Repeat C region led to visible dissociation of the RNA from the Xi (Beletskii et al. 2001; Sarma et al. 2010), indicating this region of the RNA likely plays a role in coating. Sequence inversion of a region of *Xist* that encompasses the latter half of exon 1 (Repeat D), and exons 2 and 3, results reduced Xi localization and failure of XCI in mutant carrier mice, suggesting this region may also be critical for *Xist* coating (Senner et al. 2011).

Finally, Xi coating by *Xist* is intimately linked to post-transcriptional processing of the RNA. Only spliced *Xist* coats the Xi; the intron-containing RNA does not (Sheardown et al. 1997; Panning and Jaenisch 1996). Furthermore, the induction of XCI is accompanied by an increase in the post-transcriptional stability of *Xist* and not necessarily increased rates of *Xist* transcription. *Xist* transcription rates are similar between ESCs, which do not have an *Xist*-coated Xi, and female fibroblasts, which do have one (Sheardown et al. 1997; Panning and Jaenisch 1996).

## 6 Post-Transcriptional Processing of *Xist*

A handful of factors have been identified as required for proper *Xist* processing, and through that role, a functional XCI response. *ASF/SF2*, an important component of the splicing machinery, binds *Xist* and is necessary for its processing and the initiation of XCI (Royce-Tolland et al. 2010). A SAGE-based expression screen for genes upregulated in female mouse embryos at the onset of XCI led to the discovery of *Upf1*, *Exosc10*, and *Eif1* as proteins required for *Xist* processing and XCI (Bourdet et al. 2006; Ciaudo et al. 2006). How these latter three genes are involved in *Xist* stabilization remains a mystery. *Upf1* and *Exosc10*, components of the nonsense mediated decay pathway and nuclear exosome, respectively, are typically involved in the destruction of RNA, not its stabilization (Houseley and Tollervey 2009). Similarly, *Eif1* has a documented role in the selection of start sites prior to translation initiation (Asano et al. 2000), but *Xist* is untranslated. Establishing an ordered pathway for *Xist* processing and retention on the Xi will likely yield critical insight into the mechanism of XCI.

## 7 *Xist* and the Mechanism of XCI-Induced Gene Silencing

The microscopically visible exclusion of RNA Polymerase II (Pol II) and general transcription factors from the nuclear domain occupied by *Xist* is one of the earliest observable events after the initiation of XCI (Chaumeil et al. 2006). Nevertheless, how the XCI machinery functions to inhibit Xi transcription remains a mystery. *Xist* coating is required for the accumulation of several heterochromatic marks over gene dense regions of the Xi, including H3K27me3, histone H2A ubiquitylation, histone H4-lysine20-monomethylation (H4K20me1), and incorporation of the histone variant macroH2A (Mak et al. 2002; Costanzi and Pehrson 1998; Kohlmaier et al. 2004; Plath et al. 2003; Silva et al. 2003). Induction of this heterochromatic state certainly is an important component of *Xist*-mediated gene silencing. However, both the coating of the Xi by *Xist* and the silencing of many X-linked genes are detected prior to Xi enrichment of these various heterochromatic marks, indicating they may be required to lock-in XCI-induced gene silencing rather than initiate the process. Consistent with this idea, Eed, a core component of the Polycomb Repressive Complex 2 (PRC2) that mediates deposition of H3K27me3, is only required for maintenance of XCI in differentiated extraembryonic derivatives, several cell division cycles after initiation of gene silencing (Kalantry et al. 2006). Remarkably, trophoblast stem cells (TSCs) lacking Eed lose Xi enrichment of all known heterochromatic marks, yet appear to maintain silencing of at least one X-linked locus, and still exclude chromatin modifications associated with active transcription from the genic Xi domain (Kalantry et al. 2006). These results again indicate that XCI-induced transcriptional repression can exist in the absence of enrichment for known, silencing-associated epigenetic marks.

Equally perplexing is the fact that coating of the Xi by *Xist* does not necessarily indicate the presence of a silenced X-chromosome. In human blastocysts, *Xist* coating and gene expression are co-detected at a high frequency over both X's, suggesting critical co-factors must co-localize with the RNA before gene silencing can proceed (Okamoto et al. 2012). This observation raises the intriguing possibility that some of the major players involved in the initiation of XCI during embryogenesis remain undiscovered. Similar factors would be expected to exist in mouse as well. Considering that imprinted XCI can initiate without *Xist* in certain mouse strains, but silencing is rapidly lost in *Xist's* absence (Kalantry et al. 2009), such factors might be loaded onto the mouse X concurrently with, or prior to, spread of *Xist*, but subsequently require the RNA for stabilization and immediate maintenance of silencing. In random XCI, where Xi choice is not pre-determined, loading of *Xist* onto the future Xi may be a prerequisite for recruitment of putative silencing factors.

Additional evidence indicating *Xist* coating is separable from X-linked gene silencing comes from a study of X-reactivation in the mouse blastocyst (Williams et al. 2011). As imprinted XCI nears completion during the early stages of mouse development, cells of the epiblast reactivate their Xi before re-initiating the second round of XCI, which randomly targets the paternal or maternal X for silencing. Quantitative analysis of gene expression via RNA FISH showed that re-activation could be detected on the Xi prior to loss of the *Xist* coat (Williams et al. 2011). Moreover, re-activation kinetics were not altered by overexpression of Nanog, which results in precocious loss of the *Xist* coat specifically in epiblast cells (Williams et al. 2011). Together, similar to the situation described above for human embryos, these results indicate that the transcriptional repression mediated by XCI and *Xist* coating of the Xi can be regulated separately in vivo.

A final piece of evidence indicating that *Xist* coating can be regulated separately from XCI-induced transcriptional repression comes from early transgenic studies of *Xist* itself. Systematic deletion of portions of the *Xist* cDNA in a transgenic mouse ESC model identified the Repeat A region as critical for the induction of gene silencing (Wutz et al. 2002). Although Repeat A mutant *Xist* was deficient in silencing, induced expression still led to *Xist* coating and accumulation of macroH2A, H3K27me3, and H4K20me1 over regions of the chromosome (Wutz et al. 2002; Kohlmaier et al. 2004; Plath et al. 2003). These data again support the notion that Xi coating by *Xist* and XCI-mediated transcriptional repression are separable events.

Contrary to what would be expected from Repeat A deletion in *Xist* transgenes, where mutant *Xist* coats the X without efficiently silencing genes (Wutz et al. 2002; Chaumeil et al. 2006; Kohlmaier et al. 2004; Plath et al. 2003), deletion of the Repeat A region from the endogenous *Xist* locus in the context of mouse development or in ESCs results in XCI failure due to a complete absence of *Xist* coating, and lack of properly spliced *Xist* RNA (Royce-Tolland et al. 2010; Hoki et al. 2009). Transcription of *Xist* appears unaltered in mutant cells (Royce-Tolland et al. 2010; Hoki et al. 2009). Together, these results indicate Repeat A is required for the post-transcriptional processing and stability of *Xist* RNA, in addition to its gene silencing properties. Inducible expression of wild-type or mutant *Xist* cDNAs from stably integrated transgenes appears to bypass XCI's post-transcriptional processing requirements, thus facilitating the identification of Repeat A as critical for *Xist*-mediated gene silencing (Wutz et al. 2002; Kohlmaier et al. 2004; Plath et al. 2003).

Beyond the requirement of Repeat A in *Xist*-mediated silencing, little is known about the mechanism by which XCI inhibits transcription. Early works showed that the nuclear domain occupied by *Xist* lacks nascent transcripts and is depleted of Pol II, general transcription factors, and splicing components (Clemson et al. 1996; Chaumeil et al. 2006; Clemson et al. 2006). Moreover, using DNA FISH to localize specific X-linked sequences relative to the mouse *Xist* domain, it was found that genes which escaped XCI were more frequently outside of the *Xist* domain than those that were subject to XCI (Chaumeil et al. 2006). Cot-1 DNA, which is primarily composed of LINE and SINE repetitive elements, also produced signal that overlapped with *Xist* RNA in FISH assays, in both mouse and human cells (Chaumeil et al. 2006; Clemson et al. 2006). Based on these data, it was hypothesized that XCI induces the formation of a repeat dense nuclear compartment, marked by *Xist*, which physically excludes Pol II and associated transcription machinery from its occupied area (Namekawa et al. 2010; Chow et al. 2010; Chaumeil et al. 2006; Clemson et al. 2006). In such a model, genes subject to XCI enter the repeat-dense silent compartment coincident with inactivation, whereas those that escape XCI remain exterior to it, allowing them access to transcriptional machinery (Chaumeil et al. 2006).

More recent work suggests revisions to this compartmentalized view of XCI (Calabrese et al. 2012). Site-specific DNA FISH found that LINE-dense regions of the Xi are most frequently located directly adjacent to the *Xist* coat, rather than at its center, supporting previous observations that *Xist* associates with predominantly gene-dense rather than repeat-dense Xi regions (Mak et al. 2004; Chadwick and Willard 2004; Duthie et al. 1999). Also, while genes escaping XCI were frequently found outside of the *Xist* domain, so were the X-inactivated genes situated adjacent to them. In this spatial conformation, escapers were frequently expressed, but adjacent X-inactivated genes remained silent, as assessed via RNA FISH and RNA-Seq (Calabrese et al. 2012).

This latter observation is consistent with the recently described notion of topologically associated chromatin domains (TADs). TADs are (roughly) megabase-sized genomic regions that preferentially interact within themselves over surrounding DNA (Dixon et al. 2012; Nora et al. 2012). TAD location is generally consistent across cell types and differentiation states, and is often conserved between species (Dixon et al. 2012; Nora et al. 2012). Although genes contained within TADs are frequently co-regulated, differential expression within TADs also occurs (Dixon et al. 2012; Nora et al. 2012). In regards to the Xi, the nuclear position of individual TADs might largely be dictated by genes that escape XCI, which would be expected to frequently interact with transcription factories located outside of the Xi's *Xist*-dense regions. Considering the existence of TADs, it follows that X-inactivated and escaping genes present within the same or nearby TAD would be located external to the *Xist*-dense Xi domain at similar frequencies.

The observation that X-inactivated genes are not expressed, regardless of their location relative to the microscopically detectable *Xist* cloud, supports a site-specific model for XCI, where XCI-induced gene silencing is maintained independently of a singular nuclear compartment dedicated to transcriptional silencing (Calabrese et al. 2012). A collection of prior works supports this site-specific model of XCI, showing that loci across the X differentially respond to the XCI machinery in a manner that depends on both developmental and cellular context. Examining the timing of X-inactivation for individual X-linked loci during the initiation of imprinted XCI, Patrat and colleagues found that while some genes were efficiently silenced at the 4-8 cell stage, during the onset of imprinted XCI, others remained active and were not silenced until later in development, in some cases well beyond the blastocyst stage (Patrat et al. 2009). Similarly, certain genes appear more sensitive to *Xist* lost than others during the initiation of imprinted XCI (Kalantry et al. 2009), and different subsets of X-linked genes escape XCI in different cell types (Patrat et al. 2009; Calabrese et al. 2012; Carrel and Willard 2005; Cotton et al. 2011; Yang et al. 2010). Lastly, an allele-specific analysis of Pol II distribution in human somatic cells found that while most X-inactivated genes lack Pol II association, a small number bind Pol II yet remain nontranscribed (Kucera et al. 2011). That XCI and escape can occur regardless of a gene's nuclear position, and that both processes show variability between cell types and developmental stages, suggests that the chromosome-level silencing capability of *Xist* requires some form of stably associated, developmentally regulated interface with specific regulatory sites to license the inactivation of individual loci.

Further insight into the physical mechanism by which XCI inhibits transcription has come from a quantitative analysis of chromatin states surrounding Xi regulatory elements. Recent work in F1 hybrid mouse TSCs found that X-inactivated promoters and intergenic regulatory elements maintained reduced levels of DNaseI hypersensitivity (DHS) despite excluding Pol II and other chromatin modifications associated with active transcription (Calabrese et al. 2012). This chromatin state appeared to be an epigenetic signature of XCI, as no single autosomal gene class—including autosomal Polycomb targets, lowly expressed, and nontranscribed genes—had a similar combination of DHS enrichment and Pol II exclusion. In autosomal contexts, DHS sites most frequently mark genomic locations bound by transcription factors engaged in the positive regulation of transcription (Song et al. 2011; Xi et al. 2007). The observation that X-inactivated regulatory elements still harbored detectable DHS in TSCs, albeit at reduced levels compared to the Xa, suggests they are still recognized and bound by cellular factors—these could be the transcription factors that bind cognate elements on the Xa, or unknown factors involved in XCI-induced silencing (Calabrese et al. 2012). Differentiating between these two possibilities, and determining whether cell types other than TSCs harbor similar Xi epigenetic signatures, will be important steps in understanding the mechanism of XCI.

## 8 Transcriptional Modulation of *Xist* as a Mechanism to Sense X-to-Autosome Ratios

The more X-chromosomes a cell has, the more it inactivates. Remarkably, however, the ratio between the number of Xa's per diploid autosomal complement remains at one, regardless of overall ploidy (Brown et al. 1992; Webb et al. 1992; Rastan 1994). These data suggest a mechanism must exist for cells to sense X-to-autosome ratios. Quantification of XCI status in diploid and tetraploid fusion ESC lines supported the presence of one to several activators of XCI present on the X chromosome, whose abundance relative to undefined autosomal loci dictated the likelihood that individual X's would undergo inactivation (Monkhorst et al. 2008). Subsequent BAC transgenic experiments identified the X-encoded ubiquitin ligase *Rnf12* (now called *Rlim*) as one of the major X-linked XCI activators (Fig. 1a; (Jonkers et al. 2009)). Overexpression of Rlim in male and female ESCs led to ectopic induction of XCI on one or both X's, respectively, and this induction depended on intact Rlim catalytic activity (Jonkers et al. 2009). Rlim therefore fit the proposed build of an XCI activator: the higher the ratio of Rlim-to-autosomes, the higher the odds that any given X would be inactivated (Jonkers et al. 2009). Genetic deletion of *Rlim* resulted in complete failure of XCI in some ESC lines (Barakat et al. 2011), and no defect in others, suggesting additional XCI activators may compensate for *Rlim* loss in a strain-specific manner (Shin et al. 2010). Maternal loading of Rlim into oocytes is required for imprinted XCI in the mouse, indicating the protein is the major XCI activator during this first wave of XCI (Shin et al. 2010).

Rlim activates XCI by indirectly inducing expression of *Xist*. A proteomic screen found Rlim to interact with the autosomal transcription factor Rex1, and target it for ubiquitylation and subsequent proteolytic degradation (Gontan et al. 2012). As a result, Rex1 protein levels inversely correlate with levels of Rlim. Rex1 represses *Xist* transcription by binding to its promoter. Therefore, increasing the ratio of Rlim (X-linked) to Rex1 (autosomal) is one way that cells increase expression of *Xist*; high Rlim leads to Rex1 degradation, which in turn relieves *Xist* repression (Gontan et al. 2012). Given the need for *Xist* in the establishment of an Xi, it follows that regulated expression of the RNA is a major mechanism by which cells sense X-to-autosome ratios.

The ncRNA *Jpx* is another dose-dependent activator of *Xist* expression (Fig. 1a; (Tian et al. 2010)). Deletion of a single copy of *Jpx* in female ESCs results in a ~10-fold loss of XCI induction, an effect that can be rescued by addition of exogenous *Jpx* in *trans*. *Jpx* differs from Rlim in that it appears to activate *Xist* expression directly, counteracting the repressive effects that *Tsix* has on the locus. *Jpx* expression is induced ~20-fold during ESC differentiation, suggesting a role for the RNA in maintenance of *Xist* expression after XCI induction (Tian et al. 2010). How *Jpx* induces *Xist* expression is currently unknown.

Lastly, another ncRNA, *Ftx*, may play a partially redundant role with *Jpx* in the activation of *Xist* (Fig. 1a; (Chureau et al. 2011)). Like *Jpx*, *Ftx* is located adjacent to *Xist* in the Xic, escapes XCI, and is upregulated upon ESC differentiation. The RNA is also a miRNA precursor, an observation that may provide insight into its mechanism of action. Deletion of *Ftx* in male ESCs reduces transcription at loci across the Xic, most significantly of *Xist*, but also *Tsix*, *Jpx*, and intergenic transcription between *Jpx* and *Ftx*. Whether *Ftx* exerts its transcriptional effects in a *cis*- or *trans*-mediated manner is unclear. It is also currently unclear what role the RNA plays in a functional XCI response. ESC deletion data would predict a role in the broad regulation of ncRNA expression within the Xic (Chureau et al. 2011).

## 9 Transcriptional Silencing of *Xist* by *Tsix*

Just as the stabilization of *Xist* RNA on one X-chromosome is required to form an Xi, the transcriptional silencing of *Xist* on the other is required to form an Xa. In the mouse, this silencing is achieved primarily through the action of another long ncRNA, *Tsix*. As its name implies, *Tsix* is transcribed antisense to *Xist*. Its transcription extends over the entire murine *Xist* locus, initiating about 15 kb away from *Xist*'s 3′ end, and terminating about 2 kb after *Xist*'s 5′ end (Fig. 1a; (Lee et al. 1999)). *Tsix* has exons and the RNA can be spliced, but splicing is not required for *Xist* silencing (Sado et al. 2006; Sado et al. 2001). Instead, transcription over *Xist*'s promoter appears to be the mechanism by which *Tsix* exerts its *cis*-mediated repressive effect (Luikenhuis et al. 2001; Ohhata et al. 2008). This transcription results in the deposition of DNA methylation and other repressive epigenetic modifications over *Xist*'s promoter that likely prevent its activation during differentiation (Ohhata et al. 2008; Sado et al. 2005). Notably, *Tsix* expression does not transcriptionally silence *Xist* in undifferentiated ESCs. Instead, its expression deposits histone H3-lysine4-dimethylation over the *Xist* locus, indicating *Tsix*'s repressive capacity is developmentally regulated (Navarro et al. 2005).

Through repression of *Xist* expression, *Tsix* plays a central role in determining which X-chromosome is chosen for silencing during random XCI. Deletion of a 65 kb region 3′ to *Xist* that encompasses *Tsix*'s 5′ end (Δ65 kb; Fig. 1a), or more targeted deletions that prevent *Tsix* transcription, result in nonrandom inactivation of the mutated allele in mice and ESCs (Clerc and Avner 1998; Lee and Lu 1999; Sado et al. 2001). This bias is near-absolute: *Tsix* mutant mice inactivate their mutant chromosome in 96 % of cells examined (Lee and Lu 1999; Sado et al. 2001). These studies indicate that transcription of *Tsix* plays a critical role in repressing *Xist* expression on the future Xa. Similar to the situation observed for *Xist* mutations in random XCI, *Tsix* mutants show evidence of a primary XCI defect, meaning that the mutation appears to influence choice of Xi directly, and not the maintenance of choice (Lee and Lu 1999). In the absence of *Tsix*, *Xist* expression may be more easily maintained throughout the initiation process, causing the severe inactivation bias.

In addition to its role in choice, maintained *Tsix* expression is required to prevent ectopic induction of XCI on the Xa during early mouse development. Male and female embryos with a maternally inherited *Tsix* mutation are recovered at a low frequency, between 1 and 15 % of what would be expected from normal Mendelian inheritance (Sado et al. 2001; Lee 2000). This lethality results from ectopic inactivation of the maternally inherited X in the extraembryonic lineages (Ohhata et al. 2006). ESC lines deficient in *Tsix* expression also undergo low levels of ectopic XCI upon differentiation (Luikenhuis et al. 2001; Sado et al. 2002; Morey et al. 2001; Vigneau et al. 2006). These studies suggest that continued expression of *Tsix* is required for normal Xa maintenance in both the embryonic and extraembryonic lineages. The requirement for *Tsix* in Xa maintenance, in both females and males, suggests *Xist* upregulation during the early stages of XCI is a blanket mechanism that affects all X-chromosomes lacking *Tsix* expression.

*Tsix* is not absolutely required for proper XCI. Surviving mouse embryos carrying a maternally inherited *Tsix* mutation are runted, but display expected XCI status and are fertile (Lee 2000). Similarly, in crosses between *Tsix* heterozygotes, *Tsix* homozygous females are recovered at only 4 % of the expected frequency, but are viable and display random XCI (Lee 2002). Female ESC populations homozygous for this same *Tsix* mutation also are capable of proper XCI upon differentiation, but display significantly elevated levels of cells carrying two Xi's, and have high levels of cell death upon differentiation (Lee 2005). The toxicity associated with the inheritance of nonfunctional *Tsix* alleles speaks to the importance of this ncRNA in the proper regulation of XCI. That certain cells are able to establish a proper Xa-to-Xi ratio in the absence of functional *Tsix* indicates a level of stochasticity associated with XCI that appears to confer robustness to the dosage compensation process.

## 10 Regulation of *Tsix* Expression as a Mechanism Driving Xi Choice

The transcriptional regulation of *Tsix* is a complex process that ultimately determines choice of Xi during random XCI. Beyond *Tsix*'s core promoter, several separate regulatory regions appear to be important for expression of the RNA. The most potent of these identified thus far is the *DXPas34* enhancer, a 1.2 kb CG-rich microsatellite repeat approximately 750 bp away from *Tsix*'s transcriptional start site (Fig. 1a; (Courtier et al. 1995; Heard et al. 1993)). Deletion of *DXPas34* results in reduction of *Tsix* transcription and nonrandom inactivation of the mutated allele, similar to that observed for *Tsix* promoter deletions and truncations (Vigneau et al. 2006; Cohen et al. 2007; Debrand et al. 1999). The region likely serves as a loading site for positive regulators of *Tsix* transcription, as it has been documented to recruit a host of transcriptional regulators, including CTCF, YY1, Rex1, Klf4, and c-Myc (Donohoe et al. 2007; Navarro et al. 2008). Consistent with an enhancer function for *DXPas34*, the element displays DHS, and increases basal Luciferase activity in reporter assays (Stavropoulos et al. 2005). *DXPas34* also produces small RNA from both orientations in ESCs (Cohen et al. 2007), similar to many known enhancer elements (Kim et al. 2010).

Another important player in the regulation of *Tsix* expression is *Xite*, which stands for *X*-inactivation *I*ntergenic *T*ranscription *E*lements (Fig. 1a; (Ogawa and Lee 2003)). *Xite* marks a cluster of intergenic transcription start sites that begins upstream of *Tsix*'s basal promoter and extends to the Tsx gene (Ogawa and Lee 2003). Deletion of *Xite* reduces *Tsix* expression, albeit to a lesser extent than does *DXPas34* deletion, and as a consequence, *Xite* mutants show biased inactivation of the targeted allele (Ogawa and Lee 2003). Truncation of *Xite* RNA via insertion of a splice acceptor and polyadenylation sites does not bias XCI, suggesting that the RNA *per se* does not modulate *Tsix* expression (Ogawa and Lee 2003). Rather, Xite DNA itself appears to be an important regulator of XCI, as ESCs stably transfected with extranumerary fragments of Xite fail to undergo XCI upon differentiation (Lee 2005).

Most recently, a number of potential *Tsix* regulatory sites were identified in a chromosome conformation capture screen examining the spatial organization of a 4.5 Mb region of the X-chromosome that surrounds the Xic (Nora et al. 2012).

This work found the *Tsix* locus and all of its previously known regulators to exist within a single TAD situated upstream of *Xist*'s 3′ end (TAD D, Fig. 1a). Within this TAD, several previously unknown contact sites were identified that formed significant interactions with *Tsix* or *Xite*, and showed features reminiscent of regulatory regions. Strikingly, many fell within an 80 kb transcribed region, which was termed *Linx*, for large intervening transcript in the Xic (Fig. 1a). *Linx* has features typical of a ncRNA, including nuclear retention and high levels of intron-containing transcripts. *Linx* is co-expressed with *Tsix* in the epiblast from around the time of implantation onwards, and shows frequent mono-allelism, potentially indicative of a function in XCI (Nora et al. 2012). Future experiments targeting the *Linx* locus should shed light on the potentially important biological function of this RNA.

Beyond the individual elements required for their transcription, the crucial factor driving Xi choice in random XCI is the establishment of asymmetrical expression patterns at *Xist* and *Tsix*. How this essential asymmetry is achieved is unknown. One potential clue comes from the analysis of DNA FISH patterns over the two X's in ESCs (Mlynarczyk-Evans et al. 2006). DNA FISH signals for single loci on the same chromosome can often appear as doublets due to the spatial separation of replicated alleles. Mlynarczyk-Evans and colleagues showed that, in a given ESC, the X-chromosome destined to become the Xi shows a characteristic pattern of singlets and doublets in DNA FISH assays: the Xic to be inactivated appears as a singlet, while the genic loci across the chromosome appear as doublets (Mlynarczyk-Evans et al. 2006). Remarkably, the other X, destined to become the Xa, shows the reciprocal pattern, with a doublet at the Xic and singlets across the remainder of the chromosome. These patterns depend on functional copies of *Xist* and *Tsix*, can fluctuate within the same cell, and are not the result of asynchronous DNA replication (Mlynarczyk-Evans et al. 2006). Although their physiological relevance is unclear, these DNA FISH patterns stand alone as the earliest known markers of the future Xa/Xi, differentiating the two X's prior to the induction of XCI.

Extensive microscopic analyses have revealed another physiological event with potential importance in both the sensing of X-chromosome dosage and ultimate choice of Xi: the transient homologous pairing of X-chromosomes. Shortly after induction of XCI via differentiation of ESCs, the Xic's of the two homologous X-chromosomes transiently co-localize in nuclear space (Xu et al. 2006; Bacher et al. 2006). This pairing is short-lived (about 45 min long), requires transcription and the *trans*-factors CTCF and Oct4, and can be driven by several regions within the Xic, including *Tsix*, *Xite*, and a region termed the X-paring region (Xpr, Fig. 1a; (Xu et al. 2006; Bacher et al. 2006; Donohoe et al. 2009; Xu et al. 2007; Masui et al. 2011; Augui et al. 2007)).

The exact role of pairing in XCI remains ambiguous. Loss of pairing is seen in almost every scenario where random XCI is disrupted, including when XCI is completely inhibited, when it is nonrandom, and when it is induced on both X-chromosomes. For example, both pairing and XCI induction are disrupted by increasing dosage of *Tsix* and Xite sequences via stable transfection into ESCs (Lee 2005; Xu et al. 2007). Conversely, *Tsix*/Xite deletions that result in non-random XCI also disrupt pairing (Xu et al. 2006; Bacher et al. 2006). The add-back of a 16 kb sequence that encompasses the *Tsix* promoter to these mutant cells can restore pairing but not random XCI (Bacher et al. 2006). Lastly, RNAi-mediated ablation of Oct4 results in loss of pairing with ectopic induction of *Xist* and inactivation of both X's—exactly the opposite effect of that seen in scenarios of *Tsix*/Xite overdose, and different from the nonrandom XCI observed when a single copy of *Tsix* is deleted (Donohoe et al. 2009). All together, these studies indicate an intimate link between pairing and proper execution of random XCI. However, pairing is not absolutely required for X-linked silencing, nor does the presence of pairing ensure randomness of inactivation.

In genetically normal cells, however, there is evidence to support a role for pairing in choice of Xi. Using live-cell imaging followed by fixation and RNA FISH, Masui and colleagues found that *Tsix* expression became monoallelic in differentiating ESCs shortly after release of pairing (Masui et al. 2011). Pairing may therefore play a role in the monoallelic assignment of *Tsix* transcription, and through this, choice of Xi. Considering this, and the data showing loss of pairing and XCI upon increased dosage of *Tsix* or Xite DNA (Lee 2005; Xu et al. 2006; Xu et al. 2007), pairing may be linked to a chromosomal counting process that requires the direct exchange of trans factors from one X to the other. The biological basis of pairing, and how it may impart monoallelic expression upon the *Tsix* locus, remains to be determined.

## 11 Other ncRNAs Associated with XCI

Beyond *Xist*, *Tsix*, and the ncRNAs controlling their expression within the Xic, at least three additional X-linked RNAs have potentially important roles in XCI. RepA is a 1.6 kb RNA located within the larger *Xist* that contains the Repeat A sequence (Figure S1A,B; (Zhao et al. 2008)). It was identified via immunoprecipitation of PRC2 complex components in ESCs and MEFs, followed by RT-PCR detection of associated RNA. In PRC2 immunoprecipitates, RNA from the 5′ end of *Xist*, which overlapped the Repeat A sequence, was consistently detected, but the remainder of *Xist* RNA was not. Northern blots probing with Repeat A sequence subsequently identified a 1.6 kb RNA, which was termed RepA. RepA associates with Ezh2, and induction of its expression from stably integrated autosomal loci recruits the PRC2 complex. RepA is polyadenylated and may be transcribed from its own promoter or processed from a larger *Xist* transcript. shRNA knockdown of RepA is not possible without reduction of full-length *Xist* transcripts, making it difficult to unambiguously ascribe function to the shorter RNA. Nonetheless, initial results suggest RepA is a co-factor involved in *Xist* activation and recruitment of PRC2 to the Xi (Zhao et al. 2008). It is important to note that while RepA may play an important role in both processes, redundant mechanisms are likely involved in PRC2 recruitment to the Xi; prior works have shown that overexpression of *Xist* cDNAs lacking the Repeat A region still cause H3K27me3 accumulation over the X, albeit at significantly reduced frequency relative to wild-type *Xist* (Kohlmaier et al. 2004; Plath et al. 2003).

RNA produced from full length LINE elements across the Xi may also be involved in XCI (Chow et al. 2010). RNA FISH analysis in differentiating ESCs showed a striking accumulation of LINE transcripts adjacent to, or directly overlapping with, the *Xist* domain in the early and late stages of XCI, respectively. These LINE transcripts were transcribed by Pol II and specific to the Tf- and Gf-LINE subfamilies (Ostertag and Kazazian 2001). Other classes of repetitive elements, such as SINEs, showed no such accumulation within the *Xist* domain. Furthermore, the induction of LINE transcripts was not specific to the Xi per se, but rather occurred whenever *Xist* was induced; *Xist* expression from autosomal stably integrated transgenes in male ESCs also led to localized accumulation of Gf- and Tf-LINE RNA (Chow et al. 2010).

The exact origin and function of these LINE-derived transcripts in XCI is unknown. The highly repetitive nature of full-length LINEs makes it difficult to pinpoint their expression to specific chromosomal loci. Furthermore, the induction of LINE RNA appears to occur stochastically, being detected in about ~25 % of differentiated ESCs with an *Xist* domain (Chow et al. 2010). This apparent stochasticity may be due to transient induction of LINE RNAs at a specific stage of XCI, making them difficult to detect via RNA FISH in a heterogeneous population of differentiating ESCs. LINE transcripts accumulate around the time that X-linked genes become silenced, correlating LINE expression with transcriptional silencing. Moreover, low abundance sense and antisense small RNAs were also produced from at least one LINE-adjacent locus during XCI induction, potentially linking LINE-derived transcripts to RNAi-mediated processes (Chow et al. 2010).

Most recently, a long ncRNA expressed specifically from the Xa was discovered in the analysis of RNA-seq data from human ESCs (Vallot et al. 2013). *XACT* is a striking ~252 kb in length, unspliced, polyadenylated and predominantly nuclear. Similar to *Xist*, *XACT* accumulates in a cloud-like structure over its chromosome of synthesis. Unlike *Xist*, however, *XACT* coats the Xa, and is expressed in both male and female human ESCs. *XACT* expression is restricted to pluripotent cells in humans. DNA FISH, RNA FISH, and RNA-seq failed to detect *XACT* expression in the mouse, suggesting it is a human-specific ncRNA. The role of *XACT* in dosage compensation is unknown. Given its expression pattern, it likely functions as a regulator of the process specifically in undifferentiated cells (Vallot et al. 2013). The recent identification of *XACT* serves as reminder of how little is understood about XCI in humans, and the complex roles that X-linked ncRNAs play in the process across mammals.

## 12 Conclusions

The last 20 years of XCI research has uncovered a surprisingly large number of ncRNAs that are either required for XCI or likely play as-of-yet understood roles in the process. By virtue of these discoveries, XCI has consistently proved its value as a paradigm for understanding diverse aspects of ncRNA function in nuclear cell biology. The human genome encodes thousands of ncRNAs, many of which are expressed with high levels of tissue-specificity and are conserved across mammals, and most of which have no known function (Dunham et al. 2012; Derrien et al. 2012; Cabili et al. 2011). In many ways, XCI is a microcosm of this ncRNA universe, and knowledge gained from its study will continue to have relevance across disciplines.

## References

Agrelo, R., et al. (2009). SATB1 defines the developmental context for gene silencing by Xist in lymphoma and embryonic cells. *Developmental Cell, 16*(4), 507–516.

Alvarez, J. D., et al. (2000). The MAR-binding protein SATB1 orchestrates temporal and spatial expression of multiple genes during T-cell development. *Genes and Development, 14*(5), 521–535.

Asano, K., et al. (2000). A multifactor complex of eukaryotic initiation factors, eIF1, eIF2, eIF3, eIF5, and initiator tRNA(Met) is an important translation initiation intermediate in vivo. *Genes and Development, 14*(19), 2534–2546.

Augui, S., et al. (2007). Sensing X chromosome pairs before X inactivation via a novel X-pairing region of the Xic. *Science, 318*(5856), 1632–1636.

Bacher, C. P., et al. (2006). Transient colocalization of X-inactivation centres accompanies the initiation of X inactivation. *Nature Cell Biology, 8*(3), 293–299.

Barakat, T. S., et al. (2011). RNF12 activates Xist and is essential for X chromosome inactivation. *PLoS Genetics, 7*(1), e1002001.

Beletskii, A., et al. (2001). PNA interference mapping demonstrates functional domains in the noncoding RNA Xist. *Proceedings of the National Academy of Sciences, 98*(16), 9215–9220.

Bourdet, A., et al. (2006). A SAGE approach to identifying novel trans-acting factors involved in the X inactivation process. *Cytogenetic and Genome Research, 113*(1–4), 325–335.

Brockdorff, N. (2011). Chromosome silencing mechanisms in X-chromosome inactivation: unknown unknowns. *Development, 138*(23), 5057–5065.

Brockdorff, N., et al. (1992). The product of the mouse Xist gene is a 15 kb inactive X-specific transcript containing no conserved ORF and located in the nucleus. *Cell, 71*(3), 515–526.

Brown, C. J., et al. (1991a). Localization of the X inactivation centre on the human X chromosome in Xq13. *Nature, 349*(6304), 82–84.

Brown, C. J., et al. (1991b). A gene from the region of the human X inactivation centre is expressed exclusively from the inactive X chromosome. *Nature, 349*(6304), 38–44.

Brown, C. J., et al. (1992). The human XIST gene: analysis of a 17 kb inactive X-specific RNA that contains conserved repeats and is highly localized within the nucleus. *Cell, 71*(3), 527–542.

Cabili, M. N., et al. (2011). Integrative annotation of human large intergenic noncoding RNAs reveals global properties and specific subclasses. *Genes and Development, 25*(18), 1915–1927.

Calabrese, J. M., et al. (2012). Site-specific silencing of regulatory elements as a mechanism of X inactivation. *Cell, 151*(5), 951–963.

Carrel, L., & Willard, H. F. (2005). X-inactivation profile reveals extensive variability in X-linked gene expression in females. *Nature, 434*(7031), 400–404.

Chadwick, B. P., & Willard, H. F. (2004). Multiple spatially distinct types of facultative heterochromatin on the human inactive X chromosome. *Proceedings of the National Academy of Sciences of the United States of America, 101*(50), 17450–17455.

Chaumeil, J., et al. (2006). A novel role for Xist RNA in the formation of a repressive nuclear compartment into which genes are recruited when silenced. *Genes and Development, 20*(16), 2223–2237.

Chow, J. C., et al. (2010). LINE-1 activity in facultative heterochromatin formation during X chromosome inactivation. *Cell, 141*(6), 956–969.

Chureau, C., et al. (2011). Ftx is a non-coding RNA which affects Xist expression and chromatin structure within the X-inactivation center region. *Human Molecular Genetics, 20*(4), 705–718.

Ciaudo, C., et al. (2006). Nuclear mRNA degradation pathway(s) are implicated in Xist regulation and X chromosome inactivation. *PLoS Genetics, 2*(6), e94.

Clemson, C. M., et al. (1996). XIST RNA paints the inactive X chromosome at interphase: evidence for a novel RNA involved in nuclear/chromosome structure. *Journal of Cell Biology, 132*(3), 259–275.

Clemson, C. M., et al. (2006). The X chromosome is organized into a gene-rich outer rim and an internal core containing silenced nongenic sequences. *Proceedings of the National Academy of Sciences, 103*(20), 7688–7693.

Clerc, P., & Avner, P. (1998). Role of the region 3′ to Xist exon 6 in the counting process of X-chromosome inactivation. *Nature Genetics, 19*(3), 249–253.

Cohen, D. E., et al. (2007). The DXPas34 repeat regulates random and imprinted X inactivation. *Developmental Cell, 12*(1), 57–71.

Costanzi, C., & Pehrson, J. R. (1998). Histone macroH2A1 is concentrated in the inactive X chromosome of female mammals. *Nature, 393*(6685), 599–601.

Cotton, A. M., et al. (2011). Chromosome-wide DNA methylation analysis predicts human tissue-specific X inactivation. *Human Genetics, 130*(2), 187–201.

Courtier, B., Heard, E., & Avner, P. (1995). Xce haplotypes show modified methylation in a region of the active X chromosome lying 3′ to Xist. *Proceedings of the National Academy of Sciences, 92*(8), 3531–3535.

Csankovszki, G., et al. (1999). Conditional deletion of Xist disrupts histone macroH2A localization but not maintenance of X inactivation. *Nature Genetics, 22*(4), 323–324.

de Belle, I., Cai, S., & Kohwi-Shigematsu, T. (1998). The genomic sequences bound to special AT-rich sequence-binding protein 1 (SATB1) in vivo in Jurkat T cells are tightly associated with the nuclear matrix at the bases of the chromatin loops. *Journal of Cell Biology, 141*(2), 335–348.

Debrand, E., et al. (1999). Functional analysis of the DXPas34 locus, a 3′ regulator of Xist expression. *Molecular and Cellular Biology, 19*(12), 8513–8525.

Derrien, T., et al. (2012). The GENCODE v7 catalog of human long noncoding RNAs: analysis of their gene structure, evolution, and expression. *Genome Research, 22*(9), 1775–1789.

Dixon, J. R., et al. (2012). Topological domains in mammalian genomes identified by analysis of chromatin interactions. *Nature, 485*(7398), 376–380.

Donohoe, M. E., et al. (2007). Identification of a Ctcf cofactor, Yy1, for the X chromosome binary switch. *Molecular Cell, 25*(1), 43–56.

Donohoe, M. E., et al. (2009). The pluripotency factor Oct4 interacts with Ctcf and also controls X-chromosome pairing and counting. *Nature, 460*(7251), 128–132.

Dunham, I., et al. (2012). An integrated encyclopedia of DNA elements in the human genome. *Nature, 489*(7414), 57–74.

Duthie, S. M., et al. (1999). Xist RNA exhibits a banded localization on the inactive X chromosome and is excluded from autosomal material in cis. *Human Molecular Genetics, 8*(2), 195–204.

Fujita, P. A. et al. (2010). The UCSC Genome Browser database: update 2011. *Nucleic Acids Research, 40*, D918–D923.
Gontan, C., et al. (2012). RNF12 initiates X-chromosome inactivation by targeting REX1 for degradation. *Nature, 485*(7398), 386–390.
Grant, J., et al. (2012). Rsx is a metatherian RNA with Xist-like properties in X-chromosome inactivation. *Nature, 487*(7406), 254–258.
Gribnau, J., et al. (2005). X chromosome choice occurs independently of asynchronous replication timing. *Journal of Cell Biology, 168*(3), 365–373.
Guttman, M., & Rinn, J. L. (2012). Modular regulatory principles of large non-coding RNAs. *Nature, 482*(7385), 339–346.
Hasegawa, Y., et al. (2010). The matrix protein hnRNP U is required for chromosomal localization of Xist RNA. *Developmental Cell, 19*(3), 469–476.
Heard, E., et al. (1993). Physical mapping and YAC contig analysis of the region surrounding Xist on the mouse X chromosome. *Genomics, 15*(3), 559–569.
Heard, E., et al. (1999). Xist yeast artificial chromosome transgenes function as X-inactivation centers only in multicopy arrays and not as single copies. *Molecular and Cellular Biology, 19*(4), 3156–3166.
Helbig, R., & Fackelmayer, F. O. (2003). Scaffold attachment factor A (SAF-A) is concentrated in inactive X chromosome territories through its RGG domain. *Chromosoma, 112*(4), 173–182.
Hoki, Y., et al. (2009). A proximal conserved repeat in the Xist gene is essential as a genomic element for X-inactivation in mouse. *Development, 136*(1), 139–146.
Hong, Y. K., Ontiveros, S. D., & Strauss, W. M. (2000). A revision of the human XIST gene organization and structural comparison with mouse Xist. *Mammalian Genome, 11*(3), 220–224.
Houseley, J., & Tollervey, D. (2009). The many pathways of RNA degradation. *Cell, 136*(4), 763–776.
Jeon, Y., & Lee, J. T. (2011). YY1 Tethers Xist RNA to the inactive X nucleation center. *Cell, 146*(1), 119–133.
Jonkers, I., et al. (2008). Xist RNA is confined to the nuclear territory of the silenced X chromosome throughout the cell cycle. *Molecular and Cellular Biology, 28*(18), 5583–5594.
Jonkers, I., et al. (2009). RNF12 is an X-Encoded dose-dependent activator of X chromosome inactivation. *Cell, 139*(5), 999–1011.
Kalantry, S., et al. (2006). The Polycomb group protein Eed protects the inactive X-chromosome from differentiation-induced reactivation. *Nature Cell Biology, 8*(2), 195–202.
Kalantry, S., et al. (2009). Evidence of Xist RNA-independent initiation of mouse imprinted X-chromosome inactivation. *Nature, 460*(7255), 647–651.
Kim, T. K., et al. (2010). Widespread transcription at neuronal activity-regulated enhancers. *Nature, 465*(7295), 182–187.
Kohlmaier, A., et al. (2004). A chromosomal memory triggered by Xist regulates histone methylation in X inactivation. *PLoS Biology, 2*(7), E171.
Kucera, K. S., et al. (2011). Allele-specific distribution of RNA polymerase II on female X chromosomes. *Human Molecular Genetics, 20*(20), 3964–3973.
Lahn, B. T., & Page, D. C. (1999). Four evolutionary strata on the human X chromosome. *Science, 286*(5441), 964–967.
Lee, J. T. (2000). Disruption of imprinted X inactivation by parent-of-origin effects at Tsix. *Cell, 103*(1), 17–27.
Lee, J. T. (2002). Homozygous Tsix mutant mice reveal a sex-ratio distortion and revert to random X-inactivation. *Nature Genetics, 32*(1), 195–200.
Lee, J. T. (2005). Regulation of X-chromosome counting by Tsix and Xite sequences. *Science, 309*(5735), 768–771.
Lee, J. T., & Lu, N. (1999). Targeted mutagenesis of Tsix leads to nonrandom X inactivation. *Cell, 99*(1), 47–57.

Lee, J. T., et al. (1996). A 450 kb transgene displays properties of the mammalian X-inactivation center. *Cell, 86*(1), 83–94.
Lee, J. T., Davidow, L. S., & Warshawsky, D. (1999). Tsix, a gene antisense to Xist at the X-inactivation centre. *Nature Genetics, 21*(4), 400–404.
Livernois, A. M., Graves, J. A., & Waters, P. D. (2012). The origin and evolution of vertebrate sex chromosomes and dosage compensation. *Heredity (Edinburgh), 108*(1), 50–58.
Luikenhuis, S., Wutz, A., & Jaenisch, R. (2001). Antisense transcription through the Xist locus mediates Tsix function in embryonic stem cells. *Molecular and Cellular Biology, 21*(24), 8512–8520.
Lyon, M. F. (1961). Gene action in the X-chromosome of the mouse (Mus musculus L.). *Nature, 190*, 372–373.
Lyon, M. F. (1998). X-chromosome inactivation: a repeat hypothesis. *Cytogenetics and Cell Genetics, 80*(1–4), 133–137.
Lyon, M. F., Searle A. G., & International Committee on Standardized Genetic Nomenclature for Mice (1989). *Genetic variants and strains of the laboratory mouse* (2nd ed.). New York : Oxford University Press , Stuttgart : G. Fischer Verlag, xiii, 876 p.
Ma, M., & Strauss, W. M. (2005). Analysis of the Xist RNA isoforms suggests two distinctly different forms of regulation. *Mammalian Genome, 16*(6), 391–404.
Mak, W., et al. (2002). Mitotically stable association of polycomb group proteins eed and enx1 with the inactive x chromosome in trophoblast stem cells. *Current Biology, 12*(12), 1016–1020.
Mak, W., et al. (2004). Reactivation of the paternal X chromosome in early mouse embryos. *Science, 303*(5658), 666–669.
Marahrens, Y., et al. (1997). Xist-deficient mice are defective in dosage compensation but not spermatogenesis. *Genes and Development, 11*(2), 156–166.
Masui, O., et al. (2011). Live-cell chromosome dynamics and outcome of X chromosome pairing events during ES cell differentiation. *Cell, 145*(3), 447–458.
McMahon, A., Fosten, M., & Monk, M. (1983). X-chromosome inactivation mosaicism in the three germ layers and the germ line of the mouse embryo. *Journal of Embryology and Experimental Morphology, 74*, 207–220.
Memili, E., et al. (2001). Murine Xist RNA isoforms are different at their 3′ ends: a role for differential polyadenylation. *Gene, 266*(1–2), 131–137.
Mlynarczyk-Evans, S., et al. (2006). X chromosomes alternate between two states prior to random X-inactivation. *PLoS Biology, 4*(6), e159.
Monkhorst, K., et al. (2008). X inactivation counting and choice is a stochastic process: evidence for involvement of an X-linked activator. *Cell, 132*(3), 410–421.
Montagutelli, X. (2000). Effect of the genetic background on the phenotype of mouse mutations. *Journal of the American Society of Nephrology, 11*(Suppl 16), S101–S105.
Moreira de Mello, J. C. et al. (2010). Random X inactivation and extensive mosaicism in human placenta revealed by analysis of allele-specific gene expression along the X chromosome. *PLoS ONE, 5*(6) p. e10947.
Morey, C., et al. (2001). Tsix-mediated repression of Xist accumulation is not sufficient for normal random X inactivation. *Human Molecular Genetics, 10*(13), 1403–1411.
Namekawa, S. H., et al. (2010). Two-step imprinted X inactivation: repeat versus genic silencing in the mouse. *Molecular and Cellular Biology, 30*(13), 3187–3205.
Navarro, P., et al. (2005). Tsix transcription across the Xist gene alters chromatin conformation without affecting Xist transcription: implications for X-chromosome inactivation. *Genes and Development, 19*(12), 1474–1484.
Navarro, P., et al. (2008). Molecular coupling of Xist regulation and pluripotency. *Science, 321*(5896), 1693–1695.
Nesterova, T. B., et al. (2001). Characterization of the genomic Xist locus in rodents reveals conservation of overall gene structure and tandem repeats but rapid evolution of unique sequence. *Genome Research, 11*(5), 833–849.

Newall, A. E., et al. (2001). Primary non-random X inactivation associated with disruption of Xist promoter regulation. *Human Molecular Genetics, 10*(6), 581–589.
Nora, E. P., et al. (2012). Spatial partitioning of the regulatory landscape of the X-inactivation centre. *Nature, 485*(7398), 381–385.
Ogawa, Y., & Lee, J. T. (2003). Xite, X-inactivation intergenic transcription elements that regulate the probability of choice. *Molecular Cell, 11*(3), 731–743.
Ohhata, T., et al. (2006). Tsix-deficient X chromosome does not undergo inactivation in the embryonic lineage in males: implications for Tsix-independent silencing of Xist. *Cytogenetic and Genome Research, 113*(1–4), 345–349.
Ohhata, T., et al. (2008). Crucial role of antisense transcription across the Xist promoter in Tsix-mediated Xist chromatin modification. *Development, 135*(2), 227–235.
Okamoto, I., et al. (2004). Epigenetic dynamics of imprinted X inactivation during early mouse development. *Science, 303*(5658), 644–649.
Okamoto, I., et al. (2005). Evidence for de novo imprinted X-chromosome inactivation independent of meiotic inactivation in mice. *Nature, 438*(7066), 369–373.
Okamoto, I., et al. (2012). Eutherian mammals use diverse strategies to initiate X-chromosome inactivation during development. *Nature, 472*(7343), 370–374.
Ostertag, E. M., & Kazazian, H. H, Jr. (2001). Biology of mammalian L1 retrotransposons. *Annual Review of Genetics, 35*, 501–538.
Panning, B., & Jaenisch, R. (1996). DNA hypomethylation can activate Xist expression and silence X-linked genes. *Genes and Development, 10*(16), 1991–2002.
Panning, B., Dausman, J., & Jaenisch, R. (1997). X chromosome inactivation is mediated by Xist RNA stabilization. *Cell, 90*(5), 907–916.
Patrat, C., et al. (2009). Dynamic changes in paternal X-chromosome activity during imprinted X-chromosome inactivation in mice. *Proceedings of the National Academy of Sciences of the United States of America, 106*(13), 5198–5203.
Penny, G. D., et al. (1996). Requirement for Xist in X chromosome inactivation. *Nature, 379*(6561), 131–137.
Plath, K., et al. (2003). Role of histone H3 lysine 27 methylation in X inactivation. *Science, 300*(5616), 131–135.
Popova, B. C., et al. (2006). Attenuated spread of X-inactivation in an X;autosome translocation. *Proceedings of the National Academy of Sciences, 103*(20), 7706–7711.
Powell, C. M. (2005). Sex chromosome and sex chromosome abnormalities. In S. Gersen & M. Keagle (Eds.), *The Principles of Clinical Cytogenetics, III* p. 207–246.
Pullirsch, D., et al. (2010). The trithorax group protein Ash2 l and Saf-A are recruited to the inactive X chromosome at the onset of stable X inactivation. *Development, 137*(6), 935–943.
Rastan, S. (1982). Timing of X-chromosome inactivation in postimplantation mouse embryos. *Journal of Embryology and Experimental Morphology, 71*, 11–24.
Rastan, S. (1994). X chromosome inactivation and the Xist gene. *Current Opinion in Genetics and Development, 4*(2), 292–297.
Rastan, S., & Brown, S. D. (1990). The search for the mouse X-chromosome inactivation centre. *Genetical Research, 56*(2–3), 99–106.
Royce-Tolland, M. E., et al. (2010). The A-repeat links ASF/SF2-dependent Xist RNA processing with random choice during X inactivation. *Nature Structural and Molecular Biology, 17*(8), 948–954.
Sado, T., et al. (2001). Regulation of imprinted X-chromosome inactivation in mice by Tsix. *Development, 128*(8), 1275–1286.
Sado, T., Li, E., & Sasaki, H. (2002). Effect of TSIX disruption on XIST expression in male ES cells. *Cytogenetic and Genome Research, 99*(1–4), 115–118.
Sado, T., Hoki, Y., & Sasaki, H. (2005). Tsix silences Xist through modification of chromatin structure. *Developmental Cell, 9*(1), 159–165.
Sado, T., Hoki, Y., & Sasaki, H. (2006). Tsix defective in splicing is competent to establish Xist silencing. *Development, 133*(24), 4925–4931.

Sarma, K., et al. (2010). Locked nucleic acids (LNAs) reveal sequence requirements and kinetics of Xist RNA localization to the X chromosome. *Proceedings of the National Academy of Sciences, 107*(51), 22196–22201.
Senner, C. E., et al. (2011). Disruption of a conserved region of Xist exon 1 impairs Xist RNA localisation and X-linked gene silencing during random and imprinted X chromosome inactivation. *Development, 138*(8), 1541–1550.
Sharman, G. B. (1971). Late DNA replication in the paternally derived X chromosome of female kangaroos. *Nature, 230*(5291), 231–232.
Sheardown, S. A., et al. (1997). Stabilization of Xist RNA mediates initiation of X chromosome inactivation. *Cell, 91*(1), 99–107.
Shin, J., et al. (2010). Maternal Rnf12/RLIM is required for imprinted X-chromosome inactivation in mice. *Nature, 467*(7318), 977–981.
Silva, J., et al. (2003). Establishment of histone h3 methylation on the inactive X chromosome requires transient recruitment of Eed-Enx1 polycomb group complexes. *Developmental Cell, 4*(4), 481–495.
Song, L., et al. (2011). Open chromatin defined by DNaseI and FAIRE identifies regulatory elements that shape cell-type identity. *Genome Research, 21*(10), 1757–1767.
Stavropoulos, N., Rowntree, R. K., & Lee, J. T. (2005). Identification of developmentally specific enhancers for Tsix in the regulation of X chromosome inactivation. *Molecular and Cellular Biology, 25*(7), 2757–2769.
Sugimoto, M., & Abe, K. (2007). X chromosome reactivation initiates in nascent primordial germ cells in mice. *PLoS Genetics, 3*(7), e116.
Tada, T., et al. (2000). Imprint switching for non-random X-chromosome inactivation during mouse oocyte growth. *Development, 127*(14), 3101–3105.
Takagi, N., & Sasaki, M. (1975). Preferential inactivation of the paternally derived X chromosome in the extraembryonic membranes of the mouse. *Nature, 256*(5519), 640–642.
Tang, Y. A., et al. (2010). Efficiency of Xist-mediated silencing on autosomes is linked to chromosomal domain organisation. *Epigenetics Chromatin, 3*(1), 10.
Tattermusch, A., & Brockdorff, N. (2011). A scaffold for X chromosome inactivation. *Human Genetics, 130*, pp. 247–253.
Tian, D., Sun, S., & Lee, J. T. (2010). The long noncoding RNA, Jpx, is a molecular switch for X chromosome inactivation. *Cell, 143*(3), 390–403.
Vallot, C., et al. (2013). XACT, a long noncoding transcript coating the active X chromosome in human pluripotent cells. *Nature Genetics, 45*(3), 239–241.
Vigneau, S., et al. (2006). An essential role for the DXPas34 tandem repeat and Tsix transcription in the counting process of X chromosome inactivation. *Proceedings of the National Academy of Sciences, 103*(19), 7390–7395.
Wake, N., Takagi, N., & Sasaki, M. (1976). Non-random inactivation of X chromosome in the rat yolk sac. *Nature, 262*(5569), 580–581.
Wang, X. et al. (2012). Random X inactivation in the mule and horse placenta. *Genome Research, 22*(10) pp. 1855–1863.
Webb, S., de Vries, T. J., & Kaufman, M. H. (1992). The differential staining pattern of the X chromosome in the embryonic and extraembryonic tissues of postimplantation homozygous tetraploid mouse embryos. *Genetical Research, 59*(3), 205–214.
West, J. D., et al. (1977). Preferential expression of the maternally derived X chromosome in the mouse yolk sac. *Cell, 12*(4), 873–882.
Williams, L. H., et al. (2011). Transcription precedes loss of Xist coating and depletion of H3K27me3 during X-chromosome reprogramming in the mouse inner cell mass. *Development, 138*(10), 2049–2057.
Wutz, A., & Jaenisch, R. (2000). A shift from reversible to irreversible X inactivation is triggered during ES cell differentiation. *Molecular Cell, 5*(4), 695–705.
Wutz, A., Rasmussen, T. P., & Jaenisch, R. (2002). Chromosomal silencing and localization are mediated by different domains of Xist RNA. *Nature Genetics, 30*(2), 167–174.

Xi, H., et al. (2007). Identification and characterization of cell type-specific and ubiquitous chromatin regulatory structures in the human genome. *PLoS Genetics, 3*(8), e136.

Xu, N., Tsai, C. L., & Lee, J. T. (2006). Transient homologous chromosome pairing marks the onset of X inactivation. *Science, 311*(5764), 1149–1152.

Xu, N., et al. (2007). Evidence that homologous X-chromosome pairing requires transcription and Ctcf protein. *Nature Genetics, 39*(11), 1390–1396.

Xue, F., et al. (2002). Aberrant patterns of X chromosome inactivation in bovine clones. *Nature Genetics, 31*(2), 216–220.

Yang, F., et al. (2010). Global survey of escape from X inactivation by RNA-sequencing in mouse. *Genome Research, 20*(5), 614–622.

Zhao, J., et al. (2008). Polycomb proteins targeted by a short repeat RNA to the mouse X chromosome. *Science, 322*(5902), 750–756.

# Roles of Long Non-coding RNAs in Genomic Imprinting

**Kristen Martins-Taylor and Stormy J. Chamberlain**

## 1 Introduction

The first long noncoding RNA (lncRNA) discovered is the product of a gene subject to regulation by genomic imprinting (Brannan et al. 1990; Bartolomei et al. 1991). Genomic imprinting is a phenomenon in which genes are expressed preferentially from one parental allele. For instance, a gene might be exclusively expressed from the maternally inherited allele and silenced on the paternally inherited allele. Since parent-of-origin dependent allele-specific expression requires exquisite epigenetic control of gene expression, the mechanisms underlying this process have been studied extensively as a paradigm for epigenetic gene regulation. As such, the mechanisms that imprinted lncRNAs may employ to regulate the expression of the other genes within their region may be more broadly applicable to other nonimprinted lncRNAs in nonimprinted regions. While it is apparent that no two regions are exactly alike, nearly every imprinted region has an imprinted lncRNA. Here, we detail the known functions of imprinted lncRNAs. Some common themes pertaining to how imprinted lncRNAs regulate gene expression have emerged.

### *1.1 H19*

*H19*, located on the p arm of human chromosome 11 and the distal tip of mouse chromosome 7, was the first mammalian lncRNA to be identified (Brannan et al. 1990) (Fig. 1). Brannan et al. compared the open reading frames of human and mouse *H19* genes and found that the small potential open reading frames were not

K. Martins-Taylor · S. J. Chamberlain (✉)
Department of Genetics and Developmental Biology, University of Connecticut Health Center, Farmington, CT 06030, USA
e-mail: chamberlain@uchc.edu

A. M. Khalil and J. Coller (eds.), *Molecular Biology of Long Non-coding RNAs*,
DOI: 10.1007/978-1-4614-8621-3_4, 

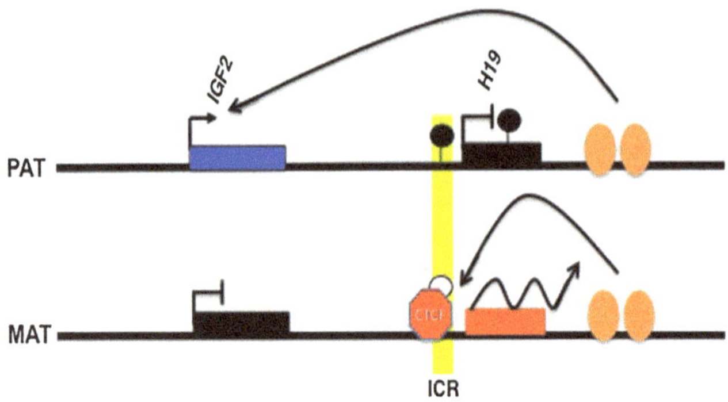


**Fig. 1** *H19/Igf2* imprinted region. Genes are represented by *colored rectangles* with *red* representing maternal allele-specific expression and *blue* representing paternal allele-specific expression. The H19 lncRNA is represented by a *wavy line*. *Ovals* represent tissue-specific enhancers and *arcs point* to the promoters they are acting on. The *yellow* highlighted region indicates the imprinting control region (ICR). *Black* and *white circles* denote differentially methylated regions (DMRs) with *black* indicating the methylated allele and *white* indicating the unmethylated allele

conserved between the species, and therefore, *H19* must function as an RNA (Brannan et al. 1990). *H19* is comprised of five exons separated by small introns, which are transcribed by RNA Polymerase II and spliced to yield a 2.3 kb RNA product. The mature RNA is capped and polyadenylated. While found in the nucleus and the cytoplasm, the bulk of *H19* RNA is localized to the cytoplasm, where it forms cytoplasmic RNA particles of unknown function (Brannan et al. 1990). *H19* is also a host transcript to the miR675 microRNA (Cai and Cullen 2007).

The *H19* lncRNA is expressed in the extraembryonic tissues from implantation onward (Poirier et al. 1991) and is highly expressed in the developing mouse embryo, especially in endodermal and mesodermal tissues (Pachnis et al. 1988). *H19* becomes silenced shortly after birth in most tissues, except for cardiac muscle, skeletal muscle, and cartilage where sustained expression may occur, in part due to the increased stability of the spliced message (Pachnis et al. 1988; Castle et al. 2010; Dudek et al. 2010).

*H19* is imprinted and almost exclusively expressed from the maternally inherited allele (Bartolomei et al. 1991). It is reciprocally imprinted with the *IGF2* gene, which is 90 kb away. Imprinted expression of *H19* is controlled by an imprinting control region (ICR). The ICR is a 2 kb sequence located approximately 2 kb upstream of the *H19* gene that consists of a differentially methylated region (DMR). The DMR within the ICR is established in the respective germlines; it becomes methylated in the paternal germline and is protected from methylation in the maternal germline. The promoter of the *H19* gene also harbors a DMR that is secondary to the ICR. *H19* and *IGF2* share distal mesodermal and endodermal enhancers. Sites within the unmethylated maternal ICR are bound by

CTCF (CCCTC binding factor), which blocks the distal enhancers from accessing the *IGF2* promoter, directing them to act on the maternal *H19* promoter. DNA methylation on the paternal ICR blocks CTCF binding allowing the distal enhancers to act preferentially on the *IGF2* promoter.

In addition to the *H19* lncRNA, a highly conserved microRNA, *miR-675*, is expressed from the gene. Additionally, an antisense transcript, known as *91H* is produced from this locus in tumor tissues.

The function of *H19* is still unknown. Imprinted genes are classically thought to regulate embryonic and placental growth, and *H19* is hypothesized to do act in this manner. A human overgrowth disorder, Beckwith-Wiedemann syndrome, is associated with hypermethylation of the ICR, resulting in reduced *H19* expression and loss of imprinted expression of *IGF2*. A reciprocal disorder, Silver-Russell syndrome, is characterized by fetal and postnatal growth retardation and is associated with hypomethylation of the ICR, resulting in increased expression of *H19* and reduced expression of *IGF2*. However, aberrant *IGF2* expression can also explain both of these syndromes. Mice lacking the *H19* gene (both the $H19^{\Delta 13mat}$ $H19^{\Delta 3mat}$ strains) are viable and fertile, but show slight placental and fetal overgrowth, perhaps also owing to aberrant *Igf2* expression (Leighton et al. 1995; Ripoche et al. 1997). Overexpression of *H19* in two different lines of transgenic mice demonstrated a role for *H19* RNA itself in the regulation of growth (Gabory et al. 2009). The *H19* transgenic mice rescued the overgrowth phenotype of the $H19^{\Delta 3mat}$ strain and demonstrated a reduced growth phenotype that was first evident at E16.5 (Gabory et al. 2009).

The mechanism by which *H19* controls growth is not completely understood. However, it is thought that *H19* RNA acts *in trans* to modulate *IGF2* expression. Furthermore, *H19* RNA modulates expression of several imprinted genes, including *Cdkn1c*, *Gnas*, *Dlk1*, *Rtl1*, and *Igf2r* as part of an imprinted gene network (IGN) (Gabory et al. 2009; Varrault et al. 2006), which together regulate embryonic growth. Future studies will reveal how *H19* elicits control on this network of genes.

## *1.2 Airn*

*Antisense Igf2r RNA noncoding* (*Airn*) is another well-studied imprinted ncRNA that is located on mouse chromosome 17 (Fig. 2). *Airn* is a 108 kb long noncoding RNA that is found in the nucleus and is polyadenylated. *Airn* is largely unspliced, with only 5 % of nascent transcripts becoming spliced. However, these spliced transcripts make up 30 % of the steady-state RNA levels. An RNA that is likely the human *AIRN* homolog has been identified in the human genome as well. Like mouse *Airn*, its promoter lies within the second intron of *Igf2r*, however, its expression has only been detected in 16–40 % of Wilms' tumors.

*Airn* is antisense to the *Igf2r* gene, and is required for its imprinted expression in mouse. Human *Igf2r* is typically not imprinted. Two other genes, *Slc22a2* and

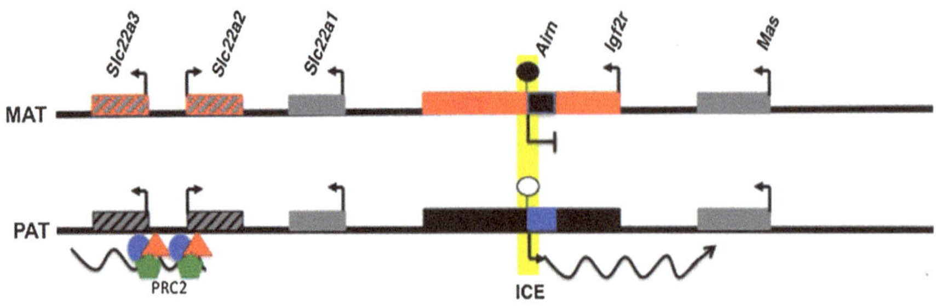


**Fig. 2** *Airn/Igf2r* imprinted region. Genes are represented by *colored rectangles* with *red* representing maternal allele-specific expression, *blue* representing paternal allele-specific expression, *gray* representing bi-allelic expression, *red/gray stripes* representing placental specific expression from the maternal allele, and *black/gray stripes* representing placental specific repression of the paternal allele. *Wavy lines* indicate lncRNAs. The *yellow* highlighted region indicates the imprinting control element (ICE). *Black* and *white circles* denote differentially methylated regions (DMRs) with *black* indicating the methylated allele and *white* indicating the unmethylated allele

*Slc22a3*, are also imprinted in mouse as a direct result of imprinted *Airn* expression. The imprinting of this locus is controlled by a 3.7 kb imprinting control element (ICE) that contains the *Airn* promoter, residing within intron two of the *Igf2r* gene (Lyle et al. 2000). The *Igf2r/Airn* ICE is differentially methylated, with methylation occurring exclusively on the maternally inherited allele. Deletion of the ICE in mice results in biallelic expression from all three genes, and truncating *Airn* at the boundary of the ICE also results in loss of imprinting for all three genes (Sleutels et al. 2002).

*Airn* acts in a bidirectional manner in *cis* to silence the paternal alleles of *Igf2r*, *Slc22a2*, and *Slc22a3* (Sleutels et al. 2002). *Airn* is transcribed from the plus strand. The *Igf2r* promoter and first exon lie distal to the *Airn* promoter, and *Igf2r* is transcribed from the minus strand. Therefore, the *Airn* transcript overlaps the first two exons and the promoter of *Igf2r* on the paternal allele. *Igf2r* is expressed preferentially from the maternal allele, being imprinted in all tissues with *Airn* expression (Sleutels et al. 2002). *Airn* was recently shown to mediate *Igf2r* imprinted expression via transcriptional interference (Latos et al. 2012). It displaces RNA Polymerase II from the *Igf2r* promoter without subsequent recruitment of H3K9me3. Since the transcription of the noncoding RNA across the *Igf2r* promoter is sufficient for its repression, neither the spliced or unspliced *Airn* products are required. Mouse embryonic stem (ES) cells that have truncated *Airn* that does not overlap the *Igf2r* promoter have biallelic *Igf2r* expression, whereas ES cells with truncated *Airn* that crossed the *Igf2r* promoter showed imprinted *Igf2r* expression (Latos et al. 2012). As a result of the *Airn*-mediated repression of paternal *Igf2r*, a CpG island at its promoter becomes paternally methylated. This second DMR is only established if *Airn* is transcribed across the *Igf2r* promoter (Santoro et al. 2013).

*Slc22a2* and *Slc22a3* lie proximal to *Airn* and are transcribed from the plus and minus strands, respectively. *Slc22a2* and *Slc22a3* are imprinted and expressed only from the maternal allele in some extraembryonic lineages (Zwart et al. 2001). Another gene, *Slc22a1*, lies proximal to *Airn* and between it and *Slc22a2*, but is not imprinted, presumably because it is not expressed in extraembryonic tissue. *Airn* does not overlap any of these genes. There are two hypotheses to explain how *Airn* represses *Slc22a2* and *Slc22a3*. One hypothesis posits that *Airn* represses *Slc22a3* by binding to the *Slc22a3* locus and recruiting EHMT2/G9a. This histone methyltransferase deposits the repressive histone modification, trimethylated histone H3, lysine 9 (H3K9me3) at the *Slc22a3* promoter (Nagano et al. 2008). The second hypothesis is that *Airn* transcription blocks the binding of transcriptional activator that establishes activating chromatin loops involving the *Slc22a2* and *Slc22a3* loci (Pauler et al. 2012). The repressed *Slc22a2/3* alleles recruit EHMT2/G9a independent of *Airn* RNA itself, leading to the H3K9me3 histone modification deposition at the *Slc22a2* and *Slc22a3* promoters. The locus then undergoes subsequent heterochromatization through the recruitment of the Polycomb group complexes 2 and 1 (PRC2 and PRC1) and deposition of their respective repressive histone modifications, H3K27me3 and H2A119u1. Finally, the heterochromatization leads to the compaction of the region, bringing *Airn* in proximity to the solute carrier genes (Pauler et al. 2012). While the precise mechanism by which *Airn* represses *Slc22a3* and *Slc22a2* is unknown, it is clear that *Airn* is in close proximity to the solute carrier genes. Using fluorescent in situ hybridization (FISH), *Airn* was shown to occupy a relatively large RNA signal that overlaps the *Slc22a3* locus in 11.5 days post-coitus (dpc) placentas, where *Slc22a3* is imprinted, suggesting that the *Airn* RNA is physically interacting with the *Slc22a3* locus (Nagano et al. 2008). By day 15.5 dpc, when *Slc22a3* is no longer imprinted, the *Airn* FISH signals in the placenta are smaller and no longer overlap *Slc22a3*. Truncated *Airn* alleles show the latter FISH signal and biallelic *Slc22a3* expression.

## *1.3 Kncq1ot1*

*Kcnq1ot1* is located on mouse chromosome 7 and human chromosome 11p15.5 (Verona et al. 2003) (Fig. 3). It is an imprinted long noncoding RNA that is transcribed by RNA Polymerase II, unspliced, and is nuclear localized. The RNA was recently reported to span 471 kb (Golding et al. 2011), although this conflicts with previous reports that estimated it to be 91 kb or 121 kb (Pandey et al. 2008; Redrup et al. 2009) and RNA-seq experiments agree with the latter estimate (Huang et al. 2011). *Kcnq1ot1* is expressed almost exclusively from the paternally inherited allele and silences genes in a bidirectional manner. The region controlled by *Kcnq1ot1* spans 1 Mb and includes 10 protein-coding genes that are maternally expressed (Paulsen et al. 1998). Like *Airn, Kcnq1ot1* controls imprinted gene expression in extraembryonic as well as embryonic tissues. *Kcnq1*, *Cdkn1c*,

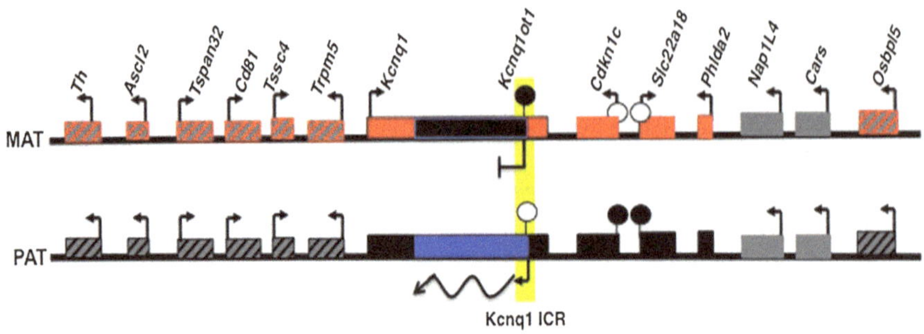


**Fig. 3** *Kcnq1ot1* imprinted region. Genes are represented by *colored rectangles* with *red* representing maternal allele-specific expression, *blue* representing paternal allele-specific expression, *gray* representing bi-allelic expression, *red/gray stripes* representing placental specific expression from the maternal allele, and *black/gray stripes* representing placental specific repression of the paternal allele. *Wavy lines* indicate lncRNAs. The *yellow* highlighted region indicates the imprinting control region (ICR). *Black* and *white circles* denote differentially methylated regions (DMRs) with *black* indicating the methylated allele and *white* indicating the unmethylated allele

*Slc22a18*, and *Phlda2* are controlled by *Kcnq1ot1* and are imprinted in both embryonic and extraembryonic tissues, while *Ascl2, Tspan32, CD81, Tssc4, Trpm5,* and *Osbpl5* are imprinted only in placenta (Umlauf et al. 2004; Shin et al. 2008; Lewis et al. 2004; Caspary et al. 1998). The *Kcnq1ot1* promoter is located in intron 10 of the *Kcnq1* gene. *Kcnq1ot1* is transcribed from the minus strand, and *Kcnq1* is transcribed from the plus strand, and thus these transcripts overlap. *Ascl2, Tspan32, CD81, Tssc4*, and *Trpm5* are proximal to the *Kcnq1ot1* promoter, while *Cdkn1c, Slc22a18, Phlda2*, and *Osbpl5* are distal to it. In addition, non-imprinted genes are interspersed with the imprinted genes; *Nap1l4* and *Cars1* show biallelic expression and are located between *Phlda2* and *Osbpl5*.

The imprinted control region for *Kcnq1ot1* is a DMR known as *Kcnq1* ICR, KvDMR, or IC2 (Engemann et al. 2000). Deletion of the *Kcnq1* ICR in mice leads to loss of imprinting at all genes in the region (Fitzpatrick et al. 2002; Mancini-Dinardo et al. 2006). Furthermore, termination of the *Kcnq1ot1* transcript 1.5 kb downstream of the transcriptional start site also leads to loss of imprinting at all loci, suggesting that either the act of transcription or the *Kcnq1ot1* RNA itself are required for establishing imprinted expression of the genes in the region (Shin et al. 2008; Fitzpatrick et al. 2002; Mancini-Dinardo et al. 2006). Using episomal vectors containing fragments of the ICR and a human placental cell line, an 890 bp region that is necessary for the silencing activity of *Kcnq1ot1* was identified (Mohammad et al. 2008). Deletion of the 890 bp silencing domain in mice led to relaxation of imprinted expression of the ubiquitously imprinted genes, *Kcnq1, Cdkn1c, Slc22a18*, and *Phlda2*, when paternally inherited. However, the relaxation of imprinting was variable. Loss of imprinted expression of *Kcnq1* was observed in both placenta and fetal liver, while loss of imprinting at the *Cdkn1c* and *Phlda2* loci was only seen in the placenta, and loss of imprinting at *Slc22a18* was only seen in the fetal liver. This loss of imprinting was accompanied by loss of DNA methylation at the *Cdkn1c* and *Slc22a18* somatic DMRs. Imprinting of the placenta-specific genes was not affected (Mohammad et al. 2008).

In addition to the silencing domain, the ICR harbors two conserved CpG islands and four conserved repeat elements. One of the CpG islands lies within the silencing domain, and the other includes the promoter of *Kcnq1ot1*. The repeat elements, termed MD, A, A1, and A2, are also located in the 5' end of the *Kcnq1ot1* RNA (Paulsen et al. 2005). The MD repeat elements are not required for silencing (Mancini-DiNardo et al. 2003), however, the A1 and A2 repeats lie within the silencing domain, and a point mutation in the A2 repeat abrogates some of the silencing function of *Kcnq1ot1*.

*Kcnq1ot1* seems to employ different mechanisms to silence the ubiquitous and placenta-specific imprinted genes. The placenta-specific imprinted genes have promoters and/or gene bodies bound by repressive histone modifications such as H3K27me3, H3K119ub1, and H3K9me3 (Umlauf et al. 2004; Lewis et al. 2004). Mice harboring mutations in *Eed, Ezh2*, and *Rnf2*, members of Polycomb Repressive Complexes 1 and 2 (PRC1 and PRC2) as well as *G9a* show loss of imprinted expression from the placenta-specific imprinted genes, but not in the ubiquitously imprinted genes (Mager et al. 2003; Terranova et al. 2008; Wagschal et al. 2008). *Kcnq1ot1* RNA has been implicated in the recruitment of repressive histone modifications to the imprinted genes at this locus. The *Kcnq1ot1* RNA can be immunoprecipitated using antibodies against *Ezh2, Suz12*, and *G9a* in placenta, but not liver (Terranova et al. 2008). Furthermore, immuno-FISH experiments show that *Ezh2* and H3K27me3 co-localize with *Kcnq1ot1* in an RNA Polymerase II-depleted domain (Terranova et al. 2008). Interestingly, the maintenance of imprinting in the placenta-specific imprinted genes does not require *Kcnqot1* RNA. Conditional depletion of the *Kcnq1ot1* RNA at 8.5 dpc did not result in the derepression of the placenta-specific genes and did not disrupt the recruitment of H3K27me3 or Ezh2 to their promoters (Mohammad et al. 2012). This suggests that *Kcnq1ot1* is required to recruit PRC2, PRC1, and presumably G9a during the establishment of imprinting in the placenta-specific genes, but it is not required for the continued recruitment of these complexes during the maintenance phase of imprinting in the placenta.

On the other hand, conditional depletion of *Kcnq1ot1* at 5.5 or at 8.5 dpc in mouse embryos led to the derepression of silenced alleles of the ubiquitously imprinted genes (Mohammad et al. 2012). This suggests that continued expression of *Kcnq1ot1* is necessary for both the establishment and maintenance of imprinted expression of the ubiquitously imprinted genes. Furthermore, DNA methylation at two somatic DMRs located at the promoters of *Cdkn1c* and *Slc22a18* is lost in both the embryonic and extraembryonic tissues in the absence of *Kcnq1ot1* expression (Mohammad et al. 2012). Therefore, *Kcnq1ot1* is also necessary for the maintenance of DNA methylation at these somatic DMRs.

Whether *Kcnq1ot1* RNA itself or the act of transcription through *Kcnq1ot1* is necessary for silencing the repressed paternal alleles is not known. However, Golding et al. showed that depletion of *Kcnq1ot1* RNA post-transcriptionally using shRNAs in embryonic stem, extraembryonic endoderm, and trophoblast stem cells had no effect on imprinted expression from the either the placenta-specific or ubiquitously imprinted genes (Golding et al. 2011). Conditional deletion of the transcript abrogates imprinted expression of the ubiquitously imprinted genes, but knockdown of the RNA post-transcriptionally in embryonic stem cells does not alter imprinted expression. This suggests that the act of transcription through the *Kcnq1ot1* locus may be more important than the RNA itself in the maintenance of imprinted expression of the ubiquitously imprinted genes. The placenta-specific imprinted genes, however, seem to lose the requirement for both *Kcnq1ot1* RNA and transcription after their imprinted expression has been established, and thus may rely on the transcriptional memory afforded by the Polycomb and/or G9a complexes for maintenance. Further experiments should help reveal whether *Kcnq1ot1* participates in chromatin silencing through transcriptional interference, recruiting repressive chromatin modification to specific compartments, or a combination of these mechanisms.

## *1.4 Gtl2*

The imprinted *Dlk1-Dio3* domain is located on mouse chromosome 12, and human chromosome 14 (Fig. 4). This 1 Mb region contains the paternally expressed protein-coding genes Delta-like homolog 1 (*Dlk1*), retrotransposon-like 1 (*Rtl1*), and iodothyronine deiodinase 3 (*Dio3*); several maternally expressed noncoding RNAs, including Gene-trap locus 2 (*Gtl2*)/maternally expressed gene 3 (*Meg3*), *Rtl1* antisense transcript (*Rtl1-as*), RNA imprinted and accumulated in nucleus (*Rian*), and Maternally expressed gene 9 (*Meg9*)/MicroRNA containing gene (*Mirg*); and a single transcript with bialleleic expression, *Dio3* antisense transcript (*Dio-as*) (Hagan et al. 2009). Within the *Dlk1-Dio3* domain, three differentially methylated regions (DMRs) have been identified, and each DMR is methylated exclusively on the paternal chromosome. An intergenic DMR (IG-DMR) is the germline DMR that regulates the allele-specific expression of all of the imprinted genes in this domain, and is located 13 kb upstream of the *Gtl2* promoter. Deletion of the IG-DMR from the maternally inherited chromosome results in the bidirectional loss of imprinting of all genes in the region, resulting in the activation of the maternally repressed imprinted genes (*Dlk1*, *Rtl1*, *Dio3*) and repression of the maternally expressed genes (*Gtl2*, *Rian*, and *Mirg*) (Buiting et al. 2003). Furthermore, the IG-DMR acts hierarchically to regulate the methylation status of the other two somatic DMRs within the region. One somatic DMR is located within the *Gtl2/Meg3* promoter. This DMR controls *Gtl2* expression as well as the expression of downstream maternally expressed genes, *Rtl1-as*, *Rian*, and *Mirg*. Disruption of the *Gtl2* DMR also alters the expression of both paternally and maternally expressed genes within the *Dlk1-Dio3* domain (Sekita et al. 2006; Takahashi et al. 2009).

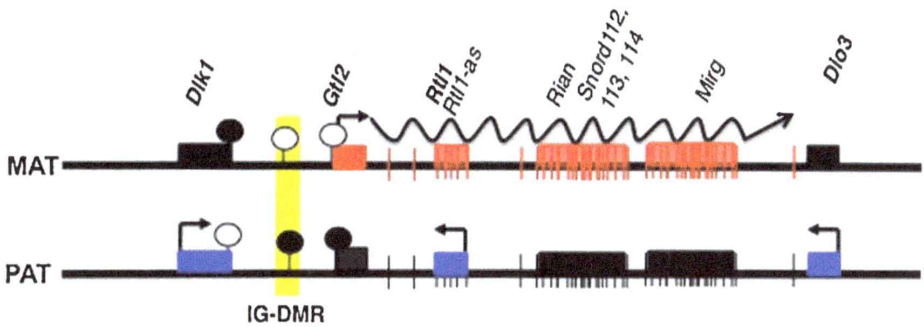


**Fig. 4** *Gtl2* imprinted region. Genes are represented by *colored rectangles* with *red* representing maternal allele-specific expression and *blue* representing paternal allele-specific expression. *Thin vertical lines* indicate miRNAs or snoRNAs. *Wavy lines* indicate lncRNAs. The *yellow* highlighted region indicates the intergenic differentially methylated region (IG-DMR). *Black* and *white circles* denote differentially methylated regions (DMRs) with *black* indicating the methylated allele and *white* indicating the unmethylated allele

The maternally expressed lncRNAS, *Gtl2/Meg3*, *Rtl1-as*, *Rian*, and *Mirg,* are all transcribed from the same strand and together span a genomic distance of approximately 200–250 kb in both mouse and human. Several pieces of evidence suggest that the maternally expressed lncRNAs might make up a larger polycistronic transcription unit. The maternal noncoding transcripts are all expressed in the same orientation as *Gtl2*, and typical promoter sequences are absent from the region (Tierling et al. 2006). Moreover, the expression of these transcripts is similar, suggesting that their expression may be coordinated (Tierling et al. 2006; Takada et al. 2000). From this putative transcription unit, 52 microRNAs (miRNAs; 42 in human) and three snoRNA clusters are produced. The individual portions of this transcriptional unit are detailed below.

*Gtl2/Meg3* encodes a long noncoding RNA (lncRNA) and is ubiquitously expressed in many tissues. It also serves as a host gene for at least one miRNA. In a genome-wide RNA-immunoprecipitation-seq experiment, *Gtl2/Meg3* was identified as an RNA that binds to the PRC2 component Ezh2. Zhao et al. then showed that post-transcriptional reduction of *Gtl2/Meg3* RNA resulted in loss of imprinted expression from the *Dlk1* gene and loss of the H3K27me3 repressive histone modification on the maternal allele (Zhao et al. 2010). Thus, the *Gtl2/Meg3* RNA itself is required for silencing the maternal allele of *Dlk1* by recruiting the PRC2 complex to the *Dlk1* promoter.

*Gtl2/Meg3* also has been shown to negatively regulate cell proliferation. In humans, hypermethylation of the *MEG3* promoter and loss of the *MEG3* RNA occurs in primary tumors and tumor cell lines, suggesting that it may be a tumor suppressor (Zhou et al. 2012; Zhao et al. 2005). Transfection of *MEG3* in human cancer cell lines increases p53 proteins levels (Zhou et al. 2007) by downregulating MDM2, an E3 ubiquitin ligase that ubiquitinates p53 and targets it for degradation by the proteosome (Zhou et al. 2007). Precisely how *MEG3* regulates MDM2 and p53 protein levels to function in tumor suppression is not known.

*Rtl1-as* is an antisense transcript to the paternally expressed *Rtl1* gene. The maternal *Rtl1-as* transcript hosts several miRNAs that are processed from hairpins that are fully complementary to the paternally transcribed *Rtl1* mRNA (Seitz et al. 2003; Davis et al. 2005). Three of the maternally expressed miRNAs hosted by the *Rtl1-as* gene have been shown to regulate the expression of *Rtl1* by mediating its degradation through RNA interference (Davis et al. 2005). To date, this is the only example of miRNA-mediated RNAi degradation involving reciprocally imprinted genes in mammals. Whether this is the sole mechanism by which *Rtl1-as* regulates *Rtl1* expression is not known.

Located ~25 kb downstream of *Rtl1* is the maternally expressed *Rian* transcript, a transcript which hosts the snoRNAs, *Snord112, Snord113,* and *Snord114.* All of these snoRNAs belong to the C/D box family of snoRNAs and are involved in the 2'-*O*-methylation of other RNA species, such as rRNA and spliceosomal RNA in the nucleolus (Bachellerie et al. 2002). In humans, these snoRNAs are arranged into two tandem arrays containing 9 and 32 paralogous copies of *SNORD113* and *SNORD114*, respectively (Bachellerie et al. 2002). The function of these snoRNAs is not well understood. The majority of C/D snoRNAs are ubiquitously expressed. However, the snoRNAs in this cluster are predominately expressed in the brain and lack complementarity to rRNA and snRNA within their sequences (Cavaille et al. 2002), suggesting that these snoRNAs may not function as canonical C/D snoRNAs and raises the possibility that these snoRNAs may have a brain-specific function.

There are two mature alternatively spliced isoforms of *Rian*: Maternally Expressed Gene 8 (*Meg8; EU434919*) and Imprinted RNA near *Meg3* (*Irm; AF498294*). *Meg8* is a 26-exon transcript that overlaps most of the *Rian* transcript (21 exons), and contains three miRNAs that are hosted within *Rian* (Hagan et al. 2009). Predicted targets of these miRNAs, include *Grb10, Lmna, Peg10,* and *Trp53* (Hagan et al. 2009). *Irm* is an 11-exon transcript that shares the last 10 exons and transcriptional termination sites with *Meg8. Irm* harbors the snoRNA clusters found within the *Rian* transcript (Hagan et al. 2009). Both *Meg8* and *Irm* are highly expressed in brain.

*Mirg* is located ~25 kb downstream of the C/D snoRNA cluster housed within *Rian* and contains ~40 miRNAs. Although some of the miRNAs in *Mirg* are single copy, the majority of the miRNAs are arranged in tandem repeats of closely related sequences (Seitz et al. 2004). These miRNAs are expressed in the embryo and placenta, but their expression is restricted to the brain in the adult (Seitz et al. 2004). The IG-DMR is located roughly 200 kb upstream from the miRNA cluster and regulates its expression. Deletion of the IG-DMR results in the loss of miRNA expression (Seitz et al. 2004), similar to that of other maternally expressed transcripts (Lin et al. 2003). The miRNA miR-134 is one of the miRNAs located within the *Mirg* transcript and is thought to negatively regulate dendritic spine size and synaptic plasticity through the inhibition of the translation of Lim-domain-containing-protein kinase 1 (Limk1) (Schratt et al. 2006). The other miRNAs found within this cluster are often dysregulated in a variety of disease pathologies (Benetatos et al. 2013).

## *1.5 Nespas*

The *Gnas/Nespas* locus is located on chromosome 2 in mouse, and on chromosome 20q13.11 in humans (Fig. 5). This complex locus includes two protein-coding transcripts, *Gnas* and *Gnasxl,* two lncRNAs, *Nespas* and *Exon 1A* (*EXON A/B* in humans), and one transcript, *Nesp* that is both a regulatory lncRNA and a protein-coding gene (Frohlich et al. 2010). Four of the transcripts, *Nesp, Gnasxl*, *Exon 1A*, and *Gnas* are transcribed from the plus strand in both human and mouse and are part of the same transcription unit. *Nesp*, *Gnasxl,* and *Exon1A* are alternatively spliced onto exon 2 of *Gnas. Nesp* has maternal-specific expression, *Gnasxl* and *Exon 1A* have paternal-specific expression, and *Gnas* has biallelic expression in most tissues. The major regulatory lncRNA in this region is *Nespas*, a paternally expressed transcript with both unspliced and spliced isoforms. *Nespas* is transcribed from the minus strand in an antisense orientation to *Nesp*. *Nespas* transcription starts approximately 2 kb upstream of the *Gnasxl* gene and ends approximately 2.5 kb past the start of the *Nesp* gene.

The imprinting control region (ICR) for this locus is a DMR located within the promoters of *Gnasxl* and *Nespas.* It has dense DNA methylation on the maternally inherited allele and is unmethylated on the paternally inherited allele. This germline methylation imprint is acquired in oocytes (Coombes et al. 2003). The *Nespas* and *Gnasxl* genes are expressed from the unmethylated paternal allele and are repressed on the methylated maternal allele. Deletion of the paternal DMR, including the *Nespas* promoter, leads to locus-wide effects on gene expression, including derepression of *Nesp* and *Gnas*, loss of *Nespas* expression, and partial repression of *Gnasxl* and *Exon 1A*. Additionally, the methylation of two other DMRs in the region is affected by loss of the paternal ICR. Methylation is lost at

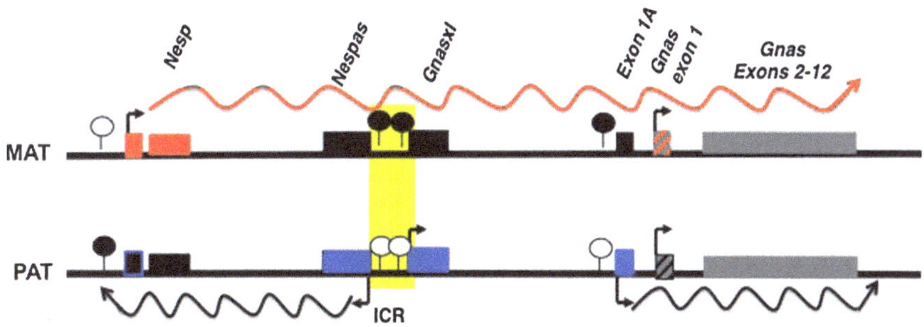


**Fig. 5** *Gnas/Nespas* imprinted region. Genes are represented by *colored rectangles* with *red* representing maternal allele-specific expression, *blue* representing paternal allele-specific expression, and *gray* representing biallelic expression. *Wavy lines* indicate lncRNAs. The *red wavy line* labels the *Nesp* noncoding RNA that is only expressed in the maternal germline. The *yellow* highlighted region indicates the imprinting control region (ICR). *Black* and *white circles* denote differentially methylated regions (DMRs) with *black* indicating the methylated allele and *white* indicating the unmethylated allele

the DMR located at the paternal promoter of *Nesp* and the DMR at the promoter of *Exon 1A* becomes partially methylated on the paternal allele (Williamson et al. 2006).

The paternally expressed *Nespas* lncRNA acts in *cis* to silence the *Nesp* gene. *Nespas* is transcribed from the unmethylated paternal allele and has an antisense orientation with respect to the *Nesp* transcript. Truncation of *Nespas* using a polyadenylation cassette inserted into exon 1 of the lncRNA results in the loss of its antisense silencing function for *Nesp* (Williamson et al. 2011). Similar to the deletion of the ICR, truncation of *Nespas* also caused loss of DNA methylation on the paternal allele of the *Nesp* DMR (Williamson et al. 2011). The paternal *Nesp* promoter is usually devoid of the activating histone mark histone H3 lysine 4 trimethylation (H3K4me3). However, increased levels of H3K4 methylation and depletion of histone H3 lysine 9 trimethylation (H3K9me3) are observed in *Nespas* truncation mutants (Williamson et al. 2011). The loss of repressive histone modifications and gain of active histone modifications is thought to prevent the recruitment or activity of DNA methyltransferases and cause loss of methylation at the *Nesp* DMR (Ooi et al. 2007 Zhang et al. 2010). It is not known whether the *Nespas* RNA or the process of its transcription is responsible for mediating the silencing of *Nesp*.

The *Nesp* coding and noncoding transcripts are maternally expressed and transcribed through the entire cluster including the *Nespas-Gnasxl* DMR and *Exon 1A* DMR. Truncation of this long transcript upstream of the two DMRs disrupts the methylation at both DMRs when the truncation is passed through the female germline (Chotalia et al. 2009). Transcription through these two germline DMRs in oocytes is thought to play an important role in establishing their maternal allele-specific methylation.

The paternally expressed *Exon 1A* lncRNA regulates the tissue-specific imprinted expression of *Gnas* (Liu et al. 2000). This lncRNA is controlled by the *Exon 1A* DMR, which is methylated on the maternally inherited allele and spans the *Exon1A* promoter. The unmethylated paternal *Exon 1A* DMR drives expression of the ubiquitously expressed *Exon 1A* lncRNA. The *Exon1A* lncRNA is transcribed across the *Gnas* promoter and is spliced onto exon 2, skipping the first exon of *Gnas*. The transcript then shares the terminal 11 exons with *Gnas*. *Gnas* is biallelically expressed in most tissues, but is paternally repressed in specific tissues (Yu et al. 1998). Expression of the *Exon 1A* lncRNA represses *Gnas* only in those specific tissues (Liu et al. 2000a; 2000b). An increase in the levels of the ncRNA *Exon 1A* is associated with the loss of *Gnas* expression and loss of methylation of the *Exon 1A* DMR (Frohlich et al. 2010; Bastepe et al. 2005), and the levels of *Exon 1A* are highest in tissues in which *Gnas* is paternally repressed (Eaton et al. 2012). Deletion of *Exon 1A* or truncation of *Exon 1A*-containting transcripts by insertion of a polyadenylation cassette within the exon resulted in the upregulation of *Gnas* due to the loss of imprinting of *Gnas* in tissues in which it is normally imprinted and repressed, but does not affect the imprinting of other transcripts in the locus (Eaton et al. 2012; Liu et al. 2005; Williamson et al. 2004). It seems most likely that the transcription of *Exon 1A* ncRNA through the *Gnas* promoter regulates the tissue-specific imprinted expression of *Gnas*, perhaps through transcriptional interference, although it is also possible that *Exon 1A* harbors a silencing or enhancer-blocking element that is disrupted in deletion or termination mutants.

## *1.6 Snrpn/Lncat/Ube3a-ats*

The *Snurf-Snrpn* gene is located in the q11-q13 region of human chromosome 15 and the central portion of mouse chromosome 7 (Fig. 6). The *Snurf-Snrpn* gene gives rise to two protein products, Snurf and Snrpn, as well as a long noncoding antisense transcript (*LNCAT*) that is also known as *Ube3a antisense transcript* (*Ube3a-ats*) (Rougeulle et al. 1998). The human *Snurf-Snrpn* transcript spans a genomic distance of approximately 600 kb, while the mouse locus is approximately 1 Mb (http://www.genome.ucsc.edu). The entire lncRNA is only made in the neurons of both species. The *Snurf-Snrpn/Lncat/Ube3a-ats* transcript is a highly spliced and transcribed by RNA polymerase II. The mouse transcript is short-lived and not polyadenylated (Meng et al. 2012) however, the human transcript is polyadenylated in nonneuronal tissues (Wevrick and Francke 1997) and is likely to be polyadenylated in neurons as well. Its half-life is not known. The *Snurf-Snrpn* transcript is expressed exclusively from the paternally inherited allele and only participates in the repression of one gene, *Ube3a* (Rougeulle et al. 1998; Meng et al. 2012; Chamberlain and Brannan 2001). *Ube3a* is expressed exclusively from the maternally inherited allele (Rougeulle et al. 1997). Another gene, *Atp10a*, is imprinted and expressed from the maternal allele in some individuals (Hogart et al. 2008). However, *Atp10a* is not imprinted in mouse (Dubose et al.

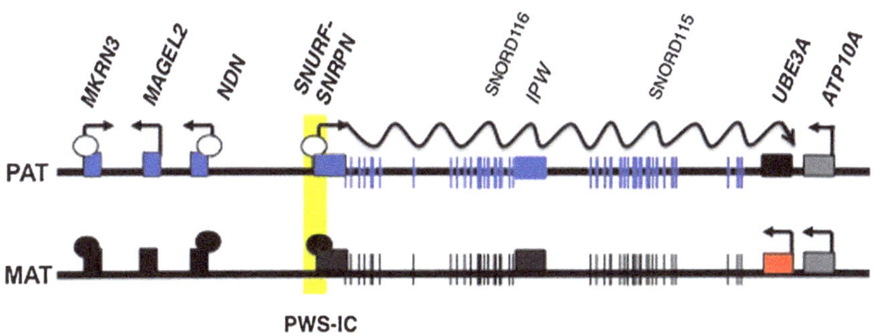


**Fig. 6** Human *SNRPN/LNCAT/UBE3A-ATS* imprinted region. Genes are represented by *colored rectangles* with *red* representing maternal allele-specific expression, *blue* representing paternal allele-specific expression, and *gray* representing biallelic expression. *Wavy lines* indicate lncRNA. The *yellow* highlighted region indicates the Prader-Willi imprinting center (PWS-IC). *Black* and *white circles* denote differentially methylated regions (DMRs) with *black* indicating the methylated allele and *white* indicating the unmethylated allele. Gene expression shown represents the neuron-specific pattern of gene expression

2009), and it is not known whether *Snurf-Snrpn* participates in allele-specific expression in humans.

*Snurf-Snrpn* is controlled by a DMR known as the Prader-Willi syndrome imprinting center (PWS-IC) (Saitoh et al. 1996). The unmethylated paternal copy of the PWS-IC acts as a promoter for *Snurf-Snrpn* and an enhancer for imprinted genes upstream of it (Brannan and Bartolomei 1999). An upstream regulatory element known as the Angelman syndrome imprinting center (AS-IC) lies upstream of the PWS-IC and seems to encompass at least one upstream, noncoding exon of *Snurf-Snrpn* (Buiting et al. 2003; Buiting et al. 2001). Together, the PWS-IC plus the AS-IC comprise the ICR for this locus. In mouse, the AS-IC functions to direct expression of the *Snurf-Snrpn* noncoding RNA through the PWS-IC in the maternal germline, which may be required to establish the maternal allele-specific methylation (Smith et al. 2011). The murine AS-IC appears to be any upstream exon of *Snurf-Snrpn* that is capable of promoting transcription through the PWS-IC. Although the act of transcription through the PWS-IC is thought to be necessary for establishing this methylation imprint, it is not known whether the noncoding RNA itself plays a role.

In mouse, only the protein-coding portions of Snurf-Snrpn are transcribed in most tissues. In brain, and more specifically, neurons alone, the entire long noncoding transcript is transcribed. Numata et al. reported that the first 25 kb of the *Ube3a* sense transcript, including the first two exons, is expressed from both alleles in brain and that the antisense transcript terminates between the second and third exons of *Ube3a* (Numata et al.2011). However, Meng et al., reported that the mouse antisense transcript spans the entire *Ube3a* locus and proceeds across the *Ube3a* promoter (Meng et al. 2012). The proximal half of the human *SNURF-SNRPN* lncRNA is produced in nonneurons (Castle et al. 2010). Transcription initiates at the major promoter of *SNURF-SNRPN* and ends at the *IPW* lncRNA. A several lncRNAs are produced as part of this transcription unit, including *PAR-SN, PAR5, HBT8*, and *IPW* (Wevrick et al. 1994; Landers et al. 2004; Chamberlain et al. 2010). Whether these RNAs have independent functions or are simply byproducts of the transcription unit is not known. In neurons, the noncoding *SNURF-SNRPN* transcript proceeds beyond *IPW* and ultimately ends within the first intron of *UBE3A* (Castle et al.2010). Additional lncRNAs, *PAR1, PAR4* and *UBE3A-ats* (Rougeulle et al.1998; Chamberlain et al. 2010; Ning et al. 1996) are produced from the neuron-specific transcript.

One major function of the *SNURF-SNRPN* lncRNA is to serve as a host gene for snoRNAs. Two major clusters of snoRNAs (*SNORD116/HBII-85 and SNORD115/HBII-52*) and five singleton snoRNAs (*SNORD107/HBII-436, SNORD64/HBII-13, SNORD108/HBII437, SNORD109A/HBII438A,* and *SNORD109B/HBII-438B)* are all produced from the *SNURF-SNRPN* lncRNA and are conserved in the mouse locus (Cavaille et al. 2000). Additionally, five sno-lncRNAs are produced from introns that harbor two individual snoRNAs and fail to use an alternatively spliced exon that lies between them (Yang et al. 2012). These sno-lncRNAs have not been described in mouse. Sno-lncRNAs are stable and long-lived than the host transcript, reaching high steady-state levels in many tissues.

The question of how the *SNURF-SNRPN* lncRNA transcript regulates neuron-specific imprinted expression of *UBE3A* remains a mystery. The transcript itself undergoes neuron-specific regulation such that it does not overlap *UBE3A* in nonneurons, and does overlap *UBE3A* in neurons. Since *UBE3A* is only imprinted in neurons, the mechanism by which *UBE3A-ATS* silences *UBE3A* is likely to require overlap between the genes. Meng et al. inserted a transcriptional termination cassette in mouse embryonic stem cells to terminate *Ube3a-ats* just before it overlapped the *Ube3a* gene (Meng et al. 2012). When these embryonic stem cells were differentiated into neurons, paternal *Ube3a* was expressed. These data support the hypothesis that *Ube3a-ats* must overlap paternal *Ube3a* to elicit repression. Future studies will reveal whether it is the act of transcription through the *Ube3a* locus or the recruitment of repressive histone modifications to the *Ube3a* promoter that initiates the repression of the paternal allele of this gene.

## 2 Perspectives

Imprinted loci make frequent use of lncRNAs to regulate allele-specific expression. In some cases, this allele-specific regulation may also be tissue-specific. Most imprinting control regions exert their main effect on an lncRNA, which is central to regulating parent-of-origin specific gene expression across the entire cluster of imprinted genes. The relative agility of lncRNAs belies their use in addition to other epigenetic modifications such as DNA methylation and covalent histone modifications. In fact, lncRNAs are at times used to establish these epigenetic modifications at imprinted loci. The lncRNAs function through various mechanisms. Common themes include the use of lncRNAs to recruit repressive histone modifications, which mainly occurs in extraembryonic lineages, and the use of lncRNAs as transcriptional templates, where the act of transcription of an lncRNA leads to the repression of transcripts that are normally transcribed from the opposite strand. Imprinted lncRNAs can also serve as host transcripts for miRNAs or snoRNAs. While imprinted lncRNAs are functioning in complex loci with allele- and tissue-specific regulatory paradigms, the basic mechanisms by which they function are undoubtedly at play in other non-imprinted loci across the genome.

## References

Brannan, C. I., Dees, E. C., Ingram, R. S., & Tilghman, S. M. (1990). The product of the H19 gene may function as an RNA. *Molecular and Cellular Biology, 10*(1), 28–36.

Bartolomei, M. S., Zemel, S., & Tilghman, S. M. (1991). Parental imprinting of the mouse H19 gene. *Nature, 351*(6322), 153–155.

Cai, X., & Cullen, B. R. (2007). The imprinted H19 noncoding RNA is a primary microRNA precursor. *RNA, 13*(3), 313–316. doi:10.1261/rna.351707.

Poirier, F., Chan, C. T., Timmons, P. M., Robertson, E. J., Evans, M. J., & Rigby, P. W. (1991). The murine H19 gene is activated during embryonic stem cell differentiation in vitro and at the time of implantation in the developing embryo. *Development, 113*(4), 1105–1114.

Pachnis, V., Brannan, C. I., & Tilghman, S. M. (1988). The structure and expression of a novel gene activated in early mouse embryogenesis. *EMBO Journal, 7*(3), 673–681.
Castle, J. C., Armour, C. D., Lower, M., Haynor, D., Biery, M., Bouzek, H., et al. (2010). Digital genome-wide ncRNA expression, including SnoRNAs, across 11 human tissues using polyA-neutral amplification. *PLoS ONE, 5*(7), e11779. doi:10.1371/journal.pone.0011779.
Dudek, K. A., Lafont, J. E., Martinez-Sanchez, A., & Murphy, C. L. (2010). Type II collagen expression is regulated by tissue-specific miR-675 in human articular chondrocytes. *Journal of Biological Chemistry, 285*(32), 24381–24387. doi:10.1074/jbc.M110.111328.
Leighton, P. A., Ingram, R. S., Eggenschwiler, J., Efstratiadis, A., & Tilghman, S. M. (1995). Disruption of imprinting caused by deletion of the H19 gene region in mice. *Nature, 375*(6526), 34–39.
Ripoche, M. A., Kress, C., Poirier, F., & Dandolo, L. (1997). Deletion of the H19 transcription unit reveals the existence of a putative imprinting control element. *Genes & Development, 11*(12), 1596–1604.
Gabory, A., Ripoche, M. A., Le Digarcher, A., Watrin, F., Ziyyat, A., Forne, T., et al. (2009). H19 acts as a trans regulator of the imprinted gene network controlling growth in mice. *Development, 136*(20), 3413–3421. doi:10.1242/dev.036061.
Varrault, A., Gueydan, C., Delalbre, A., Bellmann, A., Houssami, S., Aknin, C., et al. (2006). Zac1 regulates an imprinted gene network critically involved in the control of embryonic growth. *Developmental Cell, 11*(5), 711–722. doi:10.1016/j.devcel.2006.09.003.
Lyle, R., Watanabe, D., te Vruchte, D., Lerchner, W., Smrzka, O. W., Wutz, A., et al. (2000). The imprinted antisense RNA at the Igf2r locus overlaps but does not imprint Mas1. *Nature Genetics, 25*(1), 19–21.
Sleutels, F., Zwart, R., & Barlow, D. P. (2002). The non-coding Air RNA is required for silencing autosomal imprinted genes. *Nature, 415*(6873), 810–813.
Latos, P. A., Pauler, F. M., Koerner, M. V., Senergin, H. B., Hudson, Q. J., Stocsits, R. R., et al. (2012). Airn transcriptional overlap, but not its lncRNA products, induces imprinted Igf2r silencing. *Science, 338*(6113), 1469–1472. doi:10.1126/science.1228110.
Santoro, F., Mayer, D., Klement, R. M., Warczok, K. E., Stukalov, A., Barlow, D. P., et al. (2013). Imprinted Igf2r silencing depends on continuous Airn lncRNA expression and is not restricted to a developmental window. *Development, 140*(6), 1184–1195. doi:10.1242/dev.088849.
Zwart, R., Sleutels, F., Wutz, A., Schinkel, A. H., & Barlow, D. P. (2001). Bidirectional action of the Igf2r imprint control element on upstream and downstream imprinted genes. *Genes & Development, 15*(18), 2361–2366.
Nagano, T., Mitchell, J. A., Sanz, L. A., Pauler, F. M., Ferguson-Smith, A. C., Feil, R., et al. (2008). The Air noncoding RNA epigenetically silences transcription by targeting G9a to chromatin. *Science, 322*(5908), 1717–1720. doi:10.1126/science.1163802.
Pauler, F. M., Barlow, D. P., & Hudson, Q. J. (2012). Mechanisms of long range silencing by imprinted macro non-coding RNAs. *Current Opinion in Genetics & Development, 22*(3), 283–289. doi:10.1016/j.gde.2012.02.005.
Verona, R. I., Mann, M. R., & Bartolomei, M. S. (2003). Genomic imprinting: Intricacies of epigenetic regulation in clusters. *Annual Review of Cell and Developmental Biology, 19*, 237–259. doi:10.1146/annurev.cellbio.19.111401.092717.
Golding, M. C., Magri, L. S., Zhang, L., Lalone, S. A., Higgins, M. J., & Mann, M. R. (2011). Depletion of Kcnq1ot1 non-coding RNA does not affect imprinting maintenance in stem cells. *Development, 138*(17), 3667–3678. doi:10.1242/dev.057778.
Pandey, R. R., Mondal, T., Mohammad, F., Enroth, S., Redrup, L., Komorowski, J., et al. (2008). Kcnq1ot1 antisense noncoding RNA mediates lineage-specific transcriptional silencing through chromatin-level regulation. *Molecular Cell, 32*(2), 232–246. doi:10.1016/j.molcel.2008.08.022.
Redrup, L., Branco, M. R., Perdeaux, E. R., Krueger, C., Lewis, A., Santos, F., et al. (2009). The long noncoding RNA Kcnq1ot1 organises a lineage-specific nuclear domain for epigenetic gene silencing. *Development, 136*(4), 525–530. doi:10.1242/dev.031328.

Huang, R., Jaritz, M., Guenzl, P., Vlatkovic, I., Sommer, A., Tamir, I. M., et al. (2011). An RNA-Seq strategy to detect the complete coding and non-coding transcriptome including full-length imprinted macro ncRNAs. *PLoS ONE, 6*(11), e27288. doi:10.1371/journal.pone.0027288.
Paulsen, M., Davies, K. R., Bowden, L. M., Villar, A. J., Franck, O., Fuermann, M., et al. (1998). Syntenic organization of the mouse distal chromosome 7 imprinting cluster and the Beckwith-Wiedemann syndrome region in chromosome 11p15.5. *Human Molecular Genetics, 7*(7), 1149–1159.
Umlauf, D., Goto, Y., Cao, R., Cerqueira, F., Wagschal, A., Zhang, Y., et al. (2004). Imprinting along the Kcnq1 domain on mouse chromosome 7 involves repressive histone methylation and recruitment of Polycomb group complexes. *Nature Genetics, 36*(12), 1296–1300. doi:10.1038/ng1467.
Shin, J. Y., Fitzpatrick, G. V., & Higgins, M. J. (2008). Two distinct mechanisms of silencing by the KvDMR1 imprinting control region. *EMBO Journal, 27*(1), 168–178. doi:10.1038/sj.emboj.7601960.
Lewis, A., Mitsuya, K., Umlauf, D., Smith, P., Dean, W., Walter, J., et al. (2004). Imprinting on distal chromosome 7 in the placenta involves repressive histone methylation independent of DNA methylation. *Nature Genetics, 36*(12), 1291–1295. doi:10.1038/ng1468.
Caspary, T., Cleary, M. A., Baker, C. C., Guan, X. J., & Tilghman, S. M. (1998). Multiple mechanisms regulate imprinting of the mouse distal chromosome 7 gene cluster. *Molecular and Cellular Biology, 18*(6), 3466–3474.
Engemann, S., Strodicke, M., Paulsen, M., Franck, O., Reinhardt, R., Lane, N., et al. (2000). Sequence and functional comparison in the Beckwith-Wiedemann region: Implications for a novel imprinting centre and extended imprinting. *Human Molecular Genetics, 9*(18), 2691–2706.
Fitzpatrick G. V., Soloway P. D., Higgins M. J. (2002) Regional loss of imprinting and growth deficiency in mice with a targeted deletion of KvDMR1. *Nature Genetics,* 32,426–431.
Mancini-Dinardo, D., Steele, S. J., Levorse, J. M., Ingram, R. S., & Tilghman, S. M. (2006). Elongation of the Kcnq1ot1 transcript is required for genomic imprinting of neighboring genes. *Genes & Development, 20*(10), 1268–1282. doi:10.1101/gad.1416906.
Mohammad, F., Pandey, R. R., Nagano, T., Chakalova, L., Mondal, T., Fraser, P., et al. (2008). Kcnq1ot1/Lit1 noncoding RNA mediates transcriptional silencing by targeting to the perinucleolar region. *Molecular and Cellular Biology, 28*(11), 3713–3728. doi:10.1128/MCB.02263-07.
Paulsen, M., Khare, T., Burgard, C., Tierling, S., & Walter, J. (2005). Evolution of the Beckwith-Wiedemann syndrome region in vertebrates. *Genome Research, 15*(1), 146–153. doi:10.1101/gr.2689805.
Mancini-DiNardo, D., Steele, S. J., Ingram, R. S., & Tilghman, S. M. (2003). A differentially methylated region within the gene Kcnq1 functions as an imprinted promoter and silencer. *Human Molecular Genetics, 12*(3), 283–294.
Mager, J., Montgomery, N. D., de Villena, F. P., & Magnuson, T. (2003). Genome imprinting regulated by the mouse Polycomb group protein Eed. *Nature Genetics, 33*(4), 502–507.
Terranova, R., Yokobayashi, S., Stadler, M. B., Otte, A. P., van Lohuizen, M., Orkin, S. H., et al. (2008). Polycomb group proteins Ezh2 and Rnf2 direct genomic contraction and imprinted repression in early mouse embryos. *Developmental Cell, 15*(5), 668–679. doi:10.1016/j.devcel.2008.08.015.
Wagschal, A., Sutherland, H. G., Woodfine, K., Henckel, A., Chebli, K., Schulz, R., et al. (2008). G9a histone methyltransferase contributes to imprinting in the mouse placenta. *Molecular and Cellular Biology, 28*(3), 1104–1113. doi:10.1128/MCB.01111-07.
Mohammad, F., Pandey, G. K., Mondal, T., Enroth, S., Redrup, L., Gyllensten, U., et al. (2012). Long noncoding RNA-mediated maintenance of DNA methylation and transcriptional gene silencing. *Development, 139*(15), 2792–2803. doi:10.1242/dev.079566.
Hagan J. P., O'Neill B. L., Stewart C. L., Kozlov S. V., Croce C. M. (2009) At least ten genes define the imprinted Dlk1-Dio3 cluster on mouse chromosome 12qF1. *PLoS One* 4(2):e4352. doi: 4310.1371/journal.pone.0004352. Epub 0002009 Feb 0004355.

Buiting K., Gross S., Lich C., Gillessen-Kaesbach G., el-Maarri O., Horsthemke B. (2003) Epimutations in Prader-Willi and angelman syndromes: A molecular study of 136 patients with an imprinting defect. *The American Journal of Human Genetics* 72(3), 571–577.

Sekita, Y., Wagatsuma, H., Irie, M., Kobayashi, S., Kohda, T., Matsuda, J., et al. (2006). Aberrant regulation of imprinted gene expression in Gtl2lacZ mice. *Cytogenet Genome Res, 113*(1–4), 223–229.

Takahashi N., Okamoto A., Kobayashi R., Shirai M., Obata Y., Ogawa H., Sotomaru Y., Kono T. (2009) Deletion of Gtl2, imprinted non-coding RNA, with its differentially methylated region induces lethal parent-origin-dependent defects in mice. Human Molecular Genetics 18(10), 1879–1888. doi: 1810.1093/hmg/ddp1108. Epub 2009 Mar 1874.

Tierling S., Dalbert S., Schoppenhorst S., Tsai C. E., Oliger S., Ferguson-Smith A. C., Paulsen M., Walter J. (2006) High-resolution map and imprinting analysis of the Gtl2-Dnchc1 domain on mouse chromosome 12. *Genomics* 87(2), 225–235. Epub 2005 November 2023.

Takada, S., Tevendale, M., Baker, J., Georgiades, P., Campbell, E., Freeman, T., et al. (2000). Delta-like and gtl2 are reciprocally expressed, differentially methylated linked imprinted genes on mouse chromosome 12. *Current Biology, 10*(18), 1135–1138.

Zhao, J., Ohsumi, T. K., Kung, J. T., Ogawa, Y., Grau, D. J., Sarma, K., et al. (2010). Genome-wide identification of polycomb-associated RNAs by RIP-seq. *Molecular Cell, 40*(6), 939–953. doi:10.1016/j.molcel.2010.12.011.

Zhou Y., Zhang X., Klibanski A. (2012) MEG3 noncoding RNA: A tumor suppressor. *Journal of Molecular Endocrinology* 48(3), R45–53. doi: 10.1530/JME-1512-0008. Print 2012.

Zhao J., Dahle D., Zhou Y., Zhang X., Klibanski A. (2005) Hypermethylation of the promoter region is associated with the loss of MEG3 gene expression in human pituitary tumors. *The Journal of Clinical Endocrinology and Metabolism* 90(4), 2179–2186. Epub 2005 Jan 2111.

Zhou Y., Zhong Y., Wang Y., Zhang X., Batista D. L., Gejman R., Ansell P. J., Zhao J., Weng C., Klibanski A. (2007) Activation of p53 by MEG3 non-coding RNA. *Journal of Biological Chemistry* 282(34), 24731–24742. Epub 22007 Jun 24713.

Seitz, H., Youngson, N., Lin, S. P., Dalbert, S., Paulsen, M., Bachellerie, J. P., et al. (2003). Imprinted microRNA genes transcribed antisense to a reciprocally imprinted retrotransposon-like gene. *Nature Genetics, 34*(3), 261–262.

Davis, E., Caiment, F., Tordoir, X., Cavaille, J., Ferguson-Smith, A., Cockett, N., et al. (2005). RNAi-mediated allelic trans-interaction at the imprinted Rtl1/Peg11 locus. *Current Biology, 15*(8), 743–749.

Bachellerie, J. P., Cavaille, J., & Huttenhofer, A. (2002). The expanding snoRNA world. *Biochimie, 84*(8), 775–790.

Cavaille, J., Seitz, H., Paulsen, M., Ferguson-Smith, A. C., & Bachellerie, J. P. (2002). Identification of tandemly-repeated C/D snoRNA genes at the imprinted human 14q32 domain reminiscent of those at the Prader-Willi/Angelman syndrome region. *Human Molecular Genetics, 11*(13), 1527–1538.

Seitz H., Royo H., Bortolin M. L., Lin S. P., Ferguson-Smith A. C., Cavaille J. (2004) A large imprinted microRNA gene cluster at the mouse Dlk1-Gtl2 domain. *Genome Research* 14(9), 1741–1748. Epub 2004 August 1712.

Lin S. P., Youngson N., Takada S., Seitz H., Reik W., Paulsen M., Cavaille J., Ferguson-Smith A. C. (2003) Asymmetric regulation of imprinting on the maternal and paternal chromosomes at the Dlk1-Gtl2 imprinted cluster on mouse chromosome 12. Nature Genetics 35(1), 97–102. Epub 2003 August 2024.

Schratt, G. M., Tuebing, F., Nigh, E. A., Kane, C. G., Sabatini, M. E., Kiebler, M., et al. (2006). A brain-specific microRNA regulates dendritic spine development. *Nature, 439*(7074), 283–289.

Benetatos L., Hatzimichael E., Londin E., Vartholomatos G., Loher P., Rigoutsos I., Briasoulis E. (2013) The microRNAs within the DLK1-DIO3 genomic region: Involvement in disease pathogenesis. *Cellular and Molecular Life Sciences* 70(5), 795–814. doi: 710.1007/s00018-00012-01080-00018. Epub 02012 July 00024.

Frohlich L. F., Mrakovcic M., Steinborn R., Chung U. I., Bastepe M., Juppner H. (2010) Targeted deletion of the Nesp55 DMR defines another Gnas imprinting control region and provides a mouse model of autosomal dominant PHP-Ib. *Proceedings of the National Academy of Science of the United States of America* 107(20), 9275–9280. doi: 9210.1073/pnas.0910224107. Epub 0910222010 Apr 0910224128.

Coombes, C., Arnaud, P., Gordon, E., Dean, W., Coar, E. A., Williamson, C. M., et al. (2003). Epigenetic properties and identification of an imprint mark in the Nesp-Gnasxl domain of the mouse Gnas imprinted locus. *Molecular and Cellular Biology, 23*(16), 5475–5488.

Williamson C. M., Turner M. D., Ball S. T., Nottingham W. T., Glenister P., Fray M., Tymowska-Lalanne Z., Plagge A., Powles-Glover N., Kelsey G., Maconochie M., Peters J. (2006) Identification of an imprinting control region affecting the expression of all transcripts in the Gnas cluster. *Nature Genetics* 38(3), 350–355. Epub 2006 February 2005.

Williamson C. M., Ball S. T., Dawson C., Mehta S., Beechey C. V., Fray M., Teboul L., Dear T. N., Kelsey G., Peters J. (2011) Uncoupling antisense-mediated silencing and DNA methylation in the imprinted Gnas cluster. *PLoS Genetics* 7(3):e1001347. doi: 1001310.1001371/journal.pgen.1001347. Epub 1002011 Mar 1001324.

Ooi, S. K., Qiu, C., Bernstein, E., Li, K., Jia, D., Yang, Z., et al. (2007). DNMT3L connects unmethylated lysine 4 of histone H3 to de novo methylation of DNA. *Nature, 448*(7154), 714–717.

Zhang Y., Jurkowska R., Soeroes S., Rajavelu A., Dhayalan A., Bock I., Rathert P., Brandt O., Reinhardt R., Fischle W., Jeltsch A. (2010) Chromatin methylation activity of Dnmt3a and Dnmt3a/3L is guided by interaction of the ADD domain with the histone H3 tail. *Nucleic Acids Research* 38(13), 4246–4253. doi: 4210.1093/nar/gkq4147. Epub 2010 March 4211.

Chotalia, M., Smallwood, S. A., Ruf, N., Dawson, C., Lucifero, D., Frontera, M., et al. (2009). Transcription is required for establishment of germline methylation marks at imprinted genes. *Genes & Development, 23*(1), 105–117. doi:110.1101/gad.495809.

Liu, J., Litman, D., Rosenberg, M. J., Yu, S., Biesecker, L. G., & Weinstein, L. S. (2000a). A GNAS1 imprinting defect in pseudohypoparathyroidism type IB. *Journal of Clinical Investigation, 106*(9), 1167–1174.

Yu, S., Yu, D., Lee, E., Eckhaus, M., Lee, R., Corria, Z., et al. (1998). Variable and tissue-specific hormone resistance in heterotrimeric Gs protein alpha-subunit (Gsalpha) knockout mice is due to tissue-specific imprinting of the gsalpha gene. *Proceedings of National Academy Science of the United States of America, 95*(15), 8715–8720.

Liu, J., Yu, S., Litman, D., Chen, W., & Weinstein, L. S. (2000b). Identification of a methylation imprint mark within the mouse Gnas locus. *Molecular and Cellular Biology, 20*(16), 5808–5817.

Bastepe M., Frohlich L. F., Linglart A., Abu-Zahra H. S., Tojo K., Ward L. M., Juppner H. (2005) Deletion of the NESP55 differentially methylated region causes loss of maternal GNAS imprints and pseudohypoparathyroidism type Ib. *Nature Genetics* 37(1), 25–27. Epub 2004 December 2012.

Eaton S. A., Williamson C. M., Ball S. T., Beechey C. V., Moir L., Edwards J., Teboul L., Maconochie M., Peters J. (2012) New mutations at the imprinted Gnas cluster show gene dosage effects of Gsalpha in postnatal growth and implicate XLalphas in bone and fat metabolism but not in suckling. *Molecular and Cellular Biology* 32(5), 1017–1029. doi: 1010.1128/MCB.06174-06111. Epub 02012 Jan 06173.

Liu J., Chen M., Deng C., Bourc'his D., Nealon J. G., Erlichman B., Bestor T. H., Weinstein L. S. (2005) Identification of the control region for tissue-specific imprinting of the stimulatory G protein alpha-subunit. *Proceedings of National Academy Science of the United States of America* 102(15), 5513–5518. Epub 2005 April 5515.

Williamson C. M., Ball S. T., Nottingham W. T., Skinner J. A., Plagge A., Turner M. D., Powles N., Hough T., Papworth D., Fraser W. D., Maconochie M., Peters J. (2004) A cis-acting control region is required exclusively for the tissue-specific imprinting of Gnas. *Nature Genetics* 36(8), 894–899. Epub 2004 July 2025.

Rougeulle, C., Cardoso, C., Fontes, M., Colleaux, L., & Lalande, M. (1998). An imprinted antisense RNA overlaps UBE3A and a second maternally expressed transcript. *Nature Genetics, 19*(1), 15–16.

Meng L., Person R. E., Beaudet A. L. (2012) Ube3a-ATS is an atypical RNA polymerase II transcript that represses the paternal expression of Ube3a. *Human Molecular Genetics*. doi:10.1093/hmg/dds130.

Wevrick, R., & Francke, U. (1997). An imprinted mouse transcript homologous to the human imprinted in Prader-Willi syndrome (IPW) gene. *Human Molecular Genetics, 6*(2), 325–332.

Chamberlain, S. J., & Brannan, C. I. (2001). The Prader-Willi syndrome imprinting center activates the paternally expressed murine Ube3a antisense transcript but represses paternal Ube3a. *Genomics, 73*(3), 316–322.

Rougeulle, C., Glatt, H., & Lalande, M. (1997). The Angelman syndrome candidate gene, UBE3A/E6-AP, is imprinted in brain. *Nature Genetics, 17*(1), 14–15.

Hogart, A., Patzel, K. A., & LaSalle, J. M. (2008). Gender influences monoallelic expression of ATP10A in human brain. *Human Genetics, 124*(3), 235–242.

Dubose A. J., Johnstone K. A., Smith E. Y., Hallett R. A, Resnick J. L. (2009) Atp10a, a gene adjacent to the PWS/AS gene cluster, is not imprinted in mouse and is insensitive to the PWS-IC. *Neurogenetics,* 11(2), 145–151.

Saitoh, S., Buiting, K., Rogan, P. K., Buxton, J. L., Driscoll, D. J., Arnemann, J., et al. (1996). Minimal definition of the imprinting center and fixation of chromosome 15q11-q13 epigenotype by imprinting mutations. *Proceedings of National Academy Science of the United States of America, 93*(15), 7811–7815.

Brannan, C. I., & Bartolomei, M. S. (1999). Mechanisms of genomic imprinting. *Current Opinion in Genetics & Development, 9*(2), 164–170.

Buiting, K., Barnicoat, A., Lich, C., Pembrey, M., Malcolm, S., & Horsthemke, B. (2001). Disruption of the bipartite imprinting center in a family with Angelman syndrome. *American Journal of Human Genetics, 68*(5), 1290–1294.

Smith, E. Y., Futtner, C. R., Chamberlain, S. J., Johnstone, K. A., & Resnick, J. L. (2011). Transcription is required to establish maternal imprinting at the Prader-Willi syndrome and Angelman syndrome locus. *PLoS Genetics, 7*(12), e1002422. doi:10.1371/journal.pgen.1002422.

Numata, K., Kohama, C., Abe, K., & Kiyosawa, H. (2011). Highly parallel SNP genotyping reveals high-resolution landscape of mono-allelic Ube3a expression associated with locus-wide antisense transcription. *Nucleic Acids Research, 39*(7), 2649–2657. doi:10.1093/nar/gkq1201.

Wevrick, R., Kerns, J. A., & Francke, U. (1994). Identification of a novel paternally expressed gene in the Prader- Willi-syndrome region. *Human Molecular Genetics, 3*(10), 1877–1882.

Landers, M., Bancescu, D. L., Le Meur, E., Rougeulle, C., Glatt-Deeley, H., Brannan, C., et al. (2004). Regulation of the large (approximately 1000 kb) imprinted murine Ube3a antisense transcript by alternative exons upstream of Snurf/Snrpn. *Nucleic Acids Research, 32*(11), 3480–3492.

Chamberlain, S. J., Chen, P. F., Ng, K. Y., Bourgois-Rocha, F., Lemtiri-Chlieh, F., Levine, E. S., et al. (2010). Induced pluripotent stem cell models of the genomic imprinting disorders Angelman and Prader-Willi syndromes. *Proceedings of National Academy Science of the United States of America, 107*(41), 17668–17673.

Ning, Y., Roschke, A., Christian, S. L., Lesser, J., Sutcliffe, J. S., & Ledbetter, D. H. (1996). Identification of a novel paternally expressed transcript adjacent to snRPN in the Prader-Willi syndrome critical region. *Genome Research, 6*(8), 742–746.

Cavaille, J., Buiting, K., Kiefmann, M., Lalande, M., Brannan, C. I., Horsthemke, B., et al. (2000). Identification of brain-specific and imprinted small nucleolar RNA genes exhibiting an unusual genomic organization. *Proceedings of National Academy Science of the United States of America, 97*(26), 14311–14316.

Yin Q. -F., Yang, L., Zhang, Y., Xiang, J. -F., Wu, Y.-W., Carmichael, G. G., Chen, L. -L. (2012) Long noncoding RNAs with snoRNA ends. Molecular Cell (48(2), 219–230 In press).

# Dysregulation of Long Non-coding RNAs in Human Disease

**Nianwei Lin and Tariq M. Rana**

## 1 Background

In the past, a substantial portion of the mammalian genome was thought to contain 'junk' DNA, but recent genome-wide surveys have revealed that many regions encode long noncoding RNAs (lncRNAs) that have important biological functions (Guttman et al. 2009, 2010; Ponting et al. 2009; Zhao et al. 2010). LncRNAs are a group of RNA species greater than 200 nucleotides in length that lack protein-coding potential. Like coding mRNAs, lncRNAs are transcribed by RNA polymerase II, 5′ capped, spliced, and polyadenylated (Sone et al. 2007).

Until the emergence of next-generation sequencing technology, genome-wide identification of lncRNAs had been challenging, due in large part to their relatively low copy numbers. Two global approaches using this technology have been established to search for large intergenic ncRNAs (lincRNAs) (Guttman et al. 2009, 2010). One approach defines lincRNAs based on the chromatin state at the gene locus (Guttman et al. 2009) and takes advantage of the fact that the promoters of genes actively transcribed by RNA polymerase II are enriched in trimethylated histone 3 (H3K4me3) whereas the remainder of the gene body is marked by H3K36me3 (Mikkelsen et al. 2007). In this approach, a genome-wide search for "K4–K36 domains" is performed by chromatin immunoprecipitation followed by deep sequencing (ChIP-Seq), and any hits overlapping annotated protein-coding genes are filtered out. Hits with protein-coding potential are further eliminated based on their maximum codon substitution frequency scores. This approach has identified ~1,600 multiexonic lincRNAs in four mouse cells types (Guttman et al. 2009) and ~3,300 lincRNAs in six human cell types (Khalil et al. 2009). The second method to identify lincRNAs, developed by the same group, is called Scripture and takes an ab initio approach to reconstructing the transcriptome of

N. Lin · T. M. Rana (✉)
Program for RNA Biology, Sanford-Burnham Medical Research Institute,
10901 North Torrey Pines Road, La Jolla, CA 92037, USA
e-mail: trana@sanfordburnham.org

A. M. Khalil and J. Coller (eds.), *Molecular Biology of Long Non-coding RNAs*,
DOI: 10.1007/978-1-4614-8621-3_5, 

mammalian cells (Guttman et al. 2010). In this technique, paired-end 76 base-pair reads are generated from poly(A)$^+$ mRNA by massively parallel sequencing. The reads are then aligned to a reference genome sequence in a manner that spans spliced exon junctions and the "aligned spliced reads" are used to create a "connectivity graph" between individual base pairs. A statistical segmentation approach then identifies paths in the connectivity graph that are enriched in read coverage, and thus represent likely lncRNA transcripts. Because this approach can detect rare transcripts, hundreds of novel lincRNAs were discovered in addition to those identified by the chromatin signature method (Guttman et al. 2010). Using the most sensitive sequencing technique, the recently updated GENCODE version 7 catalog contains 15,512 lncRNA transcripts grouped in 9,640 gene loci (Derrien et al. 2012). Although this is the most complete human lncRNA annotation to date, the number of transcripts will undoubtedly increase as sequencing technologies advance.

LincRNAs show some evidence of evolutionary conservation, but the level of sequence conservation lies somewhere between that of protein-coding exons and introns (Guttman et al. 2009, 2010). These lncRNAs have been proposed to function through their secondary structure, suggesting that the low overall conservation reflects the reduced selection pressure for the RNA structure compared with the codon sequence. However, it is not yet clear how the primary sequences of lincRNAs "translate" into secondary structural motifs, or how these motifs contribute to function. Rinn et al. performed a tiling array-based systematic investigation of the human *HOX* loci, which control anatomic organization during development, and identified hundreds of *HOX* lncRNAs (Rinn et al. 2007). A comparison of the primary sequences identified some motifs that correlated with lncRNA expression along the anatomic anterior–posterior division. However, the function and significance of these motifs remain to be elucidated.

Conserved motifs in lncRNAs may serve as functional units to modulate RNA–protein or RNA–DNA interactions. The Chang laboratory developed the chromatin isolation by RNA purification (ChIRP) technique to construct genomic maps of lncRNA occupancy (Chu et al. 2011). For this, chromatin–protein–lncRNA complexes are isolated using tiling oligonucleotides that target the specific lncRNA, after which the eluted DNA is deep sequenced. ChIRP-Seq data has revealed that binding sites of the lncRNA *HOTAIR* (HOX antisense intergenic RNA) overlapped with regions of occupancy of the histone methyltransferase Polycomb Repressive Complex 2 (PRC2; EZH2; and SUZ12) and H3K27me3, which is consistent with an earlier hypothesis that *HOTAIR* enables epigenetic silencing by recruiting the PRC2 complex to target genes (Rinn et al. 2007; Tsai et al. 2010). Another lncRNA investigated in this study was *Drosophila roX2*, which binds with the protein MSL (male specific lethal) to sites on male X-linked genes and is responsible for dosage compensation. Interestingly, ChIRP-Seq analyses of *HOTAIR* and *roX2* showed that both lncRNAs preferentially bind to GA-rich motifs (Chu et al. 2011), suggesting that diverse species may use similar mechanisms to recruit lncRNA–protein complexes to the target loci. Another study sought to identify regions of the *HOTAIR* primary sequence required for RNA–protein interactions. *HOTAIR* was found to act as a modular scaffold for multiple histone modification complexes, binding PRC2 through 300 bp in the 5′ domain and the LSD1 (lysine-specific demethylase 1) complex through 646 bp at the 3′ domain (Tsai et al. 2010). It will be interesting to see whether the 5′ and 3′ domains of *HOTAIR* are evolutionarily conserved. The examples described here suggest that we will gain a better understanding of the link between the primary sequences and functions of lincRNAs by performing systematic analyses of lincRNA protein- and DNA-interacting domains.

The low sequence conservation (Guttman et al. 2010) and often very low expression of lncRNAs (Cabili et al. 2011; Ravasi et al. 2006) have prompted concern that they may simply be the products of transcriptional ‘noise’ rather than molecules with specific biological functions (Ebisuya et al. 2008; De Santa et al. 2010; Kim et al. 2010). Although some lncRNAs may indeed be functionless byproducts of transcription (Ebisuya et al. 2008; Struhl 2007), accumulating evidence suggests that lncRNAs play diverse biological roles through a range of molecular mechanisms. As will be detailed in this review, lncRNAs can regulate gene expression through epigenetic mechanisms, as well as through transcriptional and post-transcriptional control. Some examples of lncRNAs that serve as binding scaffolds for chromatin-modifying proteins are *XIST*, *Tsix*, *RepA*, and *Jpx* in X-chromosome inactivation (Lee et al. 1999; Zhao et al. 2008; Brown et al. 1991; Tian et al. 2010), *HOTAIR* in *trans*-acting gene repression (Rinn et al. 2007; Gupta et al. 2010), *HOTTIP* and *Mira* in *trans*-acting gene activation (Wang et al. 2011; Bertani et al. 2011), and *H19*, *Air*, and *KCNQ1OT1* in imprinting control (Brannan et al. 1990; Sotomaru et al. 2002; Pandey et al. 2008). Other studies have also suggested that lncRNAs may have *cis*-acting enhancer-like functions (Orom et al. 2010; Hung et al. 2011) and $ncRNA_{CCND1}$ has been shown to act as a *cis* transcriptional repressor (Wang et al. 2008). $ncRNA_{CCND1}$ recruits the RNA-binding protein TLS (translocated in liposarcoma) to the neighboring *cyclin D1* promoter, where it inhibits CREB-binding protein and p300 histone acetyltransferase activities (Wang et al. 2008). Other studies have demonstrated roles for lncRNAs via interactions with protein-coding genes in such diverse functions as control of p53 response pathways (Huarte et al. 2010) and the maintenance of pluripotency (Huarte et al. 2010; Guttman et al. 2011; Loewer et al. 2010; Sheik Mohamed et al. 2010; Dinger et al. 2008). *LincRNA-p21* can also function as an inhibitor of translation (Yoon et al. 2012). LincRNAs can also function by interacting with other ncRNAs, as shown in a recent study describing crosstalk between microRNAs and a muscle-specific lincRNA, *linc-MD1*. The authors showed that *linc-MD1* acts as a “sponge” to prevent miR-133 and miR-135 from inhibiting the muscle-specific transcription factors MAML1 and MEF2C (Cesana et al. 2011). Gong and Maquat reported an unexpected function of lncRNAs in recycling mRNA. They showed that certain lncRNAs, which they termed half-STAU1-binding site RNAs, were involved in the transactivation of STAU1-mediated mRNA decay (Gong and Maquat 2011). LncRNAs can also regulate mRNA processing. High levels of *MALAT-1* (metastasis-associated lung adenocarcinoma transcript 1) have been shown to modulate alternative splicing and are associated with cancer metastasis (Tripathi et al. 2010). Another study showed that the evolutionarily conserved ncRNA *NRON* (non-coding repressor of nuclear factor of activated T cells [NFAT]) can function as a specific regulator of NFAT nuclear trafficking (Willingham et al. 2005). Finally, using a live-cell imaging system, Mao et al. demonstrated that Men $\varepsilon/\beta$ (multiple endocrine neoplasia 1) ncRNAs are essential to initiate the *de novo* assembly of paraspeckles in the nuclei of mammalian cells (Mao et al. 2011).

This section has highlighted some of the mechanisms by which lncRNAs perform their biological functions, including *cis*- and *trans*-regulation, allosteric modification, decoy functions, chromatin remodeling, translational regulation, and post-translational processing (Wang and Chang 2011; Guttman and Rinn 2012; Mercer et al. 2009; Wapinski and Chang 2011). Here, we review recent studies linking lncRNAs to diverse human diseases, grouped according to their regulatory mechanisms.

# 2 Molecular Mechanisms of lncRNA Function in Human Diseases

## *2.1 Epigenetic Regulation and Chromatin Remodeling*

### 2.1.1 *HOTAIR* and Cancer Metastasis

*HOTAIR*, one of the best-characterized lncRNAs, is a 2.2 kb antisense transcript residing in the *HOXC* locus (Rinn et al. 2007). *HOTAIR* mediates epigenetic silencing by physically interacting with the PRC2 and LSD1-CoREST complexes via its 5′ and 3′ domains, ,respectively (Tsai et al. 2010). The EZH2 subunit of PRC2 has histone methyltransferase activity and trimethylates histone 3 at lysine 27, whereas LSD1 demethylates H3K4me2 and H3K4me1 (Fig. 1). Knockdown of *HOTAIR* decreases H3K27me3 and SUZ12 occupancy across $\sim$40 kb of the *HOXD* locus (Rinn et al. 2007), demonstrating that *HOTAIR* represses transcription in *trans* by acting as a scaffold for histone modification complexes (Rinn et al. 2007; Tsai et al. 2010).

Overexpression of *HOTAIR* has been associated with breast cancer metastasis and poor survival (Gupta et al. 2010). Enforced expression of HOTAIR in breast cancer cells promoted cancer cell invasion and metastasis in matrix invasion assays and xenograft experiments, respectively, and this was impaired by depletion of either PRC2 SUZ12 or EZH subunits. In addition, overexpression of *HOTAIR* in breast cancer cells induced global retargeting of PRC2, resulting in an occupancy pattern resembling that of embryonic fibroblasts. Specifically, alterations in chromatin methylation silenced several metastasis suppressor genes and concomitantly increased expression of genes associated with metastasis (Gupta et al. 2010). These findings suggest that *HOTAIR* expression may be a useful marker for the prediction of breast cancer metastasis and survival.

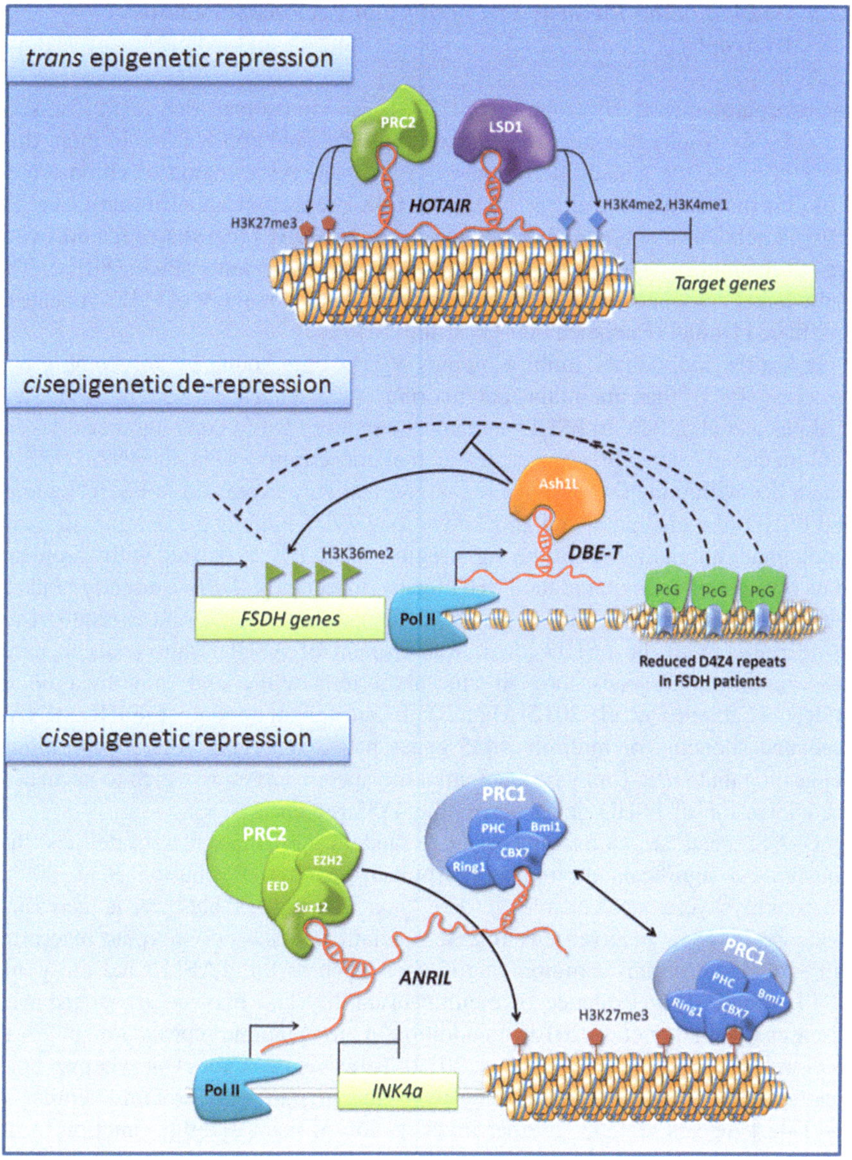


**Fig. 1** Epigenetic regulation mediated by the lncRNAs *HOTAIR*, *DBE-T*, and *ANRIL*

Several key questions about *HOTAIR* function remain to be addressed. For example, how are *HOTAIR*-associated repressive complexes recruited to their target genes? Does *HOTAIR* possess additional motifs that are required for its function? How does *HOTAIR* overexpression alter the PRC2 binding pattern? How is *HOTAIR* expression elevated in metastatic tumors? The answers to these questions could contribute to the therapeutic potential of *HOTAIR*.

### 2.1.2 D4Z4 Binding Element Transcript and Facioscapulohumeral Dystrophy

Facioscapulohumeral dystrophy (FSHD) is an autosomal dominant form of muscular dystrophy that affects the face, shoulders, and upper arms. In more than 95 % of cases, the genetic defect maps to a subtelomeric region of chromosome 4 (4q35) that contains multiple copies of D4Z4 repeats (Cabianca and Gabellini 2010). Each D4Z4 unit contains a binding element (DBE) that shares features with the *cis*-regulatory Polycomb and Trithorax response elements (PREs/TREs). The main defect associated with FSHD is a reduction in the number of D4Z4 repeats to less than 11 units (Cabianca and Gabellini 2010).

In healthy individuals, multiple copies of D4Z4 are bound by Polycomb group proteins (PcG) that maintain neighboring 4q35 genes in a repressed state (Cabianca et al. 2012). In FSHD patients, the reduced D4Z4 copy number weakens PcG-mediated silencing and results in the transcription of a lncRNA, *DBE-T*, which lies within the D4Z4 unit. *DBE-T* is selectively expressed in FSHD patients and FSHD-like disorders, and knockdown of *DBE-T* inhibits the derepression and topological reorganization of the 4q35 region. *DBE-T* is associated with chromatin at its own genomic region at the FSHD locus. Importantly, *DBE-T* directly binds to the methyltransferase Trithorax group (TrxG) protein ASH1L and is required for its recruitment to the FSHD locus. Recruitment of ASHL1 thus leads to derepression of 4q35 genes through chromatin remodeling and dimethylation of H3K36 (Cabianca et al. 2012) (Fig. 1). In an animal model of FSHD, RNAi-mediated silencing of multiple 4q35 genes had a synergistic therapeutic effect, suggesting that *DBE-T* may be a potential therapeutic target by which to normalize the expression of FSHD genes across the 4q35 region.

D4Z4 repeats are an example of large tandem repeats, or macrosatellites, that constitute a significant portion of the human genome (Warburton et al. 2008). These arrays were once considered to be "junk" sequences, but there is increasing evidence for their involvement in gene regulation. Some polymorphic macrosatellites residing within common fragile sites, such as the TAF11-Like array and SST1 arrays, show evidence of meiotic instability and may be associated with diseases (Tremblay et al. 2010). In addition, many genomic repeats are subject to PcG-mediated silencing (Leeb et al. 2010). However, it is not clear whether other macrosatellite sequences encode lncRNAs that recruit PcG proteins, similar to the D4Z4 repeats at 4q35. Further investigation of macrosatellite function in the initiation and maintenance of epigenetic repression will shed light on our understanding of these "dark boxes" in the human genome.

### 2.1.3 *ANRIL* and Tumorigenesis

The lncRNA *ANRIL* (antisense lncRNA of the *INK4* locus) is located within the *INK4b/ARF/INK4a* locus that encodes three tumor suppressor genes: *INK4b* encodes p15/CDKN2B (cyclin-dependent kinase inhibitor 2B), *ARF* encodes p14/ARF (alternative reading frame), and *INK4a* encodes p16/CDKN2A (cyclin-dependent kinase 2A) (Pasmant et al. 2007). Notably, the *INK4b/ARF/INK4a* locus is part of a 403 kb germline deletion present in the largest known family with melanoma-neural system tumor syndrome (Pasmant et al. 2007). *ANRIL* is an antisense transcript that overlaps with the *INK4b* gene and spans a genomic region of 30–40 kb, suggesting it may play a role in the pathology associated with this deletion.

A structure-guided study revealed that *ANRIL* regulates transcriptional silencing of *INK4a* by recruiting the PcG protein Chromobox 7 (CBX7) to the *INK4b/ARF/INK4a* locus (Yap et al. 2010). Interestingly, Yap et al. showed that expression levels of both *ANRIL* and CBX7 are elevated in prostate cancer tissues (Yap et al. 2010). CBX7 is a component of PRC1, which binds to the H3K27me3 repressive mark and is required for maintenance of epigenetic gene silencing (Fig. 1). *ANRIL* RNA interacts with the chromodomain of CBX7 and point mutations in this region compromise the ability of CBX7 to silence the *INK4b/ARF/INK4a* locus and disrupts cell senescence (Yap et al. 2010). However, the secondary RNA structure involved in the *ANRIL*–CBX7 interaction is not yet known.

Genome-wide association studies (GWAS) integrating single-nucleotide polymorphisms (SNPs) in common diseases and gene expression profiles have shown that the *ANRIL* gene locus is a hotspot associated with high risk of coronary heart disease (McPherson et al. 2007; Broadbent et al. 2008; Samani et al. 2008), intracranial aneurysm (Helgadottir et al. 2008; Bilguvar et al. 2008; Yasuno et al. 2010), type 2 diabetes (Zeggini et al. 2007), breast cancer (Stacey et al. 2009), nasopharyngeal carcinoma (Bei et al. 2010), basal cell carcinoma (Stacey et al. 2009), and glioma (Shete et al. 2009). Although it is very likely that *ANRIL* plays a major role in the epigenetic silencing of the *INK4b/ARF/INK4a* locus, it is unclear how mutations in *ANRIL* may contribute to the pathophysiology of these diseases.

## 2.2 *Transcriptional Control*

### 2.2.1 Growth arrest-specific transcript 5, Autoimmune Disorders, and Cancer

*Gas5* (growth arrest-specific transcript 5) is a non-coding multiple small-nucleolar RNA (snoRNA) host gene that contains a 5′ terminal oligopyrimidine tract (Smith and Steitz 1998). The transcript encoded by *Gas5* accumulates in cells undergoing growth arrest caused by growth factor or nutrient deprivation (Schneider et al. 1988). A study by Kino et al. showed that *Gas5* sensitizes cells to apoptosis by acting as a decoy to block transcription of glucocorticoid-responsive genes in response to nutrient deprivation (Kino et al. 2010).

Glucocorticoids are a family of steroid hormones that play physiological roles in diverse processes, particularly immune responses. Glucocorticoid signaling is mainly mediated by the intracellular glucocorticoid receptor (GR), which translocates from the cytoplasm to the nucleus upon ligand binding. In the nucleus, the GR binds through its DNA-binding domain to glucocorticoid-responsive elements (GREs) in target genes (Chrousos and Kino 2005). Many GR-targeted genes are apoptosis suppressors, including cellular inhibitor of apoptosis 2 (cIAP2), which under normal conditions inhibits the function of proapoptotic proteins such as caspases 3, 7, and 9. In their study of cells undergoing nutrient deprivation-induced growth arrest, Kino et al. found high levels of *Gas5* associated with the GR, which prevented binding of GR to *cIAP2* GREs (Kino et al. 2010) (Fig. 2). The repression of *cIAP2* transcription in turn resulted in increased levels of active caspases and subsequent starvation-associated cell death. The *Gas5* region interacting with GR was mapped to a sequence between nucleotides 400 and 598, which contains two conserved "GRE-mimic" sequences.

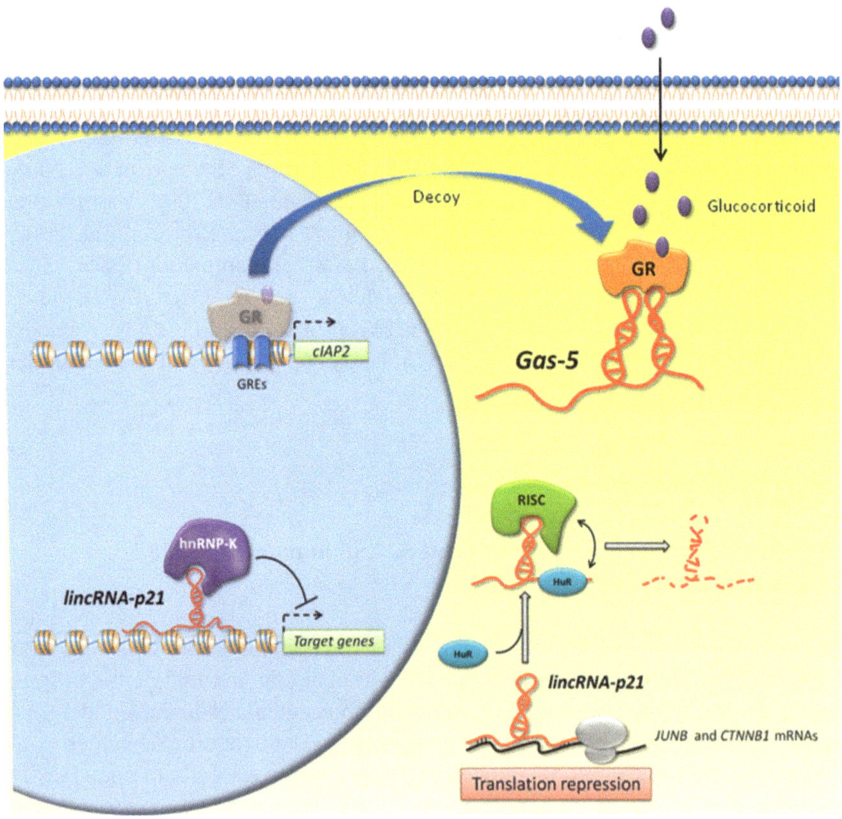


**Fig. 2** Transcriptional control mediated by lncRNA *Gas-5* and *lincRNA-p21*. LncRNA *Gas-5* functions as a molecular decoy to prevent binding of the nuclear transcription factor GR to GREs in the *cIAP2* promoter region. *LincRNA-p21* acts as a repressor at both the transcriptional and translational levels

*Gas5* has been linked to breast cancer in its capacity as an apoptosis sensitizer (Mourtada-Maarabouni et al. 2009). Overexpression of *Gas5* induced growth arrest in several human cell lines and sensitized the cells to stress-induced apoptosis. In addition, breast cancer tissues showed reduced expression of *Gas5* compared with adjacent normal breast epithelial tissues (Mourtada-Maarabouni et al. 2009). Notably, the human *Gas5* gene is located at 1q25, a region linked to susceptibility to several cancers, including melanoma (Smedley et al. 2000), prostate cancer (Nupponen and Carpten 2001), breast cancer (Stange et al. 2006; Morrison et al. 2007), and B-cell lymphoma (Nakamura et al. 2008). Collectively, these results suggest that *Gas5* may normally function as a tumor suppressor by maintaining physiologically appropriate apoptotic responses.

The *Gas5* locus has also been linked to disease susceptibility in the BXSB mouse model of systemic lupus erythematosus, suggesting it may be associated with the development of autoimmunity (Haywood et al. 2006). Endogenous glucocorticoids are important modulators of the adaptive immune response and susceptibility to autoimmunity. Therefore, dysregulation of *Gas5* might contribute to autoimmunity by disrupting glucocorticoid signaling and expression of GR-regulated target genes. Consistent with this, glucocorticoids such as prednisone are potent immunosuppressants and are standard treatments for many autoimmune diseases. More detailed studies will be necessary to confirm the putative functions of *Gas5* in autoimmunity and cancer.

### 2.2.2 *LincRNA-p21* and Cancer

*LincRNA-p21* was originally identified in multiple independent cell-based systems as a transcriptional target of the tumor suppressor p53 (Huarte et al. 2010). *LincRNA-p21* is located ~15 kb upstream of the cell-cycle regulator gene *p21/Cdkn1a*, a canonical target of p53, and encodes a 3.1 kb transcript with two exons. Doxorubicin-induced DNA damage of mouse embryonic fibroblasts and several tumor-derived cell lines (lung tumor, sarcoma, and lymphoma) induced *lincRNA-p21* transcription in a p53-dependent manner. RNA pull-down experiments indicated that *lincRNA-p21* interacts through its 5′ terminal region with heterogeneous nuclear ribonucleoprotein K (hnRNP-K). HnRNP-K binds to the promoters of genes that are co-repressed by *lincRNA-p21* and p53, and as expected, these interactions were significantly disrupted by siRNA-mediated knockdown of *lincRNA-p21* (Huarte et al. 2010). These data therefore suggest that *lincRNA-p21* acts as a transcriptional repressor of the p53 response by recruiting hnRNP-K to target gene promoters (Fig. 2).This function supports the possibility that *lincRNA-p21* is involved in cancer initiation and/or metastasis; however, the mechanism by which this might occur remains to be elucidated.

*LincRNA-p21* also acts as a translational repressor. A recent study with human cervical carcinoma HeLa cells found that *lincRNA-p21* suppresses translation of two cell growth-related genes *JUNB* (transcription factor jun-B) and *CTNNB1* (*β*-catenin) by binding to their mRNAs (Yoon et al. 2012). *LincRNA-p21* was shown to be associated with the RNA-binding protein HuR, which recruits the let-7/Ago2 microRNA/protein complex and accelerates *lincRNA-p21* degradation. When HuR levels are reduced, *lincRNA-p21* accumulates in the cytoplasm and directly interacts with *JUNB* and *CTNNB1* mRNAs at polysomes to prevent their translation. In contrast, overexpression of HuR reduces *lincRNA-p21* levels, which results in translational derepression of both proteins (Yoon et al. 2012) (Fig. 2).

Although direct evidence for the involvement of *lincRNA-p21* in disease is lacking, its effects on tumor suppressor and oncogenic pathways supports its association with various cancers. *LincRNA-p21* regulates genes at transcriptional and post-transcriptional levels, underlining its importance in maintaining homeostasis. Further studies on the mechanism of action of *lincRNA-p21* are needed to uncover the full extent of its pathophysiological function.

## *2.3 Post-Transcriptional Processing*

### 2.3.1 *MALAT-1* and Non-small Cell Lung Cancer

The lncRNA *MALAT-1* was identified in a study that used subtractive hybridization to discover predictive markers for early-stage non-small cell lung cancer (NSCLC) (Ji et al. 2003). *MALAT-1* is a ~6.5 kb lncRNA transcribed from chromosome 11q13, and is highly expressed in the lungs and several other organs. Notably, *MALAT-1* was found to be more highly expressed (threefold) in primary NSCLC tumors that went on to metastasize compared with tumors from the nonmetastatic group and was identified as a prognostic parameter for survival in stage I NSCLC.

Two independent studies showed that *MALAT-1* is retained in the nucleus where it modulates alternative splicing by recruiting serine/arginine (SR) splicing factors to the transcription site (Tripathi et al. 2010; Bernard et al. 2010). The alternative splicing pattern was shown to be tightly regulated by the cellular concentration and/or phosphorylation status of the SR proteins. Thus, although total SR factor levels were increased in *MALAT-1*-depleted cells, splicing was still affected because most of the protein was present in the dephosphorylated inactive form (Tripathi et al. 2010; Bernard et al. 2010) (Fig. 3). Further work will be necessary to determine the location of *MALAT-1* in nuclear speckles and to identify which domains or secondary structures are required for the interaction between *MALAT-1* and the SR splicing factors.

Although it is not clear how dysregulation of *MALAT-1* contributes to the pathogenesis of NSCLC, gene expression analysis in neuroblastoma cells showed that *MALAT-1* depletion affected numerous genes controlling nuclear processes and synapse function (Bernard et al. 2010). Identifying the genes and processes affected by *MALAT-1* will greatly increase our understanding of the normal physiological function of *MALAT-1* as well as how its dysregulation contributes to cancer.

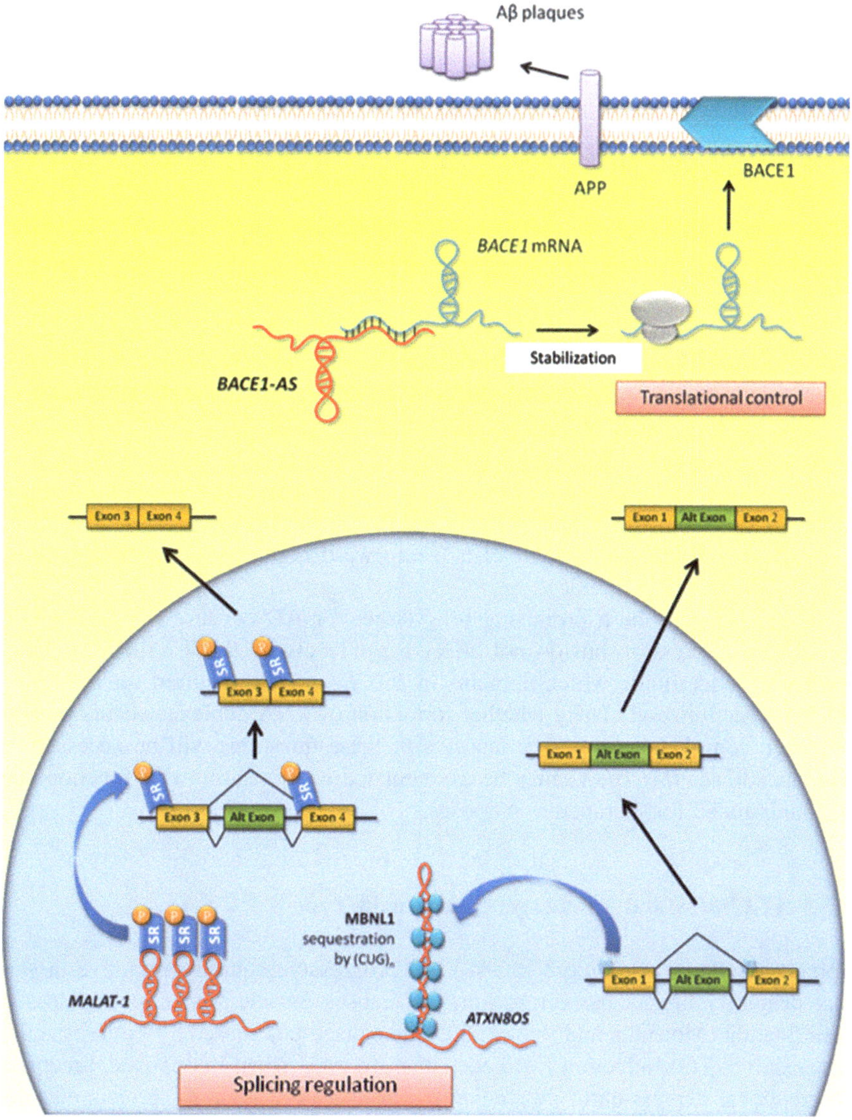


**Fig. 3** Post-transcriptional regulation mediated by lncRNAs *MALAT-1*, *BACE1-AS*, and *ATXN8OS*

### 2.3.2 *BACE1-AS* and Alzheimer's Disease

*BACE1-AS* (BACE1-antisense transcript) regulates gene expression at the translational level. BACE1 ($\beta$-secretase 1) is an aspartyl protease that cleaves amyloid precursor protein (APP) at the $\beta$ site and generates A$\beta$ (amyloid-$\beta$ peptide), a component of amyloid plaques associated with Alzheimer's disease (AD). Dysregulation of *BACE1* has been implicated in AD pathophysiology (McConlogue et al. 2007; Zhong et al. 2007), and consistent with this, gene deletion affects behavior in a mouse model of AD (Kobayashi et al. 2008).

*BACE1-AS* is transcribed from the positive strand of chromosome 11 at the *BACE1* locus (Faghihi et al. 2008) and regulates *BACE1* at both the mRNA and protein levels in vitro and in vivo. Both *BACE1-AS* and *BACE1* mRNA levels increase in cells exposed to stressors such as high temperature, serum starvation, A$\beta$ 1–42, $H_2O_2$, and high glucose concentrations. Importantly, the induction of BACE1 observed in response to cell treatment with A$\beta$ 1–42 is dependent on *BACE1-AS*.

*BACE1* and *BACE1-AS* form RNA duplexes that stabilize the mRNA. Thus, knockdown or overexpression of *BACE1-AS* decreases or increases the stability of *BACE1* mRNA, respectively. Interestingly, levels of *BACE1* and *BACE1-AS* are elevated in the brains of AD patients, and APP transgenic mice also display increased levels of *BACE1-AS*. These data indicate that lncRNA *BACE1-AS* stabilizes its sense mRNA *BACE1* and thus increases levels of BACE1 protein. In turn, APP processing is increased, A$\beta$ 1–42 accumulates, and *BACE1-AS* expression is stimulated as part of a feedforward regulation of APP processing (Fig. 3) (Faghihi et al. 2008).

*BACE1-AS* may be a promising drug target for AD because RNAi treatment reduces its expression but do not affect basal levels of *BACE1* mRNA. Future studies will determine which domains in *BACE1-AS* are required for the RNA–RNA interaction, and clarify whether formation of RNA duplexes occurs through sequence complementarity. The answers to these questions will be necessary to understand how *BACE1-AS* may be dysregulated in AD and how its function may be manipulated for therapeutic purposes.

### 2.3.3 *ATXN8OS* and Spinocerebellar Ataxia Type 8

Spinocerebellar ataxia type 8 (SCA8) is a progressive neurodegenerative disease that primarily affects the cerebellum and causes muscle weakness and loss of coordination (Mutsuddi and Rebay 2005). The disease is caused by a microsatellite expansion $(CTG)_n$ affecting two genes that are transcribed in opposite directions. Bidirectional expression of this region produces a polyglutamine protein encoded by *ATXN8* and a non-coding expansion transcript *ATXN8OS* (ataxin 8 opposite strand) (Moseley et al. 2006).

Using SCA8 transgenic mice expressing the expansion mutation, the Ranum group demonstrated that the $(CUG)_n$ expansion-associated RNA gain-of-function seems to play the most significant role in SCA8 through modulation of splicing events (Daughters et al. 2009). $(CUG)_n$ expansion-positive ribonuclear inclusions were shown to colocalize in the brain with the alternative splicing factor MBNL1 (muscleblind-like 1). Depletion of *Mbnl1* enhanced the coordination and balance deficits in SCA8 mice. Moreover, nuclear accumulation of the $(CUG)_n$ expansion-containing transcripts disrupted alternative splicing of the neurotransmitter

GABA-A transporter 4 (GAT4/Gabt4), a target gene of CUG binding protein 1-MBNL1, resulting in the loss of cerebellar GABAergic inhibition in SCA8 mice (Daughters et al. 2009) (Fig. 3). Despite the clear association between the $(CUG)_n$ expansions in *ATXN8OS*, altered splicing events, and SCA8, the mechanism by which the microsatellite repeats interact with the splicing factors remains elusive. It is unclear whether the disrupted RNA structure of *ATXN8OS* also contributes to its pathophysiological role.

Microsatellite expansion-associated RNA gain-of-function has been reported in other neuromuscular disorders. Myotonic dystrophy type 1 (DM1) is caused by CTG repeats in the 3′ UTR of the dystrophica myotonica-protein kinase (*DMPK*) gene, whereas myotonic dystrophy type 2 (DM2) is caused by 75–11,000 CCTG repeats in a *DMPK* intron (Ranum and Cooper 2006; O'Rourke and Swanson 2009). These toxic microsatellite expansions contribute to disease pathogenesis by sequestering alternative splicing factors, including MBNL1 and CELF (ETR-3–like factor). A similar mechanism may be responsible for other dominant non-coding expansion disorders, including fragile X tremor ataxia syndrome, spinocerebellar ataxia type 10, and Huntington's disease-like 2 (Ranum and Cooper 2006).

## 2.4 Others

The preceding sections have described many of the disease-associated lncRNAs that function through epigenetic remodeling and transcriptional and post-transcriptional regulation. However, numerous lncRNAs are thought to be involved in disease pathogenesis by mechanisms that are either unknown or poorly understood. In this section, we will discuss some of these lncRNAs and their potential to be used as diagnostic markers or therapeutic targets.

### 2.4.1 Genomic Imprinting Diseases

Genomic imprinting is a non mendelian genetic phenomenon occurring in diploid organisms in which genes are expressed from only one of the parental alleles. The imprinted genes are silenced through epigenetic mechanisms involving DNA methylation and histone modifications but the DNA sequence is unaffected (Koerner et al. 2009). The first imprinting-associated lncRNA, *H19*, was identified in the *Igf2* locus; since then many additional lncRNAs have been discovered in imprinted gene clusters, where they usually have *cis*-regulatory effects on flanking genes (Koerner et al. 2009).

One of the best-characterized imprinted gene clusters is the Prader-Willi syndrome (PWS) and Angelman syndrome (AS) locus. PWS and AS are distinct neurological disorders that results from aberrant gene expression regulated by the PWS imprinting center (PWS-IC) and the AS-IC. *Ube3a* (ubiquitin protein ligase E3A) is a candidate gene for AS, located adjacent to the PWS-IC (PWS imprinted center), and is maternally expressed in the brain. Its antisense non-coding transcript, *Ube3a-as*, is only expressed from the paternally derived allele. Interestingly, deletion of PWS-IC caused an upregulation of the paternal *Ube3a* allele, while suppressed the paternally expressed *Ube3a-as* gene (Chamberlain and Brannan 2001; Johnstone et al. 2006). It is likely that *Ube3a-as* plays a role in repressing the paternal *Ube3a* expression. Moreover, two paternally expressed intronic lncRNAs in PWS-AS domain accumulate near their transcription sites (Vitali et al. 2010). The allelic specific chromatin structure revealed by DNA FISH suggested that these nuclear-retained lncRNAs might regulate the spatial organization of gene expression by modulating nuclear architecture. Therefore, lncRNAs might be involved in the imprinting-associated diseases, although their function and disease connections require more detailed studies.

### 2.4.2 Neurological Disorders

The development, homeostasis, and plasticity of the central nervous system are controlled by complex signaling pathways and gene regulatory networks. Independent studies in the mouse have identified hundreds of lncRNAs with brain-specific expression patterns, suggesting these molecules play roles in regulating normal brain function (Ponjavic et al. 2009; Mercer et al. 2008). Consistent with this, a recently developed genome-wide catalog of SNP-trait associations showed that most susceptibility loci for common psychiatric and neurophysiological diseases are located in non-coding regions. These studies point to the functional importance of lncRNAs in the brain and suggest this area of lncRNA research will advance our understanding of brain development and disease (Qureshi et al. 2010).

#### Huntington's Disease

Huntington's disease (HD) is a dominant inherited neurodegenerative disorder that affects muscle coordination and cognition. HD is caused by expansion of a CAG repeat in the first exon of the *Huntingtin* (*Htt*) gene, which results in toxic gain-of-function of the mutant protein (mutHTT) (Johnson 2012). HTT normally functions in neurons by sequestering the transcriptional repressor REST (repressor element 1 silencing transcription factor) within the cytoplasm. MutHTT has a lower affinity for REST than wild type HTT, which leads to aberrant nuclear localization of REST and transcriptional repression (Zuccato et al. 2003). REST mRNA was shown to be increased in HD mice and in neuron-like cells expressing mutHTT (Ravache et al. 2010). DiGeorge syndrome-associated noncoding RNA (*DGCR5*) was identified in a genome-wide search for REST-targeted lncRNAs and is negatively regulated by REST (Johnson et al. 2009). Consistent with this finding, *DGCR5* is downregulated in the brains of HD patients. Expression of several other lncRNAs is also altered in HD brains: *MEG3* (maternally expressed 3) is downregulated, whereas *TUG1* (taurine upregulated 1) and *NEAT1* (nuclear paraspeckle assembly transcript 1) are upregulated (Johnson 2012). Although *NEAT1* and *MEG3* both contain REST binding sites, the upregulation of *MEG3* is inconsistent with the negative regulatory role of REST. More information about these lncRNAs will be necessary to understand their pathophysiological functions in HD disease.

### Alzheimer's Disease

In addition to *BACE1-AS* described above, the lncRNA *BC200* is also aberrantly expressed in AD brains. *BC200* RNA is selectively enriched in somatodendritic domains of neurons, where it is thought to modulate local protein synthesis and contribute to the maintenance of long-term synaptic plasticity (Tiedge et al. 1993; Wang and Tiedge 2004; Kondrashov et al. 2005). *BC200* expression is increased in brain areas involved in AD in proportion to the severity of the disease (Mus et al. 2007). In addition, the somatodendritic distribution of *BC200* is greatly disturbed in severe AD cases. These findings suggest that *BC200* expression may be useful as a prognostic marker in early AD (Albert 1996). Although *BC200* has been reported to inhibit protein translation in vitro through its oligo(A)-rich region (Kondrashov et al. 2005), the mechanism by which this lncRNA is involved in the pathology of AD remains unclear.

### Down's Syndrome

Down's syndrome (DS) or trisomy 21 is the most common chromosomal abnormality in humans and is caused by an extra copy of chromosome 21. DS is associated with cognitive disability and growth abnormalities. Although the transcription factor NFAT was originally described and named for its function in T cell activation, it also plays a role in the development of organs such as the heart, muscles, and nervous tissue (Hogan et al. 2003). In animal models of DS, increased expression of two genes on chromosome 21, *DSCR1* and *DYRK1A*, prevented activation of NFATc and resulted in many of the features of DS (Arron et al. 2006). The lncRNA *NRON* (noncoding repressor of NFAT) has been shown to interact with nuclear transport factors and acts as a specific regulator of NFAT nuclear trafficking (Willingham et al. 2005), suggesting that *NRON* may play a role in the pathophysiology of DS.

### Neuropsychiatric Disorders

The protein-coding gene Disrupted in schizophrenia 1 (*DISC1*) and its antisense non-coding gene *DISC2* were originally identified in a large Scottish family and are candidate susceptibility genes for schizophrenia and related psychiatric diseases (Millar et al. 2000). *DISC1* is known to be involved in normal brain development

(Brandon et al. 2009), but a number of association studies in diverse populations support a link between *DISC1* and *DISC2* and schizophrenia, bipolar disorder, major depressive disorder, and autism spectrum disorders (Millar et al. 2000; Chubb et al. 2008; Williams et al. 2009). LncRNA *DISC2* is thought to function by regulating the expression of *DISC1* (Devon et al. 2001; Ekelund et al. 2004), although the mechanism by which this might occur remains unclear. *DISC2* may also play a *DISC1*-independent role in psychiatric diseases (Chubb et al. 2008), suggesting a more complex involvement in the pathogenesis of psychiatric disorders.

### 2.4.3 Cancer

Many lncRNAs are thought to be involved in cancer initiation and/or progression through unknown mechanisms. One example is the imprinted gene *H19* located in the *Igf2* gene cluster, which is expressed only from the maternal allele. Loss of imprinting and biallelic expression of *H19* has been described in medulloblastomas and meningiomas (Albrecht et al. 1996; Muller et al. 2000). In glioblastoma-derived primary cell lines, *H19* is expressed at higher levels in CD133 + cancer stem cells than in CD133− tumor cells (Beier et al. 2007). Other studies showed that *H19* is a target of the transcription factors GLI1, p53, c-Myc, and E2F1 in various human cancers (Yoon et al. 2002; Dugimont et al. 1998; Barsyte-Lovejoy et al. 2006; Berteaux et al. 2005), raising the possibility that this lncRNA is actively involved in oncogenic and/or tumor suppressor regulatory networks. Another lncRNA demonstrating cancer-specific dysregulation is *MEG3*. This lncRNA is highly expressed in normal tissues but markedly downregulated in gall bladder, retinal, and prostate cancers (Miyoshi et al. 2000; Zhang et al. 2003; Benetatos et al. 2011), suggesting it may function as a tumor suppressor.

A recent study generated an atlas for lncRNA profiling in cancer (Gibb et al. 2011). The authors compiled 272 human SAGE (serial analysis of gene expression) libraries to construct lncRNA transcriptomes for various normal and cancer tissues (Gibb et al. 2011).This study found that, in addition to the known cancer-associated lncRNAs, numerous lncRNAs were aberrantly expressed in human cancers, including *NEAT1*, *XIST*, *SNHG6* (small-nucleolar RNA host gene 6), *SNHG5*, *SCAND2* (SCAN domain containing 2), as well as a number of novel lncRNAs (Gibb et al. 2011). This pilot investigation provides strong evidence that lncRNAs may be much more extensively involved in human cancers than previously recognized.

## 3 Perspectives

The emergence of noncoding RNAs as key regulators of diverse biological activities has challenged the "DNA to RNA to protein" central dogma of molecular biology (Mattick 2003). The most recent update of the encyclopedia of

DNA elements (ENCODE) consortium shows that ~75 % of the human genome may be transcribed, and of the more than 9640 loci classified as lncRNAs, only ~100 have well-characterized cellular roles (Derrien et al. 2012; Banfai et al. 2012; Djebali et al. 2012; Bernstein et al. 2012). Although lncRNAs and protein-coding genes are transcribed by similar pathways, and lncRNAs display canonical gene structures and histone modifications, lncRNAs are preferentially localized in the chromatin and nucleus, consistent with their major role as epigenetic regulators. LncRNAs are under weaker selective constraints than protein-coding genes, but show stronger sequence conservation than neutrally evolving sequences. Importantly, about 30 % of human lncRNA transcripts seem to be primate specific. LncRNAs are generally present at lower levels than the products of protein-coding genes but display a more tissue-specific pattern of expression. Interestingly, a high proportion of identified lncRNAs are expressed specifically in the brain, suggesting significant involvement in brain development and the pathophysiology of neurological diseases. Finally, there is a remarkable positive correlation between expression of lncRNAs and their overlapping antisense mRNAs (Derrien et al. 2012).

Our understanding of the molecular mechanisms by which lncRNAs function remains poor, although it is clear they can interact with DNA, RNA, or proteins (Mattick 2003). These interactions might occur through complementary base pairing or specific secondary structures. However, the relatively low sequence conservation of lncRNAs makes it difficult to identify common functional motifs in the primary sequences. Further investigations will be necessary to link sequence and structural features of lncRNAs with their biological functions.

The discovery that the human genome contains vast numbers of lncRNAs that play diverse biological roles has ushered in a new era of molecular genetics. This rapidly evolving field is not only identifying new mechanisms of gene regulation but also uncovering novel links to human diseases. Further advances will undoubtedly pave the way for the development of innovative therapeutic strategies using lncRNA-based drugs.

## References


Albert, M. S. (1996). Cognitive and neurobiologic markers of early Alzheimer disease. *Proceedings of the National Academy of Sciences of the United States of America, 93*(24), 13547–13551.

Albrecht, S., et al. (1996). Variable imprinting of H19 and IGF2 in fetal cerebellum and medulloblastoma. *Journal of Neuropathology and Experimental Neurology, 55*(12), 1270–1276.

Arron, J. R., et al. (2006). NFAT dysregulation by increased dosage of DSCR and DYRK1A on chromosome 21. *Nature, 441*(7093), 595–600.

Banfai, B., et al. (2012). Long noncoding RNAs are rarely translated in two human cell lines. *Genome Research, 22*(9), 1646–1657.

Barsyte-Lovejoy, D., et al. (2006). The c-Myc oncogene directly induces the H19 noncoding RNA by allele-specific binding to potentiate tumorigenesis. *Cancer Research, 66*(10), 5330–5337.

Bei, J. X., et al. (2010). A genome-wide association study of nasopharyngeal carcinoma identifies three new susceptibility loci. *Nature Genetics, 42*(7), 599–603.

Beier, D., et al. (2007). CD133(+) and CD133(−) glioblastoma-derived cancer stem cells show differential growth characteristics and molecular profiles. *Cancer Research, 67*(9), 4010–4015.

Benetatos, L., Vartholomatos, G., & Hatzimichael, E. (2011). MEG3 imprinted gene contribution in tumorigenesis. *International Journal of Cancer, 129*(4), 773–779.

Bernard, D., et al. (2010). A long nuclear-retained non-coding RNA regulates synaptogenesis by modulating gene expression. *EMBO Journal, 29*(18), 3082–3093.

Bernstein, B. E., et al. (2012). An integrated encyclopedia of DNA elements in the human genome. *Nature, 489*(7414), 57–74.

Bertani, S., et al. (2011). The noncoding RNA Mistral activates Hoxa6 and Hoxa7 expression and stem cell differentiation by recruiting MLL1 to chromatin. *Molecular Cell, 43*(6), 1040–1046.

Berteaux, N., et al. (2005). H19 mRNA-like noncoding RNA promotes breast cancer cell proliferation through positive control by E2F1. *Journal of Biological Chemistry, 280*(33), 29625–29636.

Bilguvar, K., et al. (2008). Susceptibility loci for intracranial aneurysm in European and Japanese populations. *Nature Genetics, 40*(12), 1472–1477.

Brandon, N. J., et al. (2009). Understanding the role of DISC in psychiatric disease and during normal development. *Journal of Neuroscience, 29*(41), 12768–12775.

Brannan, C. I., et al. (1990). The product of the H19 gene may function as an RNA. *Molecular and Cellular Biology, 10*(1), 28–36.

Broadbent, H. M., et al. (2008). Susceptibility to coronary artery disease and diabetes is encoded by distinct, tightly linked SNPs in the ANRIL locus on chromosome 9p. *Human Molecular Genetics, 17*(6), 806–814.

Brown, C. J., et al. (1991). A gene from the region of the human X inactivation centre is expressed exclusively from the inactive X chromosome. *Nature, 349*(6304), 38–44.

Cabianca, D. S., & Gabellini, D. (2010). The cell biology of disease: FSHD: Copy number variations on the theme of muscular dystrophy. *Journal of Cell Biology, 191*(6), 1049–1060.

Cabianca, D. S., et al. (2012). A long ncRNA links copy number variation to a polycomb/trithorax epigenetic switch in FSHD muscular dystrophy. *Cell, 149*(4), 819–831.

Cabili, M. N., et al. (2011). Integrative annotation of human large intergenic noncoding RNAs reveals global properties and specific subclasses. *Genes & Development, 25*(18), 1915–1927.

Cesana, M., et al. (2011). A long noncoding RNA controls muscle differentiation by functioning as a competing endogenous RNA. *Cell, 147*(2), 358–369.

Chamberlain, S. J., & Brannan, C. I. (2001). The Prader-Willi syndrome imprinting center activates the paternally expressed murine Ube3a antisense transcript but represses paternal Ube3a. *Genomics, 73*(3), 316–322.

Chrousos, G. P., & Kino, T. (2005). Intracellular glucocorticoid signaling: A formerly simple system turns stochastic. *Science STKE, 2005*(304), pe48.

Chu, C., et al. (2011). Genomic maps of long noncoding RNA occupancy reveal principles of RNA-chromatin interactions. *Molecular Cell, 44*(4), 667–678.

Chubb, J. E., et al. (2008). The DISC locus in psychiatric illness. *Molecular Psychiatry, 13*(1), 36–64.

Daughters, R. S., et al. (2009). RNA gain-of-function in spinocerebellar ataxia type. *PLoS Genet, 5*(8), e1000600.

De Santa, F., et al. (2010). A large fraction of extragenic RNA pol II transcription sites overlap enhancers. *PLoS Biology, 8*(5), e1000384.

Derrien, T., et al. (2012). The GENCODE v7 catalog of human long noncoding RNAs: Analysis of their gene structure, evolution, and expression. *Genome Research, 22*(9), 1775–1789.

Devon, R. S., et al. (2001). Identification of polymorphisms within Disrupted in Schizophrenia 1 and Disrupted in Schizophrenia 2, and an investigation of their association with schizophrenia and bipolar affective disorder. *Psychiatric Genetics, 11*(2), 71–78.

Dinger, M. E., et al. (2008). Long noncoding RNAs in mouse embryonic stem cell pluripotency and differentiation. *Genome Research, 18*(9), 1433–1445.
Djebali, S., et al. (2012). Landscape of transcription in human cells. *Nature, 489*(7414), 101–108.
Dugimont, T., et al. (1998). The HTATA-less promoter is efficiently repressed by wild-type tumor suppressor gene product p53. *Oncogene, 16*(18), 2395–2401.
Ebisuya, M., et al. (2008). Ripples from neighbouring transcription. *Nature Cell Biology, 10*(9), 1106–1113.
Ekelund, J., et al. (2004). Replication of 1q42 linkage in Finnish schizophrenia pedigrees. *Molecular Psychiatry, 9*(11), 1037–1041.
Faghihi, M. A., et al. (2008). Expression of a noncoding RNA is elevated in Alzheimer's disease and drives rapid feed-forward regulation of beta-secretase. *Nature Medicine, 14*(7), 723–730.
Gibb, E. A., et al. (2011). Human cancer long non-coding RNA transcriptomes. *PLoS ONE, 6*(10), e25915.
Gong, C., & Maquat, L. E. (2011). lncRNAs transactivate STAU1-mediated mRNA decay by duplexing with 3′ UTRs via Alu elements. *Nature, 470*(7333), 284–288.
Gupta, R. A., et al. (2010). Long non-coding RNA HOTAIR reprograms chromatin state to promote cancer metastasis. *Nature, 464*(7291), 1071–1076.
Guttman, M., & Rinn, J. L. (2012). Modular regulatory principles of large non-coding RNAs. *Nature, 482*(7385), 339–346.
Guttman, M., et al. (2009). Chromatin signature reveals over a thousand highly conserved large non-coding RNAs in mammals. *Nature, 458*(7235), 223–227.
Guttman, M., et al. (2010). Ab initio reconstruction of cell type-specific transcriptomes in mouse reveals the conserved multi-exonic structure of lincRNAs. *Nature Biotechnology, 28*(5), 503–510.
Guttman, M., et al. (2011). LincRNAs act in the circuitry controlling pluripotency and differentiation. *Nature, 477*(7364), 295–300.
Haywood, M. E., et al. (2006). Overlapping BXSB congenic intervals, in combination with microarray gene expression, reveal novel lupus candidate genes. *Genes and Immunity, 7*(3), 250–263.
Helgadottir, A., et al. (2008). The same sequence variant on 9p21 associates with myocardial infarction, abdominal aortic aneurysm and intracranial aneurysm. *Nature Genetics, 40*(2), 217–224.
Hogan, P. G., et al. (2003). Transcriptional regulation by calcium, calcineurin, and NFAT. *Genes & Development, 17*(18), 2205–2232.
Huarte, M., et al. (2010). A large intergenic noncoding RNA induced by p53 mediates global gene repression in the p53 response. *Cell, 142*(3), 409–419.
Hung, T., et al. (2011). Extensive and coordinated transcription of noncoding RNAs within cell-cycle promoters. *Nature Genetics, 43*(7), 621–629.
Ji, P., et al. (2003). MALAT-1, a novel noncoding RNA, and thymosin beta4 predict metastasis and survival in early-stage non-small cell lung cancer. *Oncogene, 22*(39), 8031–8041.
Johnson, R. (2012). Long non-coding RNAs in Huntington's disease neurodegeneration. *Neurobiology of Diseases, 46*(2), 245–254.
Johnson, R., et al. (2009). Regulation of neural macroRNAs by the transcriptional repressor REST. *RNA, 15*(1), 85–96.
Johnstone, K. A., et al. (2006). A human imprinting centre demonstrates conserved acquisition but diverged maintenance of imprinting in a mouse model for Angelman syndrome imprinting defects. *Human Molecular Genetics, 15*(3), 393–404.
Khalil, A. M., et al. (2009). Many human large intergenic noncoding RNAs associate with chromatin-modifying complexes and affect gene expression. *Proceedings of the National Academy of Sciences of the United States of America, 106*(28), 11667–11672.
Kim, T. K., et al. (2010). Widespread transcription at neuronal activity-regulated enhancers. *Nature, 465*(7295), 182–187.
Kino, T., et al. (2010). Noncoding RNA gas5 is a growth arrest- and starvation-associated repressor of the glucocorticoid receptor. *Science Signal, 3*(107), ra8.

Kobayashi, D., et al. (2008). BACE gene deletion: Impact on behavioral function in a model of Alzheimer's disease. *Neurobiology of Aging, 29*(6), 861–873.
Koerner, M. V., et al. (2009). The function of non-coding RNAs in genomic imprinting. *Development, 136*(11), 1771–1783.
Kondrashov, A. V., et al. (2005). Inhibitory effect of naked neural BC1 RNA or BC200 RNA on eukaryotic in vitro translation systems is reversed by poly(A)-binding protein (PABP). *Journal of Molecular Biology, 353*(1), 88–103.
Lee, J. T., Davidow, L. S., & Warshawsky, D. (1999). Tsix, a gene antisense to Xist at the X-inactivation centre. *Nature Genetics, 21*(4), 400–404.
Leeb, M., et al. (2010). Polycomb complexes act redundantly to repress genomic repeats and genes. *Genes & Development, 24*(3), 265–276.
Loewer, S., et al. (2010). Large intergenic non-coding RNA-RoR modulates reprogramming of human induced pluripotent stem cells. *Nature Genetics, 42*(12), 1113–1117.
Mao, Y. S., et al. (2011). Direct visualization of the co-transcriptional assembly of a nuclear body by noncoding RNAs. *Nature Cell Biology, 13*(1), 95–101.
Mattick, J. S. (2003). Challenging the dogma: The hidden layer of non-protein-coding RNAs in complex organisms. *Bioessays, 25*(10), 930–939.
McConlogue, L., et al. (2007). Partial reduction of BACE has dramatic effects on Alzheimer plaque and synaptic pathology in APP Transgenic Mice. *Journal of Biological Chemistry, 282*(36), 26326–26334.
McPherson, R., et al. (2007). A common allele on chromosome associated with coronary heart disease. *Science, 316*(5830), 1488–1491.
Mercer, T. R., et al. (2008). Specific expression of long noncoding RNAs in the mouse brain. *Proceedings of the National Academy of Sciences of the United States of America, 105*(2), 716–721.
Mercer, T. R., Dinger, M. E., & Mattick, J. S. (2009). Long non-coding RNAs: Insights into functions. *Nature Reviews Genetics, 10*(3), 155–159.
Mikkelsen, T. S., et al. (2007). Genome-wide maps of chromatin state in pluripotent and lineage-committed cells. *Nature, 448*(7153), 553–560.
Millar, J. K., et al. (2000). Disruption of two novel genes by a translocation co-segregating with schizophrenia. *Human Molecular Genetics, 9*(9), 1415–1423.
Miyoshi, N., et al. (2000). Identification of an imprinted gene, Meg3/Gtl2 and its human homologue MEG3, first mapped on mouse distal chromosome 12 and human chromosome 14q. *Genes to Cells, 5*(3), 211–220.
Morrison, L. E., et al. (2007). Effects of ERBB2 amplicon size and genomic alterations of chromosomes 1, 3, and 10 on patient response to trastuzumab in metastatic breast cancer. *Genes, Chromosomes and Cancer, 46*(4), 397–405.
Moseley, M. L., et al. (2006). Bidirectional expression of CUG and CAG expansion transcripts and intranuclear polyglutamine inclusions in spinocerebellar ataxia type 8. *Nature Genetics, 38*(7), 758–769.
Mourtada-Maarabouni, M., et al. (2009). GAS5, a non-protein-coding RNA, controls apoptosis and is downregulated in breast cancer. *Oncogene, 28*(2), 195–208.
Muller, S., et al. (2000). Genomic imprinting of IGF and H19 in human meningiomas. *European Journal of Cancer, 36*(5), 651–655.
Mus, E., Hof, P. R., & Tiedge, H. (2007). Dendritic BC200 RNA in aging and in Alzheimer's disease. *Proceedings of the National Academy of Sciences of the United States of America, 104*(25), 10679–10684.
Mutsuddi, M., & Rebay, I. (2005). Molecular genetics of spinocerebellar ataxia type 8 (SCA8). *RNA Biology, 2*(2), 49–52.
Nakamura, Y., et al. (2008). The GAS5 (growth arrest-specific transcript 5) gene fuses to BCL6 as a result of t(1;3)(q25;q27) in a patient with B-cell lymphoma. *Cancer Genetics and Cytogenetics, 182*(2), 144–149.
Nupponen, N. N., & Carpten, J. D. (2001). Prostate cancer susceptibility genes: Many studies, many results, no answers. *Cancer and Metastasis Reviews, 20*(3–4), 155–164.

Orom, U. A., et al. (2010). Long noncoding RNAs with enhancer-like function in human cells. *Cell, 143*(1), 46–58.

O'Rourke, J. R., & Swanson, M. S. (2009). Mechanisms of RNA-mediated disease. *Journal of Biological Chemistry, 284*(12), 7419–7423.

Pandey, R. R., et al. (2008). Kcnq1ot1 antisense noncoding RNA mediates lineage-specific transcriptional silencing through chromatin-level regulation. *Molecular Cell, 32*(2), 232–246.

Pasmant, E., et al. (2007). Characterization of a germ-line deletion, including the entire INK4/ARF locus, in a melanoma-neural system tumor family: Identification of ANRIL, an antisense noncoding RNA whose expression coclusters with ARF. *Cancer Research, 67*(8), 3963–3969.

Ponjavic, J., et al. (2009). Genomic and transcriptional co-localization of protein-coding and long non-coding RNA pairs in the developing brain. *PLoS Genetics, 5*(8), e1000617.

Ponting, C. P., Oliver, P. L., & Reik, W. (2009). Evolution and functions of long noncoding RNAs. *Cell, 136*(4), 629–641.

Qureshi, I. A., Mattick, J. S., & Mehler, M. F. (2010). Long non-coding RNAs in nervous system function and disease. *Brain Research, 1338*, 20–35.

Ranum, L. P., & Cooper, T. A. (2006). RNA-mediated neuromuscular disorders. *Annual Review of Neuroscience, 29*, 259–277.

Ravache, M., et al. (2010). Transcriptional activation of REST by Spin Huntington's disease models. *PLoS ONE, 5*(12), e14311.

Ravasi, T., et al. (2006). Experimental validation of the regulated expression of large numbers of non-coding RNAs from the mouse genome. *Genome Research, 16*(1), 11–19.

Rinn, J. L., et al. (2007). Functional demarcation of active and silent chromatin domains in human HOX loci by noncoding RNAs. *Cell, 129*(7), 1311–1323.

Samani, N. J., et al. (2008). Coronary artery disease-associated locus on chromosome 9p and early markers of atherosclerosis. *Arteriosclerosis, Thrombosis, and Vascular Biology, 28*(9), 1679–1683.

Schneider, C., King, R. M., & Philipson, L. (1988). Genes specifically expressed at growth arrest of mammalian cells. *Cell, 54*(6), 787–793.

Sheik Mohamed, J., et al. (2010). Conserved long noncoding RNAs transcriptionally regulated by Octand Nanog modulate pluripotency in mouse embryonic stem cells. *RNA, 16*(2), 324–337.

Shete, S., et al. (2009). Genome-wide association study identifies five susceptibility loci for glioma. *Nature Genetics, 41*(8), 899–904.

Smedley, D., et al. (2000). Characterization of chromosome abnormalities in malignant melanomas. *Genes, Chromosomes and Cancer, 28*(1), 121–125.

Smith, C. M., & Steitz, J. A. (1998). Classification of gas5 as a multi-small-nucleolar-RNA (snoRNA) host gene and a member of the 5′-terminal oligopyrimidine gene family reveals common features of snoRNA host genes. *Molecular and Cellular Biology, 18*(12), 6897–6909.

Sone, M., et al. (2007). The mRNA-like noncoding RNA Gomafu constitutes a novel nuclear domain in a subset of neurons. *Journal of Cell Science, 120*(Pt 15), 2498–2506.

Sotomaru, Y., et al. (2002). Unregulated expression of the imprinted genes H19 and Igf2r in mouse uniparental fetuses. *Journal of Biological Chemistry, 277*(14), 12474–12478.

Stacey, S. N., et al. (2009). New common variants affecting susceptibility to basal cell carcinoma. *Nature Genetics, 41*(8), 909–914.

Stange, D. E., et al. (2006). High-resolution genomic profiling reveals association of chromosomal aberrations on 1q and 16p with histologic and genetic subgroups of invasive breast cancer. *Clinical Cancer Research, 12*(2), 345–352.

Struhl, K. (2007). Transcriptional noise and the fidelity of initiation by RNA polymerase II. *Nature Structural & Molecular Biology, 14*(2), 103–105.

Tian, D., Sun, S., & Lee, J. T. (2010). The long noncoding RNA, Jpx, is a molecular switch for X chromosome inactivation. *Cell, 143*(3), 390–403.

Tiedge, H., Chen, W., & Brosius, J. (1993). Primary structure, neural-specific expression, and dendritic location of human BC200 RNA. *Journal of Neuroscience, 13*(6), 2382–2390.

Tremblay, D. C., et al. (2010). Expression, tandem repeat copy number variation and stability of four macrosatellite arrays in the human genome. *BMC Genomics, 11*, 632.

Tripathi, V., et al. (2010). The nuclear-retained noncoding RNA MALAT1 regulates alternative splicing by modulating SR splicing factor phosphorylation. *Molecular Cell, 39*(6), 925–938.

Tsai, M. C., et al. (2010). Long noncoding RNA as modular scaffold of histone modification complexes. *Science, 329*(5992), 689–693.

Vitali, P., et al. (2010). Long nuclear-retained non-coding RNAs and allele-specific higher-order chromatin organization at imprinted snoRNA gene arrays. *Journal of Cell Science, 123*(Pt 1), 70–83.

Wang, K. C., & Chang, H. Y. (2011). Molecular mechanisms of long noncoding RNAs. *Molecular Cell, 43*(6), 904–914.

Wang, H., & Tiedge, H. (2004). Translational control at the synapse. *Neuroscientist, 10*(5), 456–466.

Wang, X., et al. (2008). Induced ncRNAs allosterically modify RNA-binding proteins in cis to inhibit transcription. *Nature, 454*(7200), 126–130.

Wang, K. C., et al. (2011). A long noncoding RNA maintains active chromatin to coordinate homeotic gene expression. *Nature, 472*(7341), 120–124.

Wapinski, O., & Chang, H. Y. (2011). Long noncoding RNAs and human disease. *Trends in Cell Biology, 21*(6), 354–361.

Warburton, P. E., et al. (2008). Analysis of the largest tandemly repeated DNA families in the human genome. *BMC Genomics, 9*, 533.

Williams, J. M., et al. (2009). A 1q42 deletion involving DISC1, DISC2, and TSNAX in an autism spectrum disorder. *American Journal of Medical Genetics Part A, 149A*(8), 1758–1762.

Willingham, A. T., et al. (2005). A strategy for probing the function of noncoding RNAs finds a repressor of NFAT. *Science, 309*(5740), 1570–1573.

Yap, K. L., et al. (2010). Molecular interplay of the noncoding RNA ANRIL and methylated histone H3 lysine 27 by polycomb CBX7 in transcriptional silencing of INK4a. *Molecular Cell, 38*(5), 662–674.

Yasuno, K., et al. (2010). Genome-wide association study of intracranial aneurysm identifies three new risk loci. *Nature Genetics, 42*(5), 420–425.

Yoon, J. W., et al. (2002). Gene expression profiling leads to identification of GLI1-binding elements in target genes and a role for multiple downstream pathways in GLI1-induced cell transformation. *Journal of Biological Chemistry, 277*(7), 5548–5555.

Yoon, J. H., et al. (2012). LincRNA-p21 suppresses target mRNA translation. *Molecular Cell, 47*(4), 648–655.

Zeggini, E., et al. (2007). Replication of genome-wide association signals in UK samples reveals risk loci for type diabetes. *Science, 316*(5829), 1336–1341.

Zhang, X., et al. (2003). A pituitary-derived MEG3 isoform functions as a growth suppressor in tumor cells. *Journal of Clinical Endocrinology and Metabolism, 88*(11), 5119–5126.

Zhao, J., et al. (2008). Polycomb proteins targeted by a short repeat RNA to the mouse X chromosome. *Science, 322*(5902), 750–756.

Zhao, J., et al. (2010). Genome-wide identification of polycomb-associated RNAs by RIP-seq. *Molecular Cell, 40*(6), 939–953.

Zhong, Z., et al. (2007). Levels of beta-secretase (BACE1) in cerebrospinal fluid as a predictor of risk in mild cognitive impairment. *Archives of General Psychiatry, 64*(6), 718–726.

Zuccato, C., et al. (2003). Huntingtin interacts with REST/NRSF to modulate the transcription of NRSE-controlled neuronal genes. *Nature Genetics, 35*(1), 76–83.

# Functions of Long Non-Coding RNAs in Non-mammalian Systems

**Alex Tuck and David Tollervey**

## 1 LncRNA Origins and Expression

The discovery of long non-coding RNAs (lncRNAs) in *Saccharomyces cerevisiae* followed the development of technologies able to simultaneously detect enormous numbers of transcripts, such as genome-wide tiling arrays (David et al. 2006; Xu et al. 2009) and high-throughput sequencing (Nagalakshmi et al. 2008; Yassour et al. 2010). Similar studies have since identified numerous lncRNAs in *Arabidopsis* (Liu et al. 2012), zebrafish (Pauli et al. 2012), *Caenorhabditis elegans* (Nam and Bartel 2012) and *Drosophila* (Young et al. 2012). LncRNAs can arise where the transcriptional machinery "hijacks" the nucleosome-free regions (NFRs) associated with protein-coding genes. A distinct pre-initiation complex (PIC) can then assemble at the upstream boundary of the 5′ NFR (Churchman and Weissman 2011; Rhee and Pugh 2012) to generate divergent (antisense), promoter-associated, lncRNAs (Neil et al. 2009; Xu et al. 2009). Similarly, PICs can assemble at 3′ NFRs to generate lncRNAs in either orientation (Murray et al. 2012; Rhee and Pugh 2012). However, not all antisense lncRNAs can be linked to the 5′ or 3′ NFR of an associated gene (Yassour et al. 2010) and long intergenic ncRNAs (lincRNAs) can be separated by several kb from genes and transcribed in gene-poor regions (Young et al. 2012). This indicates that lncRNAs also arise from dedicated promoters. In such cases, lncRNA transcription can be driven by canonical transcription factors (Bird et al. 2006; Houseley et al. 2008; Pinskaya et al. 2009; Xu et al. 2011), suggesting that their expression is actively regulated. Within protein-coding genes, transcription of lncRNAs is repressed by a refractory nucleosome organization, directed by the histone deacetylases Set3C (Kim et al. 2012) and Rpd3S (Churchman and Weissman 2011). This suppresses spurious intragenic transcription initiation, while permitting elongation by polymerase II (Pol II) assisted by histone chaperones and chromatin remodelers that disassemble

A. Tuck · D. Tollervey (✉)
Wellcome Trust Centre for Cell Biology, The University of Edinburgh, Edinburgh, Scotland
e-mail: d.tollervey@ed.ac.uk

A. M. Khalil and J. Coller (eds.), *Molecular Biology of Long Non-coding RNAs*,
DOI: 10.1007/978-1-4614-8621-3_6, 

nucleosomes ahead of Pol II and correctly reassemble them in its wake (Cheung et al. 2008). Additionally, the juxtaposition of promoters and terminators of protein-coding genes to form gene loops helps maintain transcription directionality and suppress antisense initiation of lncRNAs (Tan-Wong et al. 2012).

Although some lncRNAs are detectable in wild-type cells, many more are apparent when the RNA surveillance machinery is inactivated in the nucleus (Gudipati et al. 2012; Neil et al. 2009; Xu et al. 2009) or cytoplasm (van Dijk et al. 2011). The stability of lncRNAs is therefore heterogeneous, with some stable species accumulating to high levels while many others are turned over rapidly. Targeting for rapid turnover is probably initiated co-transcriptionally, as the exosome nuclease complex associates with chromatin (Hessle et al. 2012) and lncRNA termination factors (Vasiljeva and Buratowski 2006).

Analyses of transcripts associated with ribosomes (Brar et al. 2012) and cross-species alignments to assess coding potential (Lin et al. 2011) confirm that most lncRNAs do not encode proteins. However, in transcriptome-wide analyses numerous lncRNAs show strongly regulated expression, suggesting functions for lncRNAs in diverse cellular processes. As examples, transcriptome profiling has identified lncRNAs expressed during specific stages of the cell cycle (Granovskaia et al. 2010; Lardenois et al. 2011), in specific tissues (Liu et al. 2012) or sub-cellular compartments (Pauli et al. 2012), in response to the circadian rhythm (Hazen et al. 2009), nutrient conditions (Xu et al. 2009) or stress (Yassour et al. 2010), in aging cells (Camblong et al. 2007) and in subpopulations of genetically identical yeast (Bumgarner et al. 2012). Furthermore, correlations have emerged between the expression of lncRNAs and neighboring mRNAs, such as reciprocal relationships between sense:antisense pairs (van Dijk et al. 2011; Xu et al. 2009; Yassour et al. 2010), suggesting that the changes in lncRNA expression impact on adjacent genes.

A growing body of evidence from individual case studies now supports the functional importance of specific lncRNAs, with diverse regulatory roles. These can be split into two major classes based upon whether the transcript itself is functional, or whether it is the act of transcription that plays the key role.

## 2 Competitive Transcription

In *S. cerevisiae*, the promoters of protein-coding genes typically comprise a ~180 bp NFR flanked by upstream (−1) and downstream (+1) nucleosomes. Transcription begins with the assembly of a PIC, containing Pol II and basal transcription factors, immediately upstream of or just within the +1 nucleosome (Rhee and Pugh 2012). The DNA duplex within the PIC then melts, generating a Pol II—open promoter complex that scans a short distance downstream to a transcription start site (TSS), where transcription of the gene is initiated. There are several points at which this process can be subverted, diverting transcription into lncRNA production and thus reducing the mRNA output from the locus (Fig. 1a–c).

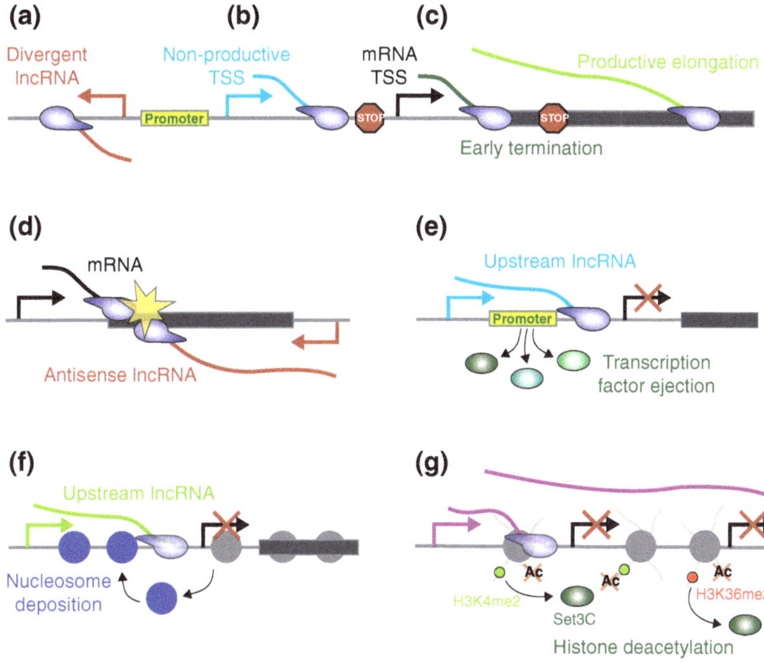


**Fig. 1** LncRNA functions attributed to the act of transcription. **a** Divergent lncRNA transcription opposes mRNA production at bidirectional promoters by competing for pre-initiation complex (PIC) components. **b** Scanning PICs can initiate transcription at non-productive start sites to produce an attenuated, non-coding transcript instead of a full length mRNA. **c** Polymerases that have initiated transcription at a canonical transcription start site can be subject to early termination. **d** Polymerases transcribing antisense lncRNAs can collide with sense-oriented polymerases and trigger their ejection from the template and/or degradation. **e** LncRNA transcription across promoters disrupts the binding of transcription factors. **f** LncRNA transcription can also promote nucleosome eviction or, as depicted here, deposition. **g** LncRNA transcription is accompanied by H3K4 dimethylation in promoter-proximal regions, and H3K36 methylation downstream. These marks are bound by Set3C and Rpd3S histone deacetylases, respectively, which direct assembly of a chromatin state refractory to transcription initiation

Firstly, a PIC can competitively assemble at the upstream border of the NFR to initiate transcription of a divergent lncRNA (Fig. 1a). In the case of yeast *TPI1*, this PIC apparently competes with the sense-oriented PIC for the local pool of basal transcription factors, recruited by shared transcriptional activators (Neil et al. 2009). The transcriptional apparatus can also be diverted after PIC formation, via initiation at an alternative, non-productive TSS (Fig. 1b). This mechanism is employed to regulate genes encoding nucleotide biogenesis factors. For example, yeast Imd2 participates in GTP biosynthesis, and in GTP replete conditions *IMD2* initiation shifts to a promoter-proximal TSS, upstream of the productive (promoter-distal) mRNA TSS. This switch is dependent on an initiating GG dinucleotide at the upstream TSS, rendering it sensitive to GTP concentration. The region between the two TSSs includes a recognition site for the Nrd1–Nab3 complex, which triggers early termination coupled to exosome-mediated decay. Transcription from the upstream TSS therefore produces unstable, attenuated lncRNAs (Jenks et al. 2008; Kuehner and Brow 2008). These alternative TSSs exhibit a reciprocal pattern of initiation, suggesting that they compete for scanning PICs from the shared promoter. Upstream, non-coding transcription thus enables a high rate of PIC assembly to be maintained in GTP replete conditions, perhaps ensuring that there is spare capacity to rapidly upregulate transcriptional output in response to GTP depletion. A similar mechanism occurs at the *URA2* gene to regulate UTP synthesis (Thiebaut et al. 2008). However, in this case increased initiation from the promoter-distal, productive TSS is not accompanied by reduced initiation from the upstream TSS, suggesting that non-productive transcription exerts a constant level of repression to minimize the basal level of transcription. The use of alternative TSSs has been documented for a number of additional genes, indicating that it may be relatively common (Creamer et al. 2011; Kim et al. 2010; Thiebaut et al. 2008).

Switching between productive and non-productive transcription can also occur at genes with a single TSS, via premature termination. Here, regulation occurs downstream of initiation, rather than during PIC scanning (Fig. 1c). The nuclear RNA-binding proteins Hrp1 (Kuehner and Brow 2008), Nrd1 (Arigo et al. 2006) and Pcf11 (Creamer et al. 2011) each participate in, and are subject to, Nrd1-dependent early termination. This constitutes an autoregulatory mechanism, since increased accumulation of Nrd1, Pcf11 or Hrp1 increases the proportion of unstable, attenuated transcripts. Nrd1 also binds toward the 5′ end of many transcripts that are highly expressed during logarithmic growth (e.g. *CLN3*, involved in cell cycle progression), and can contribute to their downregulation in response to starvation or stress (Creamer et al. 2011). In nutrient rich conditions, Nrd1-dependent premature termination is opposed by Ras signaling. Conversely, *FKS2*, encoding an enzyme that synthesizes a structural polysaccharide component of the cell wall, is upregulated during stress via alleviation of Nrd1-dependent attenuation. Cell wall stressors trigger a signaling cascade that culminates in activation of the kinase Mkp1, which associates with a central regulator of transcription (the Pol II-associated complex, Paf1C) to promote *FKS2* elongation (Kim and Levin 2011).

In general, regulation via switching between productive and non-coding transcription downstream of PIC assembly is most widely employed at promoters that must respond rapidly to changes in environmental or physiological conditions. Faced with limited nucleotide availability, cells must immediately transcribe genes encoding nucleotide biosynthetic factors, before nucleotides drop to levels that inhibit transcription. Similarly, during stress cells must conserve resources to mount a stress response. Promoter-proximal non-productive transcription enables a promoter to maintain a reserve of engaged polymerases and thus be poised for rapid upregulation, but also to rapidly downregulate transcription, even if it has already initiated. This mechanism is reminiscent of promoter-proximal pausing of Pol II, which is evident from transcription run-on data (Rodriguez-Gil et al. 2010), analyses of nascent transcripts (Churchman and Weissman 2011) and the presence of $\sim$18 nt promoter-proximal fragments in *Drosophila* apparently protected by

Pol II (Taft et al. 2011). The relationship between pausing and promoter-proximal termination is unclear, but perhaps a proportion of paused polymerases fail to reinitiate and are resolved by termination.

The partitioning of chromatin into isolated domains and the ability of genes to form loops (O'Sullivan et al. 2004) suggest that even transcription from opposite ends of a gene may compete for a shared pool of transcription factors. In consequence, lncRNA transcription initiating near the 3′ end of protein-coding genes might reduce productive mRNA output from the locus. Consistent with this model, genes with antisense partners generally have lower basal expression levels (Xu et al. 2011) and show stress-induced increases in mRNA production that exceed the total increase in total Pol II occupancy (Kim et al. 2011). This suggests that Pol II can be redistributed from lncRNA to mRNA transcription to activate protein expression, although this interpretation must be treated with caution as changes in mRNA stability can also explain discrepancies between transcription rate and mRNA abundance (Garcia-Martinez et al. 2012).

## 3 Chromatin Resurfacing

In addition to downregulating mRNA synthesis by diverting Pol II, lncRNA transcription can play a more active role, disrupting the association of Pol II, transcription factors and nucleosomes with both promoters and transcribed regions.

Head on collisions between polymerases transcribing opposite DNA strands are inhibitory. The elongating polymerases cannot bypass each other, so one must be ejected via ubiquitylation-directed proteolysis (Hobson et al. 2012) (Fig. 1d). Direct Pol II collisions might, therefore, explain the inhibitory effect of some antisense lncRNAs. For example, expression of the *S. cerevisiae* inducer of meiosis Ime4 is suppressed in haploid cells by an antisense lncRNA, RME2. Transcription of *IME4* mRNA and *RME2* is mutually exclusive, so upregulation of either one represses the other (Hongay et al. 2006). RME2 can only act in *cis*, at the locus from which it is transcribed, leading to a model in which polymerases transcribing the antisense strand collide with, and inhibit, sense-oriented polymerases. Intriguingly, repression requires a 450 bp promoter-proximal tract in *IME4*, suggesting that this region might be particularly sensitive to Pol II collisions (Gelfand et al. 2011). Inverting this region abolishes lncRNA-mediated repression, resulting in expression of *IME4* in haploid cells and a reduction in RME2. Conceivably, this region may contain elements that affect the orientation of the "winning" polymerase (i.e., the polymerase that remains on the template, when the other is displaced), perhaps directing preferential ejection of sense transcribing polymerases over antisense.

# 4 Transcription Factor Ejection

Transcriptional interference can also occur upstream of the mRNA TSS, where lncRNA transcription can disrupt protein interactions at the promoter (Fig. 1e). In *Drosophila*, this is epitomized by the regulation of the Bithorax complex (BX-C) homeotic genes *Ubx*, *abd-A* and *abd-B*. Sense-oriented lncRNAs are transcribed from the *bxd* region upstream of *Ubx* and extend across the *Ubx* promoter. *Ubx* and *bxd* show reciprocal expression patterns in RNA fluorescence in situ hybridization (FISH) studies of embryonic tissues and in sorted nuclei (Petruk et al. 2006). Reciprocally expressed lncRNAs are also transcribed upstream of *abd-A* (Petruk et al. 2007), suggesting that upstream lncRNA transcription interferes with both the *Ubx* and *abd-A* promoters.

Studies in *S. cerevisiae* provide insights into the mechanisms by which lncRNAs interfere with mRNA promoter function. For example, *ADH1* encodes a zinc-dependent alcohol dehydrogenase and its promoter is repressed by lncRNA transcription during zinc deficiency. The lncRNA is transcribed across the binding site for the transcriptional activator Rap1 and chemical modification data suggest that this results in transient displacement of Rap1 (Bird et al. 2006).

LncRNA transcription also functions to displace transcription factors from the promoter of *FLO11*, which encodes a cell wall glycoprotein conferring adhesion properties. In this case, regulation involves a pair of *cis*-interfering lncRNAs, *ICR1* and *PWR1*, which form the basis of an epigenetic toggle (Bumgarner et al. 2009). RNA FISH analyses enabled the number of *ICR1*, *PWR1*, and *FLO11* transcripts to be counted in individual cells, revealing an inverse correlation between the two lncRNAs and identifying three previously predicted *FLO11* promoter states (silenced, basal and active) (Bumgarner et al. 2012; Octavio et al. 2009). The ratio of active to basal cells was sensitive to *ICR1* expression, the transcriptional activator Flo8 and the transcriptional repressor Sfl1. ChIP analyses revealed that *ICR1* transcription ejects Flo8 and Sfl1 from the *FLO11* promoter. Together, this leads to a model in which Flo8 and Sfl1 compete for binding to the basal state of the *FLO11* promoter and induce activation or silencing, respectively. Flo8 promotes *PWR1* expression, which inhibits *ICR1* transcription and facilitates binding of additional activators. However, infrequent Sfl1-promoted *ICR1* expression can eject transcriptional activators and inhibit *PWR1* transcription, resetting the promoter to its basal state and enabling Sfl1 and Flo8 to compete for binding anew. The stochastic nature of this toggle results in variegated *FLO11* expression and thus heterogeneous adhesive properties within an isogenic population of yeast, with some individuals adhering to local surfaces and others forming filaments or washing away. This might ensure that new nutrient sources are located before the local supply is exhausted. In general, phenotypic heterogeneity helps a population anticipate a change in environmental conditions by maintaining distinct subpopulations equipped to deal with various scenarios. Many genes involved in signaling, metabolism and stress responses are associated with lncRNAs (Yassour et al. 2010). This suggests that lncRNA-dependent variegated expression might be a prevalent source of phenotypic heterogeneity within populations of genetically identical microorganisms.

LncRNAs are well suited to providing toggle functions, as they can provide a digital output (being either transcribed or not) but are themselves regulated by multiple analog or digital inputs. They therefore potentially integrate information from a variety of sources in the region upstream of a gene and transmit a binary decision to the proximal promoter. The *FLO11* toggle also illustrates the versatility of lncRNA transcription, which functions here in a slow (less than once per cell division) epigenetic toggle (Halme et al. 2004), but elsewhere in rapid responses to starvation or stress.

# 5 Nucleosome Remodeling

Transcription factors often promote chromatin modifications, so their ejection by lncRNA transcription indirectly affects chromatin structure. However, lncRNA transcription can also directly trigger chromatin rearrangements, because nucleosomes are partially disassembled ahead of Pol II and reassembled in its wake during the normal transcription cycle (Fig. 1f).

*SRG1* is a sense-oriented lncRNA transcribed across the promoter of *SER3*, which encodes a component of the serine biosynthetic pathway. High serine levels trigger *SRG1* expression via the serine-responsive activator Cha4, leading to *SER3* repression (Martens et al. 2005). *SRG1* was proposed to act via ejection of transcription factors, as it represses *SER3* in *cis* and can disrupt binding of the transcriptional activator Gal4 to ectopic binding sites placed at the *SER3* locus (Martens et al. 2004). However, *SRG1* transcription also generates a broad region of micrococcal nuclease protection, indicative of the presence of nucleosomes across the *SER3* promoter and in contrast to the NFR generally located at promoters (Hainer and Martens 2011). Histone turnover in the *SER3* upstream region occurs at a high rate, suggesting that continuous nucleosome reassembly is required (Thebault et al. 2011). Mutations in the histone chaperones Spt6 and Spt16 (Hainer et al. 2011) or the HMG-like protein Spt2 (Thebault et al. 2011) result in *SER3* derepression and loss of nucleosomes across the *SER3* promoter, even though *SRG1* transcription is maintained. Spt6, Spt16 and Spt2 contribute to nucleosome reassembly behind Pol II (Thebault et al. 2011), suggesting that *SRG1* transcription continually directs the deposition of nucleosomes over the *SER3* promoter.

LncRNA transcription can also displace nucleosomes to promote transcriptional activation. For example, antisense transcription across the *PHO5* locus in *S. cerevisiae* facilitates rapid eviction of four positioned nucleosomes in the *PHO5* promoter (Uhler et al. 2007). Similarly, induction of *fbp1+* in *Schizosaccharomyces pombe* is accompanied by a cascade of lncRNA transcription that progressively disrupts chromatin across the promoter and enables activators to bind (Hirota et al. 2008).

## 6 Histone Modifications

In addition to histone chaperones, Pol II associates with chromatin modifying enzymes at specific stages during elongation, dependent on the phosphorylation status of heptad repeats within the C-terminal domain (CTD) of the large subunit of Pol II. At the promoter, the Pol II CTD is phosphorylated at serine 5 (Ser5P) and bound by the Set1 histone methyltransferase, which methylates histone H3 residue K4 (H3K4) to produce H3K4me3 in promoter–proximal regions, H3K4me2 slightly further downstream and H3K4me1 across the gene body. After initiation, the Pol II CTD is progressively phosphorylated at Ser2P and the H3K36 methyl-transferase Set2 is recruited to the doubly phosphorylated Pol II CTD (Ser2P,5P) (Kizer et al. 2005). This leads to H3K36 di- and trimethylation in the mid and 3′ regions of the gene. H3K4me2 and H3K36me2/me3 are recognized by the histone deacetylases Set3C (Kim and Buratowski 2009) and Rpd3S (Li et al. 2009), respectively, which promote the assembly of chromatin with a less accessible state (Fig. 1g).

In the case of *SER3* (Hainer and Martens 2011; Thebault et al. 2011) and *PHO5* (Uhler et al. 2007), lncRNA-dependent chromatin remodeling does not require Set1 or Set2, indicating that nucleosome assembly/disassembly is a direct consequence of lncRNA transcription. At other loci, however, histone modification is a key step in lncRNA-dependent regulation. For example, an antisense lncRNA initiating within the *GAL10* coding region is transcribed across the *GAL1-10* promoter, resulting in reduced induction of *GAL1* and *GAL10* at low galactose concentrations (Houseley et al. 2008; Pinskaya et al. 2009). Transcription of GAL10as directs methylation of both H3K4 and H3K36, and repression is dependent on histone deacetylation by the Rpd3S complex, recruited either via its Eaf3 subunit binding to H3K36me2/me3 (Houseley et al. 2008), or the Rco1 subunit binding to H3K4me2/me3 (Pinskaya et al. 2009). Repression of *GAL1* is alleviated when binding sites for Reb1, an activator of GAL10as transcription, are mutated. However, accumulation of GAL10as in mutants with defective nuclear surveillance has no effect on *GAL1* expression. These observations are consistent with transcription of the lncRNA, rather than the transcript itself, repressing *GAL1*. The GAL10as lncRNA also accumulates in mutants lacking the decapping protein Dcp2 or 5′-to-3′ exonuclease Rat1, but in this case *GAL1* induction is delayed (Geisler et al. 2012). Rat1 participates in transcription termination, suggesting that decapping of GAL10as and degradation by Rat1 might occur co-transcriptionally, leading to transcription termination before repressive histone marks are deposited. Given the disruptive nature of non-coding transcription and its ability to pervade even silenced regions of the genome, a rapid termination pathway might be valuable in protecting against spurious chromatin disruption.

LncRNA transcription can also recruit the Set3C histone deacetylase complex, via Set1-dependent H3K4 dimethylation (Kim et al. 2012; van Werven et al. 2012). Expression of the lncRNA IRT1 invokes this mechanism, together with H3K36me3-dependent Rpd3S recruitment, to silence the promoter region of *IME1* (van Werven et al. 2012). Ime1 and Ime4 (see above) are the central inducers of meiosis in *S. cerevisiae* and both are repressed by non-coding transcription, so gametogenesis in yeast is primarily controlled by lncRNAs.

The promoters of many other genes overlap with the H3K4 dimethylation "zones" of adjacent lncRNAs and are subject to Set3C-dependent repression, suggesting that this mechanism is widespread (Kim et al. 2012). Notably, Set3C-dependent repression is most apparent during transition periods, such as a galactose induction. Together with the role of the GAL10as lncRNA in modulating *GAL1-10* induction, this suggests that lncRNA transcription is particularly important in regulating the kinetics of gene induction or repression. non-coding transcription acts in *cis* and, as the underlying sequence is relatively unimportant, the rate of transcription can be tuned rapidly via evolution. LncRNAs are therefore ideally suited to offering a layer of autonomous regulation, fine-tuning the expression levels of individual genes against a backdrop of general signaling.

In summary, the transcription of lncRNAs can facilitate nucleosome reorganization, eject transcription factors and direct histone modifications. In many cases, it is difficult to establish at precisely which stage non-coding transcription acts, as the pathways of chromatin regulation are interwoven. Collectively, however, non-coding transcription enables old marks to be removed and new ones laid down, effectively "resurfacing" chromatin. We suggest that this is an important mechanism to ensure that genes remain responsive to incoming signals, rather than irreversibly committing to a particular state.

## 7 Regulatory LncRNAs

The functions of the lncRNAs discussed above can largely be attributed to the act of transcription, but in many cases the transcript is itself functional. In this event, experimental intervention to manipulate lncRNA abundance or sequence can give insights into its functions, and some lncRNAs can operate when expressed ectopically from a plasmid or distant genetic locus. The heterogeneous nature of lncRNAs provides scope for a broad variety of regulatory mechanisms.

## 8 Regulators of Protein Activity

LncRNAs can modulate the activity of proteins in various ways. This has been best characterized in human cells, where lncRNAs can block interactions between PIC components (Martianov et al. 2007), modify the affinity of proteins for various histone modifications (Yang et al. 2011), or allosterically activate transcriptional co-repressors (Wang et al. 2008). LncRNAs are likely to perform similar roles in other organisms. For example, in diverse eukaryotes, including humans and *S. cerevisiae*, telomeric repeats are transcribed into telomeric repeat-containing RNA

(TERRA) (Luke et al. 2008). TERRA inhibits the human telomerase in vitro through interactions with both RNA and protein components (Redon et al. 2010). In *S. cerevisiae*, the accumulation of TERRA transcripts in RNA surveillance mutants results in defective telomere elongation, suggesting that TERRA similarly inhibits yeast telomerase activity. However, *S. cerevisiae* TERRA also interacts with the Ku protein complex, an inhibitor of the 5′ to 3′ DNA exonuclease Exo1, resulting in nuclease activation and telomere degradation (Fig. 2a) (Pfeiffer and Lingner 2012). TERRA-induced telomere shortening can occur in the absence of telomerase and is suppressed by Exo1 deletion, suggesting that this is the major mechanism of telomere length control in yeast. In both pathways, modulation of protein activity by TERRA is central.

Yeast lncRNAs are also implicated in copy number control of the TY1 retrotransposon. The presence of an elevated number of TY1 elements results in increased expression of TY1 antisense lncRNAs, which can suppress TY1 mobility. Regulation can occur post-transcriptionally, since the levels of mature integrase (IN) and reverse transcriptase (RT) proteins are reduced, whereas the abundance of the polycistronic TY1 mRNA is not affected (Matsuda and Garfinkel 2009). Reverse transcription occurs within cytoplasmic virus-like particles (VLPs), into which the TY1-as RNAs are packaged together with TY1 mRNA. However, structure probing (SHAPE) analyses did not detect interactions between these RNAs (Purzycka et al. 2012). This suggests that the TY1-as lncRNA does not inhibit translation of the IN and RT proteins, but might instead destabilize them or prevent their excision from the precursor polyprotein.

LncRNAs can also influence nuclear protein localization. *Drosophila hsrω-n* is a nuclear lncRNA that interacts with RNA processing factors and co-localizes with them in nuclear foci (Fig. 2b). Following heat shock, *hsrω-n* is upregulated and these foci coalesce into a single region at the *hsrω-n* gene locus (Prasanth et al. 2000). *hsrω-n* is required both for the integrity and dynamics of these foci (Lakhotia et al. 2012) and its abundance negatively correlates with global protein synthesis (Johnson et al. 2009). Sequestration of pre-mRNA associated proteins by *hsrω-n* may regulate their availability to function in pre-mRNA processing and export.

## 9 Assembly of Nuclear Bodies

Other lncRNAs contribute to the formation of nuclear structures and this has been extensively characterized in *Drosophila* dosage compensation. *Drosophila* males have a single X chromosome, which is transcribed at twice the level of each of the two female X chromosomes. In males, a dosage compensation complex (DCC), containing two functionally redundant lncRNAs (roX1 and roX2) (Meller and Rattner 2002), binds X-linked genes and doubles their expression (Fig. 2c). The DCC proteins bind specific loci (chromatin entry sites) containing a GA-rich sequence motif and then spread to flanking sites within active genes (Alekseyenko et al. 2008;

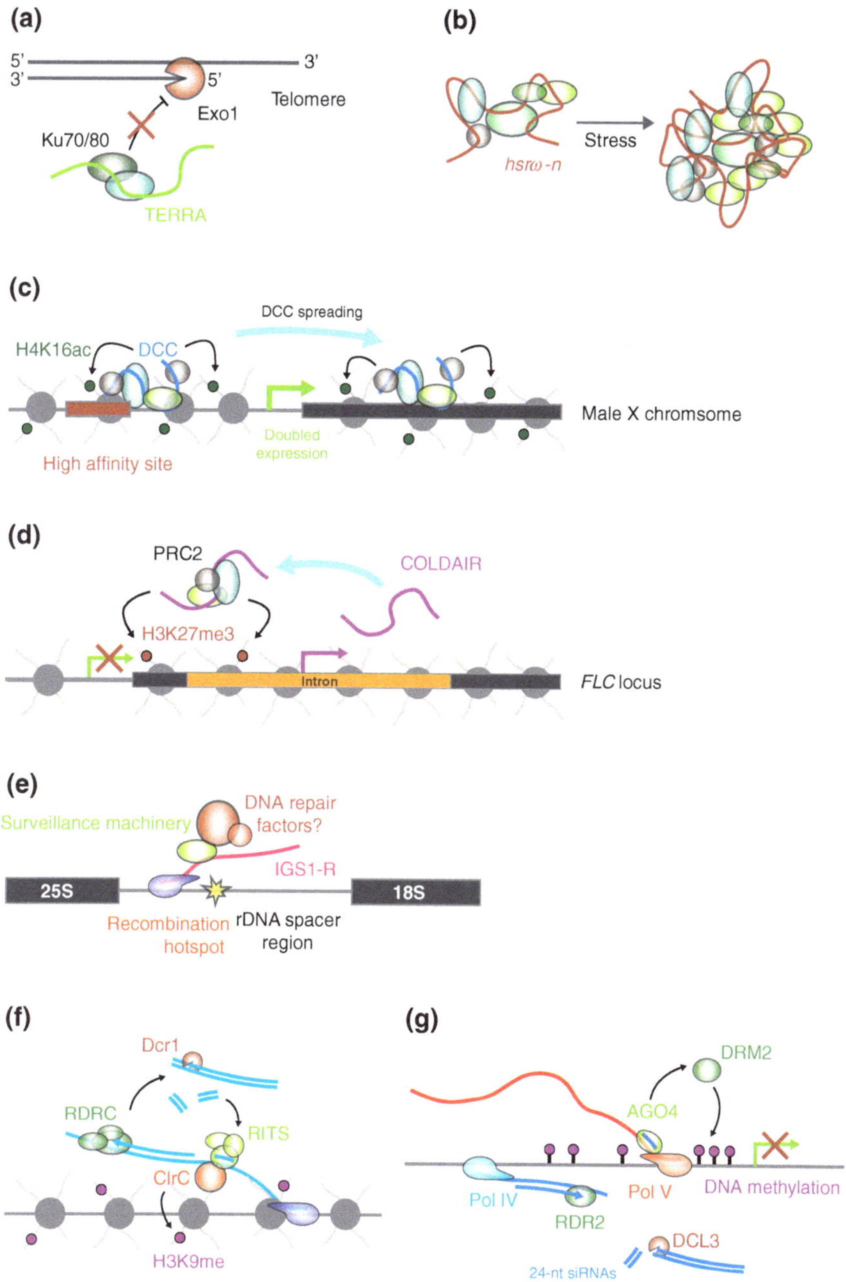

(a)
5'
3'
3'
5'
Telomere
Exo1
Ku70/80
TERRA
(b)
hsrω-n
Stress
(c)
DCC spreading
H4K16ac
DCC
Male X chromsome
Doubled expression
High affinity site
(d)
PRC2
COLDAIR
H3K27me3
Intron
FLC locus
(e)
DNA repair factors?
Surveillance machinery
IGS1-R
25S
18S
Recombination hotspot
rDNA spacer region
(f)
Dcr1
RDRC
RITS
ClrC
H3K9me
(g)
DRM2
AGO4
Pol IV
RDR2
Pol V
DNA methylation
DCL3
24-nt siRNAs

◀ **Fig. 2** LncRNA functions dependent on the transcript. **a** The telomeric repeat-containing RNA (TERRA) interacts with and inhibits the Ku protein complex, enabling the 5′-to-3′ exonuclease Exo1 to degrade telomeric DNA. **b** The *hsrω-n* lncRNA sequesters RNA-binding proteins, which disrupts RNA processing and export, particularly when *hsrω-n* expression is increased following stress. **c** Gene expression is doubled on the single X chromosome in *Drosophila* males by the dosage compensation complex (DCC), which contains roX lncRNAs. The DCC binds high affinity sites on the X chromosome then spreads to flanking regions, directing H4K16 hyper-acetylation to establish a chromosome-wide activated domain. **d** Upon prolonged exposure to cold, the *Arabidopsis* COLDAIR intronic lncRNA recruits a modified PRC2 complex to *FLC* and this complex establishes silencing via H3K27 methylation. **e** Recognition of lncRNAs such as IGS1-R by the nuclear surveillance machinery might recruit DNA repair or silencing factors. **f** Heterochromatin formation in *S. pombe* is primarily directed via an siRNA-dependent mechanism, in which long non-coding transcripts are converted into dsRNA by the RNA-directed RNA polymerase complex (RDRC) then processed into siRNAs by the Dcr1 endonuclease. These siRNAs target the Ago1-containing RNA-induced transcriptional silencing (RITS) complex by hybridizing to nascent lncRNAs, which thus act as both tethers and siRNA precursors. **g** In *Arabidopsis*, precursor lncRNAs transcribed by Pol IV are processed into 24 nt siRNAs by the RNA-dependent RNA polymerase RDR2 and the endonuclease DCL3. These siRNAs are loaded onto AGO4, which they direct to specific targets by base pairing with scaffold lncRNAs transcribed by Pol V. This culminates in DNA methylation by DRM2. In some cases, the Pol V-transcribed scaffolds are also processed into siRNAs, resulting in amplification

Conrad et al. 2012) and the roX2 lncRNA shows a similar distribution (Chu et al. 2011; Simon et al. 2011). DCC binding induces H4K16 hyper-acetylation across the gene bodies, and Pol II ChIP analyses reveal the upregulation of both transcription initiation (Conrad et al. 2012) and elongation (Larschan et al. 2011).

The roX lncRNAs effectively coat the X chromosome and in this they resemble the *Xist* lncRNA, which coats and inactivates one X chromosome in female mammals. LncRNAs can thus establish chromosome-wide domains, in which expression is repressed (by *Xist*) or upregulated (by roX1/roX2). Notably, the roX2 binding sites lack significant complementarity to roX2 (Simon et al. 2011), suggesting that, as for *Xist* (Hasegawa et al. 2010; Jeon and Lee 2011), bridging proteins link the lncRNAs to the X chromosome.

Several other roles have been reported for lncRNAs in large-scale chromatin organization. In *Arabidopsis*, repeat-rich regions such as the centromeres and ribosomal DNA arrays are assembled into heterochromatic "chromocenters." Although the majority of *Arabidopsis* repeat silencing occurs via RNAi-based mechanisms, chromocenter formation is dependent on the activity of a specialized polymerase, Pol V, which transcribes these regions into lncRNAs and acts to silence some classes of repeats independently of RNAi (Pontes et al. 2009). Furthermore, chromocenters are dispersed by RNase A treatment, leading to the suggestion that Pol V-transcribed lncRNAs act as structural components.

LncRNAs might also assist in pairing homologous chromosomes during meiosis. In *S. pombe*, non-coding transcription of the *sme2* locus on both chromosome II homologues is required for efficient pairing during meiosis I (Ding et al. 2012), perhaps imparting chromosome-specificity upon the pairing process. Extrapolating from this result, the authors suggest that each chromosome might be associated with specific lncRNA-containing complexes, enabling homologues to be matched.

In support of this hypothesis, pairing occurs at recombination hotspots, and these are typically associated with non-coding transcription (Wahls et al. 2008).

LncRNAs are therefore widely employed as architectural components to organize the genome into domains, enabling specific regions to be paired and stretches of chromatin (perhaps entire chromosomes) to be partitioned into distinct bodies subject to communal regulation. On a smaller scale, lncRNAs can facilitate contacts between two loci, such as an enhancer and its distal target.

## 10 Recruitment of Chromatin-Modifying Factors

Within nuclear bodies lncRNAs have dual functions, both acting as architectural components and recruiting chromatin-modifying enzymes. However, lncRNAs are also widely employed to target chromatin modifications outside of nuclear bodies. Nascent lncRNAs can act as chromatin-anchored tethers in *cis*, whereas *trans*-acting lncRNAs can direct chromatin-modifying enzymes to distal loci. Furthermore, through simultaneous interactions with two or more histone modifying enzymes, lncRNAs can facilitate collaboration and integration between activities.

Many studies have reported roles for lncRNAs in the targeting of Polycomb group (PcG) proteins. These assemble into various repressive complexes such as PRC1 and PRC2, which contribute to gene silencing via catalyzing H2A mono-ubiquitylation and H3K27 methylation, respectively. In *Arabidopsis*, flowering is accelerated following prolonged exposure to cold, a process known as vernalization. This occurs via silencing of the flowering repressor *FLC* by a modified PRC2 complex that methylates H3K27 at a promoter-proximal “nucleation site” (Fig. 2d). An intronic sense-oriented *FLC* lncRNA, COLDAIR, is expressed after a prolonged exposure to cold and binds PRC2 (Heo and Sung 2011). COLDAIR is required for PRC2 complex recruitment and presumably targets it to *FLC*. Mammalian PRC2 directly recognizes a double stem-loop motif in lncRNAs (Zhao et al. 2010) and binds many different lncRNAs, including HOTAIR that directs PRC2 to hundreds of genomic loci *in trans* (Chu et al. 2011; Gupta et al. 2010). In addition, mammalian PRC2 can be recruited by binding short, promoter-proximal RNAs (Kanhere et al. 2010). *Drosophila* PRC1 is also enriched at promoters with small, promoter-proximal RNAs that may arise from Pol II stalling (Enderle et al. 2011; Nechaev et al. 2010) but it remains unclear whether PRC1 directly binds ncRNAs.

In *S. cerevisiae*, several lncRNAs are reported to direct histone modifications. For example, the TY1-as lncRNA can repress *TY1* mRNA transcription in *trans* (Berretta et al. 2008; Matsuda and Garfinkel 2009), in addition to the post-transcriptional roles described above. Accumulation of TY1-as in strains lacking the 5′ exonuclease Xrn1 results in reduced Pol II occupancy at the *TY1* locus. Genetic analyses reveal that *TY1* silencing requires Set1-dependent H3K4 methyl-ation and histone deacetylation. It is, however, unclear whether the TY1-as lncRNA directly recruits histone-modifying enzymes, or acts as a silencing factor in response to these modifications. The *PHO84* locus is also regulated by an antisense lncRNA, in this case functioning in *cis* to direct histone deacetylation at the *PHO84* promoter by the activity of Hda1/2/3 (Camblong et al. 2007). The PHO84-as lncRNA can also repress *PHO84* when expressed ectopically from a plasmid, but this activity is independent of Hda1/2/3 (Camblong et al. 2009). The 5′ and 3′ regions of PHO84-as are both required for *trans*-repression and the 3′ region is homologous to the upstream activating sequence (UAS) of the *PHO84* promoter. Conceivably, PHO84-as might hybridize to the UAS via its 3′ end and recruit silencing factors via its 5′ end.

Many lncRNA are targets for nuclear surveillance pathways and recognition of the lncRNA may be responsible for some downstream functions. In *S. cerevisiae*, the IGS1-R lncRNA is transcribed from the region between tandem rDNA repeats, and is degraded by the exosome nuclease complex, assisted by the TRAMP poly-adenylation complex (Houseley et al. 2007) (Fig. 2e). Double-strand breaks occur at a hotspot within *IGS1* and can lead to recombination-based repeat expansion or loss. Notably, in several different mutants with hyper-recombination phenotypes, where the rDNA repeat number rapidly fluctuates, the deletion of *TRF4* results in a dramatic loss of rDNA repeats (repeat instability). Transcription of the lncRNA might promote recombination, by increasing chromatin accessibility and thus susceptibility to damage (Vasiljeva et al. 2008). However, this cannot explain the synthetic defect seen upon *TRF4* deletion, as *IGS1-R* transcription is not affected. Instead, Trf4 recruitment is suggested to contribute to DNA repair mechanisms, a model supported by multiple synthetic-lethal interactions between Trf4 and DNA repair factors (Houseley and Tollervey 2008; Houseley and Tollervey 2011). In *trf4Δ* strains, other repair mechanisms biased toward repeat loss dominate and the rDNA repeat number collapses.

Although there is little direct evidence supporting surveillance-based recruitment of DNA repair or chromatin modifying factors, it seems conceivable that the recognition of aberrant transcripts by the surveillance machinery acts as a proxy to detect the state of the underlying chromatin. In this model, damaged or silenced chromatin would produce aberrant transcripts that are recognized by the surveillance machinery, which would degrade the transcripts, but also recruit repair and/or silencing factors to remedy or silence the locus. Support for such a dual role of the surveillance machinery is provided by studies of heterochromatin formation in *S. pombe*. This is classically directed by small interfering RNA (siRNA)-dependent methylation of H3K9 by the ClrC complex, but ClrC can still mediate H3K9 methylation and silencing in strains deleted for the siRNA-binding protein Ago1 (Shanker et al. 2010). This is abolished in strains lacking the exosome-associated nuclease Rrp6 (Reyes-Turcu et al. 2011). In mitotic cells, meiotic genes undergo siRNA-independent, ClrC-dependent silencing that also requires Rrp6, which interacts with the mRNAs in a complex with the surveillance factors Mmi1 and Red1 (Zofall et al. 2012). These results suggest that the nuclear exosome contributes to transcriptional silencing, in addition to its role in degrading heterochromatin-derived transcripts.

From the results described above, it is clear that ncRNAs can target chromatin-modifying enzymes to specific loci via several distinct mechanisms. Where the lncRNA acts only on the locus from which it was transcribed, it might function as a nascent transcript recruiting either specific proteins or the nuclear RNA surveillance system, or remain tethered after transcription. LncRNAs acting at distant, homologous loci might anneal to the nascent transcripts or associate with the DNA; e.g. via R-loops, in which the lncRNA invades the DNA duplex, or by triplex formation as has been reported for the mammalian rDNA locus (Schmitz et al. 2010). Finally, some lncRNAs such as roX2 can apparently target many loci over very large chromatin domains with little or no homology, perhaps acting via protein bridges or binding with low affinity.

## 11 RNA Intersections

Numerous reports have described interactions between lncRNAs and other RNA species via base pairing. For example, in *S. cerevisiae* the lncRNA KCS1-as is induced by Pho4-dependent, low phosphate signaling and acts in *trans* to direct production of a truncated Kcs1 protein (Nishizawa et al. 2008). When both KCS1 mRNA and KCS1-as are present the region of complementarity is protected from RNase digestion, suggesting that KCS1 mRNA and KCS1-as form a duplex, which might interfere with translation. However, the ability of lncRNAs to hybridize with other transcripts is most extensively characterized in cases where lncRNAs impact upon small RNA regulatory systems.

## 12 LncRNAs Act as Precursors or Tethers for Small RNAs

In diverse eukaryotes, with the notable exception of *S. cerevisiae*, lncRNAs function alongside very small ($\sim$21–25 nt) microRNAs (miRNAs) and siRNAs. The Dicer family of endonucleases processes siRNAs from extended RNA duplexes, whereas miRNAs are excised from shorter pre-miRNA hairpins themselves derived from primary miRNAs (pri-miRNAs) (Czech and Hannon 2011). The siRNAs and miRNAs bind Argonaute family proteins and direct them to specific RNA targets via hybridization. This can result in transcriptional gene silencing (TGS) or post-transcriptional gene silencing (PTGS), by mRNA destabilization or translational repression. There is extensive crosstalk between long and small ncRNA systems, with lncRNAs acting as precursors, tethers or competitors.

In *Drosophila*, endogenous siRNAs (endo-siRNAs) are excised by Dicer 2 (DCR2) from (i) transcripts containing inverted repeats that fold into hairpins, (ii) *cis*-acting natural antisense transcripts (cis-NATs) produced from overlapping, oppositely oriented genes, and (iii) repetitive elements such as retrotransposons (Czech et al. 2008; Ghildiyal et al. 2008; Kawamura et al. 2008; Okamura et al. 2008a, b). In some cis-NAT pairs one of the partners is an antisense lncRNA (Czech et al. 2008), and antisense lncRNAs are suggested to provide the complementary strand where siRNAs are generated from repetitive elements. Like siRNAs derived from exogenous sources (exo-siRNAs), endo-siRNAs in *Drosophila* direct AGO2-dependent PTGS. However, DCR2 and AGO2 can also associate with euchromatic loci from which siRNAs are generated, including heat shock protein (HSP) genes, and promote Pol II pausing (Cernilogar et al. 2011). As HSP genes are associated with antisense lncRNAs, these siRNAs might arise from lncRNA:mRNA duplexes. In *Drosophila*, therefore, lncRNAs potentially provide the second strand required for a dsRNA Dicer substrate, facilitating siRNA-dependent TGS and PTGS. In other organisms, endo-siRNAs are predominantly reported to direct silencing at the transcriptional level, suggesting that in contrast to exo-siRNAs and miRNAs, TGS is the major effector mechanism for endo-siRNAs.

TGS is extensively characterized in *S. pombe* centromeric heterochromatin formation, where the combination of lncRNAs and siRNAs plays a central role (Fig. 2f). Pol II transcribes pericentromeric repeats to generate lncRNAs that are processed by Dcr1 into siRNAs (Kato et al. 2005). In the current model, these are loaded onto Ago1 within the RNA-induced transcriptional silencing (RITS) effector complex and target it to heterochromatic regions via base pairing with nascent lncRNAs (Buhler et al. 2006; Motamedi et al. 2004). RITS associates with ClrC, an E3 ubiquitin ligase complex that also contains a histone methyltransferase, thus directing repressive H3K9 histone methylation. LncRNAs therefore act as both precursors and tethers in TGS. Notably, this requires the action of the RNA-directed RNA polymerase complex (RDRC), which contains an RNA-dependent RNA polymerase (Rdp1), to convert the lncRNAs into dsRNA for Dcr1-dependent cleavage. However, RDRC recruitment occurs downstream of siRNA-programmed RITS binding, so it is unclear how the initial siRNAs can be generated. Small RNAs have been detected bound to Ago1 in the absence of Dcr1 or RDRC (Halic and Moazed 2010). These contain untemplated nucleotides at the 3′ end, indicating that they have been targeted by the exosome and TRAMP RNA surveillance complexes. These so-called primal RNAs might be degradation fragments from pervasive transcripts that load onto Ago1 after trimming by the nuclear surveillance machinery and act as the initial trigger for TGS. Pervasive transcription therefore plays an important role in the formation of double-stranded Dicer substrates, either providing both strands directly, or assisting in the recruitment of RNA-dependent RNA polymerase complexes to generate the complementary strand.

## 13 Specialized Polymerases in *Arabidopsis* Transcribe siRNA Precursors and Tethers

In *Arabidopsis*, TGS is also directed by small RNAs. Here, the RNA-dependent RNA polymerase RDR2 converts single-stranded precursor lncRNAs into dsRNAs, which are processed into siRNAs by the Dicer protein DCL3. siRNA-programmed AGO4 then guides the *de novo* DNA methyltransferase DRM2 to specific sites (Wierzbicki et al. 2012), resulting in cytosine methylation, mainly in the context of

CHH motifs (where H is A.T or C) (Fig. 2g). This process is collectively referred to as RNA-directed DNA methylation (RdDM) and is a form of TGS. As in *S. pombe*, lncRNAs act as precursors and scaffolds, but in *Arabidopsis* they are transcribed by dedicated polymerases, Pol IV and Pol V (Herr et al. 2005; Kanno et al. 2005; Onodera et al. 2005; Pontier et al. 2005). Functional differences between Pol IV and Pol V have provided insight into the individual contributions of distinct steps in siRNA-dependent TGS.

Pol IV is required for >90 % of siRNA generation, as well as the silencing and methylation of repetitive elements such as transposons and 5S rDNA (Mosher et al. 2008; Wierzbicki et al. 2012). Pol V is also required for methylation and silencing at many of these loci, but only acts to reinforce or amplify siRNA levels (Huettel et al. 2006; Kanno et al. 2005; Mosher et al. 2008; Pontes et al. 2006; Pontier et al. 2005; Wierzbicki et al. 2008, 2012). Pol IV-dependent lncRNAs have not been detected, but as Pol IV associates with RDR2, these lncRNAs might only exist fleetingly, before being processed into small RNAs (Haag Jeremy et al. 2012). In contrast, Pol V transcripts are more stable (Wierzbicki et al. 2008), and both Pol V-transcribed lncRNAs (Wierzbicki et al. 2009) and the Pol V protein are reported to bind AGO4 (El-Shami et al. 2007; Li et al. 2006). This may be reinforced by KTF1 which binds both ssRNA and AGO4 (He et al. 2009). Pol V is required for the association of AGO4 with most target genes (Zheng et al. 2012), and the location of AGO4 binding sites correlates more closely with Pol V binding sites than with small RNA generating loci (Zheng et al. 2012). Together, these data suggest that Pol V-dependent lncRNAs act as tethers for AGO4 when programmed with siRNAs produced by Pol IV. AGO4 can then recruit RDM2 for DNA methylation, and at some loci, Pol V lncRNAs might be cleaved by AGO4 to stimulate secondary siRNA generation. This model is supported by immunolocalization studies probing the order of assembly of nuclear foci in which RdDM is suggested to take place (Pontes et al. 2006). Pol V-transcribed lncRNAs also interact with the SWI/SNF chromatin-remodeling complex, via the RNA-binding protein IDN2 (Zhu et al. 2012). Nucleosome positioning by SWI/SNF contributes directly to transcriptional silencing and facilitates DNA methylation. LncRNA and small RNA systems therefore collaborate at multiple steps in RdDM.

Genome-wide ChIP and sequencing of small RNAs indicate that Pol IV, Pol V, and AGO4 act predominantly at rDNA repeats and loci overlapping with transposable elements in pericentromeric regions, but also bind intergenic and promoter regions within euchromatin (Mosher et al. 2008; Wierzbicki et al. 2012; Zheng et al. 2012; Zhong et al. 2012). There is some evidence that Pol IV and V target different types of transposable elements (Lee et al. 2012), though at many loci they appear to act together. Thus, two non-coding RNA pathways cooperate to precisely target RdDM to loci at which they intersect.

## 14 Regulating LncRNA Entry into the siRNA Pathway

Entry of lncRNAs into the siRNA pathway is not always desirable. For example, in fission yeast lacking a subunit of the TRAMP complex, the Ago1-bound pool of small RNAs is perturbed. Small RNA fragments derived from rRNA and tRNA bind Ago1 at the expense of *bona fide* siRNAs (Buhler et al. 2008), and this disrupts pericentromeric silencing (Bühler et al. 2007). This indicates that there are two fates available to lncRNAs, either processing into siRNAs or turnover by the nuclear surveillance machinery, and disturbing the balance between these pathways can be deleterious. At heterochromatic loci and some protein-coding genes, these pathways act in parallel to suppress expression at both the transcriptional (siRNA-mediated) and post-transcriptional (exosome-mediated) level (Buhler 2009; Yamanaka et al. 2012). In other situations, lncRNAs are selectively channeled into one or the other pathway. Several studies suggest that RNA-binding proteins act as gatekeepers to maintain the appropriate balance between these two fates. For example, exosome-dependent turnover of meiotic transcripts in mitotic fission yeast cells is promoted by the canonical poly(A) polymerase Pla1, the poly(A) binding protein Pab2 and the surveillance factor Red1 (which associates with Rrp6 and Pla1) (Sugiyama and Sugioka-Sugiyama 2011; Yamanaka et al. 2010). Moreover, deletion analyses suggest that Red1 and Pla1 act upstream of both exosome-mediated turnover and small RNA-dependent TGS at developmentally regulated genes and retrotransposons (Yamanaka et al. 2012). Additionally, Mlo3, an mRNA export factor, binds both TRAMP and ClrC and is required for full silencing of antisense RNAs in euchromatic regions that are targeted by both RNAi-dependent TGS and exosome-dependent turnover (Zhang et al. 2011). These observations suggest that RNA processing and surveillance factors such as Mlo3, Pla1 and Red1 are recruited to lncRNAs and both stimulate and regulate turnover and processing into small RNAs. The recruitment of surveillance factors therefore constitutes an important step during heterochromatin formation in siRNA-dependent pathways, in addition to the siRNA-independent mechanisms apparent in strains lacking Ago1. Overall, lncRNAs can compete with, or contribute to, RNAi-based silencing mechanisms, and the trafficking of lncRNAs into RNAi versus turnover pathways is highly regulated.

## 15 Concluding Remarks

Analyses of eukaryotic transcription are revealing a bewildering number of short and long ncRNAs. From the data surveyed here it will be clear that the relatively small numbers of lncRNAs that have been characterized to date have already revealed a wide range of functions, mechanisms and targets; and there seems every reason to think that many more remain to be identified. The complexity and heterogeneity of lncRNAs offer enormous numbers of possibilities for both site-specific and global regulation of gene expression. LncRNAs significantly expand the repertoire of regulatory and architectural molecules available to the cell, and interactions between lncRNAs, small RNAs and/or proteins enable them to collaborate in regulatory circuits, which exploit the unique capabilities of each class. The pervasive and disruptive nature of lncRNA transcription and the ability of lncRNAs to impact upon diverse cellular processes can also pose a threat to the cell, so lncRNAs must be appropriately managed. An important future challenge is therefore to understand the regulation, processing and turnover of lncRNAs, which will help reveal the mechanisms by which they act and perhaps also present novel experimental and therapeutic opportunities.

**Acknowledgments** This work was funded by Wellcome Trust [086587 and 077248]. Work in the Wellcome Trust Centre for Cell Biology is supported by Wellcome Trust core funding [092076].

## References

Alekseyenko, A. A., Peng, S., Larschan, E., Gorchakov, A. A., Lee, O. K., Kharchenko, P., et al. (2008). A sequence motif within chromatin entry sites directs MSL establishment on the *Drosophila* X chromosome. *Cell, 134*(4), 599–609.

Arigo, J. T., Carroll, K. L., Ames, J. M., & Corden, J. L. (2006). Regulation of yeast NRD1 expression by premature transcription termination. *Molecular Cell, 21*(5), 641–651.

Berretta, J., Pinskaya, M., & Morillon, A. (2008). A cryptic unstable transcript mediates transcriptional trans-silencing of the Ty1 retrotransposon in *Saccharomyces cerevisiae*. *Genes & Development, 22*(5), 615–626.

Bird, A. J., Gordon, M., Eide, D. J., & Winge, D. R. (2006). Repression of ADH1 and ADH3 during zinc deficiency by Zap1-induced intergenic RNA transcripts. *EMBO Journal, 25*(24), 5726–5734.

Brar, G. A., Yassour, M., Friedman, N., Regev, A., Ingolia, N. T., & Weissman, J. S. (2012). High-resolution view of the yeast meiotic program revealed by ribosome profiling. *Science, 335*(6068), 552–557.

Buhler, M. (2009). RNA turnover and chromatin-dependent gene silencing. *Chromosoma, 118*(2), 141–151.

Buhler, M., Verdel, A., & Moazed, D. (2006). Tethering RITS to a nascent transcript initiates RNAi- and heterochromatin-dependent gene silencing. *Cell, 125*(5), 873–886.

Buhler, M., Spies, N., Bartel, D. P., & Moazed, D. (2008). TRAMP-mediated RNA surveillance prevents spurious entry of RNAs into the *Schizosaccharomyces pombe* siRNA pathway. *Nature Structural & Molecular Biology, 15*(10), 1015–1023.

Bühler, M., Haas, W., Gygi, S. P., & Moazed, D. (2007). RNAi-dependent and -independent RNA turnover mechanisms contribute to heterochromatic gene silencing. *Cell, 129*(4), 707–721.

Bumgarner, S. L., Dowell, R. D., Grisafi, P., Gifford, D. K., & Fink, G. R. (2009). Toggle involving cis-interfering non-coding RNAs controls variegated gene expression in yeast. *Proceedings of the National Academy of Sciences of the United States of America, 106*(43), 18321–18326.

Bumgarner, S. L., Neuert, G., Voight, F. R., Symbor-Nagrabska, A., Grisafi, P., van Oudenaarden, A., et al. (2012). Single-cell analysis reveals that non-coding RNAs contribute to clonal heterogeneity by modulating transcription factor recruitment. *Molecular Cell, 45*, 1–13.

Camblong, J., Iglesias, N., Fickentscher, C., Dieppois, G., & Stutz, F. (2007). Antisense RNA stabilization induces transcriptional gene silencing via histone deacetylation in *Saccharomyces cerevisiae*. *Cell, 131*(4), 706–717.
Camblong, J., Beyrouthy, N., Guffanti, E., Schlaepfer, G., Steinmetz, L. M., & Stutz, F. (2009). Trans-acting antisense RNAs mediate transcriptional gene cosuppression in *S. cerevisiae*. *Genes & Development, 23*(13), 1534–1545.
Cernilogar, F. M., Onorati, M. C., Kothe, G. O., Burroughs, A. M., Parsi, K. M., Breiling, A., et al. (2011). Chromatin-associated RNA interference components contribute to transcriptional regulation in *Drosophila*. *Nature, 480*(7377), 391–395.
Cheung, V., Chua, G., Batada, N. N., Landry, C. R., Michnick, S. W., Hughes, T. R., et al. (2008). Chromatin- and transcription-related factors repress transcription from within coding regions throughout the *Saccharomyces cerevisiae* genome. *PLoS Biology, 6*(11), e277.
Chu, C., Qu, K., Zhong Franklin, L., Artandi Steven, E., & Chang Howard, Y. (2011). Genomic maps of long non-coding RNA occupancy reveal principles of RNA-chromatin interactions. *Molecular Cell, 44*(4), 667–678.
Churchman, L. S., & Weissman, J. S. (2011). Nascent transcript sequencing visualizes transcription at nucleotide resolution. *Nature, 469*(7330), 368–373.
Conrad, T., Cavalli, F. M. G., Vaquerizas, J. M., Luscombe, N. M., & Akhtar, A. (2012). *Drosophila* dosage compensation involves enhanced Pol II recruitment to male X-linked promoters. *Science, 337*(6095), 742–746.
Creamer, T. J., Darby, M. M., Jamonnak, N., Schaughency, P., Hao, H., Wheelan, S. J., et al. (2011). Transcriptome-wide binding sites for components of the *Saccharomyces cerevisiae* non-poly(A) termination pathway: Nrd1, Nab3, and Sen1. *PLoS Genetics, 7*(10), e1002329.
Czech, B., & Hannon, G. J. (2011). Small RNA sorting: matchmaking for Argonautes. *Nature Reviews Genetics, 12*(1), 19–31.
Czech, B., Malone, C. D., Zhou, R., Stark, A., Schlingeheyde, C., Dus, M., et al. (2008). An endogenous small interfering RNA pathway in *Drosophila*. *Nature, 453*(7196), 798–802.
David, L., Huber, W., Granovskaia, M., Toedling, J., Palm, C. J., Bofkin, L., et al. (2006). A high-resolution map of transcription in the yeast genome. *Proceedings of the National Academy of Sciences of the United States of America, 103*(14), 5320–5325.
Ding, D.-Q., Okamasa, K., Yamane, M., Tsutsumi, C., Haraguchi, T., Yamamoto, M., et al. (2012). Meiosis-specific non-coding RNA mediates robust pairing of homologous chromosomes in meiosis. *Science, 336*(6082), 732–736.
El-Shami, M., Pontier, D., Lahmy, S., Braun, L., Picart, C., Vega, D., et al. (2007). Reiterated WG/GW motifs form functionally and evolutionarily conserved ARGONAUTE-binding platforms in RNAi-related components. *Genes & Development, 21*(20), 2539–2544.
Enderle, D., Beisel, C., Stadler, M. B., Gerstung, M., Athri, P., & Paro, R. (2011). Polycomb preferentially targets stalled promoters of coding and non-coding transcripts. *Genome Research, 21*(2), 216–226.
Garcia-Martinez, J., Ayala, G., Pelechano, V., Chavez, S., Herrero, E., & Perez-Ortin, J. E. (2012). The relative importance of transcription rate, cryptic transcription and mRNA stability on shaping stress responses in yeast. *Transcription, 3*(1), 39–44.
Geisler, S., Lojek, L., Khalil, A. M., Baker, K. E., & Coller, J. (2012). Decapping of long non-coding RNAs regulates inducible genes. *Molecular Cell, 45*(3), 279–291.
Gelfand, B., Mead, J., Bruning, A., Apostolopoulos, N., Tadigotla, V., Nagaraj, V., et al. (2011). Regulated antisense transcription controls expression of cell-type-specific genes in yeast. *Molecular and Cellular Biology, 31*(8), 1701–1709.
Ghildiyal, M., Seitz, H., Horwich, M. D., Li, C., Du, T., Lee, S., et al. (2008). Endogenous siRNAs derived from transposons and mRNAs in *Drosophila* somatic cells. *Science, 320*(5879), 1077–1081.
Granovskaia, M., Jensen, L., Ritchie, M., Toedling, J., Ning, Y., Bork, P., et al. (2010). High-resolution transcription atlas of the mitotic cell cycle in budding yeast. *Genome Biology, 11*(3), R24.

Gudipati, R. K., Xu, Z., Lebreton, A., Seraphin, B., Steinmetz, L. M., Jacquier, A., et al. (2012). Extensive degradation of RNA precursors by the exosome in wild-type cells. *Molecular Cell, 48*(3), 409–421.

Gupta, R. A., Shah, N., Wang, K. C., Kim, J., Horlings, H. M., Wong, D. J., et al. (2010). Long non-coding RNA HOTAIR reprograms chromatin state to promote cancer metastasis. *Nature, 464*(7291), 1071–1076.

Haag, J. R., Ream, T. S., Marasco, M., Nicora, C. D., Norbeck, A. D., Pasa-Tolic, L., Pikaard, C. S. (2012). In vitro transcription activities of Pol IV, Pol V, and RDR2 reveal coupling of Pol IV and RDR2 for dsRNA synthesis in plant RNA silencing. *Molecular Cell*, 48, 811–818.

Hainer, S. J., & Martens, J. A. (2011). Identification of histone mutants that are defective for transcription-coupled nucleosome occupancy. *Molecular and Cellular Biology, 31*(17), 3557–3568.

Hainer, S. J., Pruneski, J. A., Mitchell, R. D., Monteverde, R. M., & Martens, J. A. (2011). Intergenic transcription causes repression by directing nucleosome assembly. *Genes & Development, 25*(1), 29–40.

Halic, M., & Moazed, D. (2010). Dicer-independent primal RNAs trigger RNAi and heterochromatin formation. *Cell, 140*(4), 504–516.

Halme, A., Bumgarner, S., Styles, C., & Fink, G. R. (2004). Genetic and epigenetic regulation of the FLO gene family generates cell-surface variation in yeast. *Cell, 116*(3), 405–415.

Hasegawa, Y., Brockdorff, N., Kawano, S., Tsutui, K., Tsutui, K., & Nakagawa, S. (2010). The matrix protein hnRNP U is required for chromosomal localization of Xist RNA. *Developmental Cell, 19*(3), 469–476.

Hazen, S., Naef, F., Quisel, T., Gendron, J., Chen, H., Ecker, J., et al. (2009). Exploring the transcriptional landscape of plant circadian rhythms using genome tiling arrays. *Genome Biology, 10*(2), R17.

He, X.-J., Hsu, Y.-F., Zhu, S., Wierzbicki, A. T., Pontes, O., Pikaard, C. S., et al. (2009). An effector of RNA-directed DNA methylation in *Arabidopsis* is an ARGONAUTE 4- and RNA-binding protein. *Cell, 137*(3), 498–508.

Heo, J. B., & Sung, S. (2011). Vernalization-mediated epigenetic silencing by a long intronic non-coding RNA. *Science, 331*(6013), 76–79.

Herr, A. J., Jensen, M. B., Dalmay, T., & Baulcombe, D. C. (2005). RNA polymerase IV directs silencing of endogenous DNA. *Science, 308*(5718), 118–120.

Hessle, V., von Euler, A., González de Valdivia, E., & Visa, N. (2012). Rrp6 is recruited to transcribed genes and accompanies the spliced mRNA to the nuclear pore. *RNA, 18*(8), 1466–1474.

Hirota, K., Miyoshi, T., Kugou, K., Hoffman, C. S., Shibata, T., & Ohta, K. (2008). Stepwise chromatin remodelling by a cascade of transcription initiation of non-coding RNAs. *Nature, 456*(7218), 130–134.

Hobson, D. J., Wei, W., Steinmetz, L. M., & Svejstrup, J. Q. (2012). RNA polymerase II collision interrupts convergent transcription. *Molecular Cell, 48*(3), 365–374.

Hongay, C. F., Grisafi, P. L., Galitski, T., & Fink, G. R. (2006). Antisense transcription controls cell fate in *Saccharomyces cerevisiae*. *Cell, 127*(4), 735–745.

Houseley, J., & Tollervey, D. (2008). The nuclear RNA surveillance machinery: The link between ncRNAs and genome structure in budding yeast? *Biochimica et Biophysica Acta (BBA): Gene Regulatory Mechanisms, 1779*(4), 239–246.

Houseley, J., & Tollervey, D. (2011). Repeat expansion in the budding yeast ribosomal DNA can occur independently of the canonical homologous recombination machinery. *Nucleic Acids Research, 39*(20), 8778–8791.

Houseley, J., Kotovic, K., El Hage, A., & Tollervey, D. (2007). Trf4 targets ncRNAs from telomeric and rDNA spacer regions and functions in rDNA copy number control. *EMBO Journal, 26*(24), 4996–5006.

Houseley, J., Rubbi, L., Grunstein, M., Tollervey, D., & Vogelauer, M. (2008). A ncRNA modulates histone modification and mRNA induction in the yeast *GAL* gene cluster. *Molecular Cell, 32*(5), 685–695.

Huettel, B., Kanno, T., Daxinger, L., Aufsatz, W., Matzke, A. J., & Matzke, M. (2006). Endogenous targets of RNA-directed DNA methylation and Pol IV in *Arabidopsis*. *EMBO Journal, 25*(12), 2828–2836.

Jenks, M. H., O'Rourke, T. W., & Reines, D. (2008). Properties of an intergenic terminator and start site switch that regulate IMD2 transcription in yeast. *Molecular and Cellular Biology, 28*(12), 3883–3893.

Jeon, Y., & Lee Jeannie, T. (2011). YY1 tethers Xist RNA to the inactive X nucleation center. *Cell, 146*(1), 119–133.

Johnson, T. K., Carrington, L. B., Hallas, R. J., & McKechnie, S. W. (2009). Protein synthesis rates in *Drosophila* associate with levels of the hsr-omega nuclear transcript. *Cell Stress Chaperones, 14*(6), 569–577.

Kanhere, A., Viiri, K., Araújo, C. C., Rasaiyaah, J., Bouwman, R. D., Whyte, W. A., et al. (2010). Short RNAs are transcribed from repressed polycomb target genes and interact with polycomb repressive complex-2. *Molecular Cell, 38*(5), 675–688.

Kanno, T., Huettel, B., Mette, M. F., Aufsatz, W., Jaligot, E., Daxinger, L., et al. (2005). Atypical RNA polymerase subunits required for RNA-directed DNA methylation. *Nature Genetics, 37*(7), 761–765.

Kato, H., Goto, D. B., Martienssen, R. A., Urano, T., Furukawa, K., & Murakami, Y. (2005). RNA polymerase II is required for RNAi-dependent heterochromatin assembly. *Science, 309*(5733), 467–469.

Kawamura, Y., Saito, K., Kin, T., Ono, Y., Asai, K., Sunohara, T., et al. (2008). *Drosophila* endogenous small RNAs bind to Argonaute-2 in somatic cells. *Nature, 453*(7196), 793–797.

Kim, T., & Buratowski, S. (2009). Dimethylation of H3K4 by Set1 recruits the Set3 histone deacetylase complex to 5′ transcribed regions. *Cell, 137*(2), 259–272.

Kim, K.-Y., & Levin, D. (2011). Mpk1 MAPK association with the Paf1 complex blocks Sen1-mediated premature transcription termination. *Cell, 144*(5), 745–756.

Kim, H., Erickson, B., Luo, W., Seward, D., Graber, J. H., Pollock, D. D., et al. (2010). Gene-specific RNA polymerase II phosphorylation and the CTD code. *Nature Structural & Molecular Biology, 17*(10), 1279–1286.

Kim, T. S., Liu, C. L., Yassour, M., Holik, J., Friedman, N., Buratowski, S., et al. (2011). RNA polymerase mapping during stress responses reveals widespread nonproductive transcription in yeast. *Genome Biology, 11*(7), R75.

Kim, T., Xu, Z., Clauder-Münster, S., Steinmetz Lars, M., & Buratowski, S. (2012). Set3 HDAC mediates effects of overlapping non-coding transcription on gene induction kinetics. *Cell, 150*(6), 1158–1169.

Kizer, K. O., Phatnani, H. P., Shibata, Y., Hall, H., Greenleaf, A. L., & Strahl, B. D. (2005). A novel domain in Set2 mediates RNA polymerase II interaction and couples histone H3 K36 methylation with transcript elongation. *Molecular and Cellular Biology, 25*(8), 3305–3316.

Kuehner, J. N., & Brow, D. A. (2008). Regulation of a eukaryotic gene by GTP-dependent start site selection and transcription attenuation. *Molecular Cell, 31*(2), 201–211.

Lakhotia, S. C., Mallik, M., Singh, A. K., & Ray, M. (2012). The large non-coding hsromega-n transcripts are essential for thermotolerance and remobilization of hnRNPs, HP1 and RNA polymerase II during recovery from heat shock in *Drosophila*. *Chromosoma, 121*(1), 49–70.

Lardenois, A., Liu, Y., Walther, T., Chalmel, F., Evrard, B., Granovskaia, M., et al. (2011). Execution of the meiotic non-coding RNA expression program and the onset of gametogenesis in yeast require the conserved exosome subunit Rrp6. *Proceedings of the National Academy of Sciences, 108*(3), 1058–1063.

Larschan, E., Bishop, E. P., Kharchenko, P. V., Core, L. J., Lis, J. T., Park, P. J., et al. (2011). X chromosome dosage compensation via enhanced transcriptional elongation in *Drosophila*. *Nature, 471*(7336), 115–118.

Lee, T. F., Gurazada, S. G., Zhai, J., Li, S., Simon, S. A., Matzke, M. A., et al. (2012). RNA polymerase V-dependent small RNAs in *Arabidopsis* originate from small, intergenic loci including most SINE repeats. *Epigenetics, 7*(7), 781–795.

Li, C. F., Pontes, O., El-Shami, M., Henderson, I. R., Bernatavichute, Y. V., Chan, S. W. L., et al. (2006). An ARGONAUTE4-containing nuclear processing center colocalized with Cajal bodies in *Arabidopsis thaliana*. *Cell, 126*(1), 93–106.

Li, B., Jackson, J., Simon, M. D., Fleharty, B., Gogol, M., Seidel, C., et al. (2009). Histone H3 lysine 36 dimethylation (H3K36me2) is sufficient to recruit the Rpd3s histone deacetylase complex and to repress spurious transcription. *Journal of Biological Chemistry, 284*(12), 7970–7976.

Lin, M. F., Jungreis, I., & Kellis, M. (2011). PhyloCSF: a comparative genomics method to distinguish protein coding and non-coding regions. *Bioinformatics, 27*(13), i275–i282.

Liu, J., Jung, C., Xu, J., Wang, H., Deng, S., Bernad, L., Arenas-Huertero, C., Chua, N. H. (2012). Genome-wide analysis uncovers regulation of long intergenic non-coding RNAs in *Arabidopsis*. *Plant Cell*, 24, 4333-4345.

Luke, B., Panza, A., Redon, S., Iglesias, N., Li, Z., & Lingner, J. (2008). The Rat1p 5′ to 3′ exonuclease degrades telomeric repeat-containing RNA and promotes telomere elongation in *Saccharomyces cerevisiae*. *Molecular Cell, 32*(4), 465–477.

Martens, J. A., Laprade, L., & Winston, F. (2004). Intergenic transcription is required to repress the *Saccharomyces cerevisiae SER3* gene. *Nature, 429*(6991), 571–574.

Martens, J. A., Wu, P. Y., & Winston, F. (2005). Regulation of an intergenic transcript controls adjacent gene transcription in *Saccharomyces cerevisiae*. *Genes & Development, 19*(22), 2695–2704.

Martianov, I., Ramadass, A., Serra Barros, A., Chow, N., & Akoulitchev, A. (2007). Repression of the human dihydrofolate reductase gene by a non-coding interfering transcript. *Nature, 445*(7128), 666–670.

Matsuda, E., & Garfinkel, D. J. (2009). Posttranslational interference of Ty1 retrotransposition by antisense RNAs. *Proceedings of the National Academy of Sciences, 106*(37), 15657–15662.

Meller, V. H., & Rattner, B. P. (2002). The roX genes encode redundant male-specific lethal transcripts required for targeting of the MSL complex. *EMBO Journal, 21*(5), 1084–1091.

Mosher, R. A., Schwach, F., Studholme, D., & Baulcombe, D. C. (2008). PolIVb influences RNA-directed DNA methylation independently of its role in siRNA biogenesis. *Proceedings of the National Academy of Sciences of the United States of America, 105*(8), 3145–3150.

Motamedi, M. R., Verdel, A., Colmenares, S. U., Gerber, S. A., Gygi, S. P., & Moazed, D. (2004). Two RNAi complexes, RITS and RDRC, physically interact and localize to non-coding centromeric RNAs. *Cell, 119*(6), 789–802.

Murray, S. C., Serra Barros, A., Brown, D. A., Dudek, P., Ayling, J., & Mellor, J. (2012). A pre-initiation complex at the 3′-end of genes drives antisense transcription independent of divergent sense transcription. *Nucleic Acids Research, 40*(6), 2432–2444.

Nagalakshmi, U., Wang, Z., Waern, K., Shou, C., Raha, D., Gerstein, M., et al. (2008). The transcriptional landscape of the yeast genome defined by RNA sequencing. *Science, 320*(5881), 1344–1349.

Nam, J. W., & Bartel, D. P. (2012). Long non-coding RNAs in *C. elegans*. *Genome Research, 22*, 2529–2540.

Nechaev, S., Fargo, D. C., dos Santos, G., Liu, L., Gao, Y., & Adelman, K. (2010). Global analysis of short RNAs reveals widespread promoter-proximal stalling and arrest of Pol II in *Drosophila*. *Science, 327*(5963), 335–338.

Neil, H., Malabat, C., d'Aubenton-Carafa, Y., Xu, Z., Steinmetz, L. M., & Jacquier, A. (2009). Widespread bidirectional promoters are the major source of cryptic transcripts in yeast. *Nature, 457*(7232), 1038–1042.

Nishizawa, M., Komai, T., Katou, Y., Shirahige, K., Ito, T., & Toh-e, A. (2008). Nutrient-regulated antisense and intragenic RNAs modulate a signal transduction pathway in yeast. *PLoS Biology, 6*(12), e326.

Octavio, L. M., Gedeon, K., & Maheshri, N. (2009). Epigenetic and conventional regulation is distributed among activators of FLO11 allowing tuning of population-level heterogeneity in its expression. *PLoS Genetics, 5*(10), e1000673.

Okamura, K., Balla, S., Martin, R., Liu, N., & Lai, E. C. (2008a). Two distinct mechanisms generate endogenous siRNAs from bidirectional transcription in *Drosophila melanogaster*. *Nature Structural & Molecular Biology, 15*(6), 581–590.

Okamura, K., Chung, W.-J., Ruby, J. G., Guo, H., Bartel, D. P., & Lai, E. C. (2008b). The *Drosophila* hairpin RNA pathway generates endogenous short interfering RNAs. *Nature, 453*(7196), 803–806.

Onodera, Y., Haag, J. R., Ream, T., Costa Nunes, P., Pontes, O., & Pikaard, C. S. (2005). Plant nuclear RNA polymerase IV mediates siRNA and DNA methylation-dependent heterochromatin formation. *Cell, 120*(5), 613–622.

O'Sullivan, J. M., Tan-Wong, S. M., Morillon, A., Lee, B., Coles, J., Mellor, J., et al. (2004). Gene loops juxtapose promoters and terminators in yeast. *Nature Genetics, 36*(9), 1014–1018.

Pauli, A., Valen, E., Lin, M. F., Garber, M., Vastenhouw, N. L., Levin, J. Z., et al. (2012). Systematic identification of long non-coding RNAs expressed during zebrafish embryogenesis. *Genome Research, 22*(3), 577–591.

Petruk, S., Sedkov, Y., Riley, K. M., Hodgson, J., Schweisguth, F., Hirose, S., et al. (2006). Transcription of bxd non-coding RNAs promoted by trithorax represses Ubx in cis by transcriptional interference. *Cell, 127*(6), 1209–1221.

Petruk, S., Sedkov, Y., Brock, H. W., & Mazo, A. (2007). A model for initiation of mosaic HOX gene expression patterns by non-coding RNAs in early embryos. *RNA Biology, 4*(1), 1–6.

Pfeiffer, V., & Lingner, J. (2012). TERRA promotes telomere shortening through exonuclease 1-mediated resection of chromosome ends. *PLoS Genetics, 8*(6), e1002747.

Pinskaya, M., Gourvennec, S., & Morillon, A. (2009). H3 lysine 4 di- and tri-methylation deposited by cryptic transcription attenuates promoter activation. *EMBO Journal, 28*, 1697–1707.

Pontes, O., Li, C. F., Costa Nunes, P., Haag, J., Ream, T., Vitins, A., et al. (2006). The *Arabidopsis* chromatin-modifying nuclear siRNA pathway involves a nucleolar RNA processing center. *Cell, 126*(1), 79–92.

Pontes, O., Costa-Nunes, P., Vithayathil, P., & Pikaard, C. S. (2009). RNA polymerase V functions in *Arabidopsis* interphase heterochromatin organization independently of the 24-nt siRNA-directed DNA methylation pathway. *Molecular Plant, 2*(4), 700–710.

Pontier, D., Yahubyan, G., Vega, D., Bulski, A., Saez-Vasquez, J., Hakimi, M. A., et al. (2005). Reinforcement of silencing at transposons and highly repeated sequences requires the concerted action of two distinct RNA polymerases IV in *Arabidopsis*. *Genes & Development, 19*(17), 2030–2040.

Prasanth, K. V., Rajendra, T. K., Lal, A. K., & Lakhotia, S. C. (2000). Omega speckles—a novel class of nuclear speckles containing hnRNPs associated with non-coding hsr-omega RNA in *Drosophila*. *Journal of Cell Science, 113*(19), 3485–3497.

Purzycka, K. J., Legiewicz, M., Matsuda, E., Eizentstat, L. D., Lusvarghi, S., Saha, A., Le Grice, S. F. J., Garfinkel, D. J. (2012). Exploring Ty1 retrotransposon RNA structure within virus-like particles. *Nucleic Acids Research*, 41, 463-473.

Redon, S., Reichenbach, P., & Lingner, J. (2010). The non-coding RNA TERRA is a natural ligand and direct inhibitor of human telomerase. *Nucleic Acids Research, 38*(17), 5797–5806.

Reyes-Turcu, F. E., Zhang, K., Zofall, M., Chen, E., & Grewal, S. I. (2011). Defects in RNA quality control factors reveal RNAi-independent nucleation of heterochromatin. *Nature Structural & Molecular Biology, 18*(10), 1132–1138.

Rhee, H. S., & Pugh, B. F. (2012). Genome-wide structure and organization of eukaryotic pre-initiation complexes. *Nature, 483*(7389), 295–301.

Rodriguez-Gil, A., Garcia-Martinez, J., Pelechano, V., Munoz-Centeno Mde, L., Geli, V., Perez-Ortin, J. E., et al. (2010). The distribution of active RNA polymerase II along the transcribed region is gene-specific and controlled by elongation factors. *Nucleic Acids Research, 38*(14), 4651–4664.

Schmitz, K.-M., Mayer, C., Postepska, A., & Grummt, I. (2010). Interaction of non-coding RNA with the rDNA promoter mediates recruitment of DNMT3b and silencing of rRNA genes. *Genes & Development, 24*(20), 2264–2269.

Shanker, S., Job, G., George, O. L., Creamer, K. M., Shaban, A., & Partridge, J. F. (2010). Continuous requirement for the Clr4 complex but not RNAi for centromeric heterochromatin assembly in fission yeast harboring a disrupted RITS complex. *PLoS Genetics, 6*(10), e1001174.

Simon, M. D., Wang, C. I., Kharchenko, P. V., West, J. A., Chapman, B. A., Alekseyenko, A. A., et al. (2011). The genomic binding sites of a non-coding RNA. *Proceedings of the National Academy of Sciences, 108*(51), 20497–20502.

Sugiyama, T., & Sugioka-Sugiyama, R. (2011). Red1 promotes the elimination of meiosis-specific mRNAs in vegetatively growing fission yeast. *EMBO Journal, 30*(6), 1027–1039.

Taft, R. J., Simons, C., Nahkuri, S., Oey, H., Korbie, D. J., Mercer, T. R., et al. (2011). Nuclear-localized tiny RNAs are associated with transcription initiation and splice sites in metazoans. *Nature Structural & Molecular Biology, 17*(8), 1030–1034.

Tan-Wong, S. M., Zaugg, J. B., Camblong, J., Xu, Z., Zhang, D. W., Mischo, H. E., et al. (2012). Gene loops enhance transcriptional directionality. *Science, 338*, 671–675.

Thebault, P., Boutin, G., Bhat, W., Rufiange, A., Martens, J., Nourani, A. (2011). Transcription regulation by the non-coding RNA SRG1 requires Spt2-dependent chromatin deposition in the wake of RNAP II. *Molecular and Cellular Biology*, doi:10.1128/MCB.01083-01010.

Thiebaut, M., Colin, J., Neil, H., Jacquier, A., Seraphin, B., Lacroute, F., et al. (2008). Futile cycle of transcription initiation and termination modulates the response to nucleotide shortage in *S. cerevisiae*. *Molecular Cell, 31*(5), 671–682.

Uhler, J. P., Hertel, C., & Svejstrup, J. Q. (2007). A role for non-coding transcription in activation of the yeast PHO5 gene. *Proceedings of the National Academy of Sciences of the United States of America, 104*(19), 8011–8016.

van Dijk, E. L., Chen, C. L., d/'Aubenton-Carafa, Y., Gourvennec, S., Kwapisz, M., Roche, V., et al. (2011). XUTs are a class of Xrn1-sensitive antisense regulatory non-coding RNA in yeast. *Nature, 475*(7354), 114–117.

van Werven Folkert, J., Neuert, G., Hendrick, N., Lardenois, A., Buratowski, S., van Oudenaarden, A., et al. (2012). Transcription of two long non-coding RNAs mediates mating-type control of gametogenesis in budding yeast. *Cell, 150*(6), 1170–1181.

Vasiljeva, L., & Buratowski, S. (2006). Nrd1 interacts with the nuclear exosome for 3′ processing of RNA polymerase II transcripts. *Molecular Cell, 21*(2), 239–248.

Vasiljeva, L., Kim, M., Terzi, N., Soares, L. M., & Buratowski, S. (2008). Transcription termination and RNA degradation contribute to silencing of RNA polymerase II transcription within heterochromatin. *Molecular Cell, 29*(3), 313–323.

Wahls, W. P., Siegel, E. R., & Davidson, M. K. (2008). Meiotic recombination hotspots of fission yeast are directed to loci that express non-coding RNA. *PLoS ONE, 3*(8), e2887.

Wang, X., Arai, S., Song, X., Reichart, D., Du, K., Pascual, G., et al. (2008). Induced ncRNAs allosterically modify RNA-binding proteins in cis to inhibit transcription. *Nature, 454*(7200), 126–130.

Wierzbicki, A. T., Haag, J. R., & Pikaard, C. S. (2008). Non-coding transcription by RNA polymerase Pol IVb/Pol V mediates transcriptional silencing of overlapping and adjacent genes. *Cell, 135*(4), 635–648.

Wierzbicki, A. T., Ream, T. S., Haag, J. R., & Pikaard, C. S. (2009). RNA polymerase V transcription guides ARGONAUTE4 to chromatin. *Nature Genetics, 41*(5), 630–634.

Wierzbicki, A. T., Cocklin, R., Mayampurath, A., Lister, R., Rowley, M. J., Gregory, B. D., et al. (2012). Spatial and functional relationships among Pol V-associated loci, Pol IV-dependent siRNAs, and cytosine methylation in the *Arabidopsis* epigenome. *Genes & Development, 26*(16), 1825–1836.

Xu, Z., Wei, W., Gagneur, J., Perocchi, F., Clauder-Munster, S., Camblong, J., et al. (2009). Bidirectional promoters generate pervasive transcription in yeast. *Nature, 457*(7232), 1033–1037.

Xu, Z., Wei, W., Gagneur, J., Clauder-Munster, S., Smolik, M., Huber, W., et al. (2011). Antisense expression increases gene expression variability and locus interdependency. *Molecular Systems Biology, 7*, 1–10.

Yamanaka, S., Yamashita, A., Harigaya, Y., Iwata, R., & Yamamoto, M. (2010). Importance of polyadenylation in the selective elimination of meiotic mRNAs in growing *S. pombe* cells. *EMBO Journal, 29*(13), 2173–2181.

Yamanaka, S., Mehta, S., Reyes-Turcu, F. E., Zhuang, F., Fuchs, R.T., Rong, Y., Robb, G. B., Grewal, S. I. (2012). RNAi triggered by specialized machinery silences developmental genes and retrotransposons. *Nature*, 493, 557-560.

Yang, L., Lin, C., Liu, W., Zhang, J., Ohgi Kenneth, A., Grinstein Jonathan, D., et al. (2011). ncRNA- and Pc2 methylation-dependent gene relocation between nuclear structures mediates gene activation programs. *Cell, 147*(4), 773–788.

Yassour, M., Pfiffner, J., Levin, J., Adiconis, X., Gnirke, A., Nusbaum, C., et al. (2010). Strand-specific RNA sequencing reveals extensive regulated long antisense transcripts that are conserved across yeast species. *Genome Biology, 11*(8), R87.

Young, R. S., Marques, A. C., Tibbit, C., Haerty, W., Bassett, A. R., Liu, J. L., et al. (2012). Identification and properties of 1,119 candidate lincRNA loci in the *Drosophila melanogaster* genome. *Genome Biology and Evolution, 4*(4), 427–442.

Zhang, K., Fischer, T., Porter, R. L., Dhakshnamoorthy, J., Zofall, M., Zhou, M., et al. (2011). Clr4/Suv39 and RNA quality control factors cooperate to trigger RNAi and suppress antisense RNA. *Science, 331*(6024), 1624–1627.

Zhao, J., Ohsumi, T. K., Kung, J. T., Ogawa, Y., Grau, D. J., Sarma, K., et al. (2010). Genome-wide identification of polycomb-associated RNAs by RIP-seq. *Molecular Cell, 40*(6), 939–953.

Zheng, Q., Rowley, M. J., Bohmdorfer, G., Sandhu, D., Gregory, B. D., Wierzbicki, A. T. (2012). RNA polymerase V targets transcriptional silencing components to promoters of protein-coding genes. *Plant Journal*, 10.1111/tpj.12034.

Zhong, X., Hale, C. J., Law, J. A., Johnson, L. M., Feng, S., Tu, A., et al. (2012). DDR complex facilitates global association of RNA polymerase V to promoters and evolutionarily young transposons. *Nature Structural & Molecular Biology, 19*(9), 870–875.

Zhu, Y., Rowley, M. J., Böhmdorfer, G., Wierzbicki Andrzej, T. (2012). A SWI/SNF chromatin-remodeling complex acts in non-coding RNA-mediated transcriptional silencing. *Molecular Cell*, 49, 298-309.

Zofall, M., Yamanaka, S., Reyes-Turcu, F. E., Zhang, K., Rubin, C., & Grewal, S. I. (2012). RNA elimination machinery targeting meiotic mRNAs promotes facultative heterochromatin formation. *Science, 335*(6064), 96–100.

# Emerging Technologies to Study Long Non-coding RNAs

**Fereshteh Jahaniani, Varsha Rao, Stephanie Nevins, Damek Spacek, Neal Bharadwaj, Jason Reuter and Michael Snyder**

## 1 Introduction

It has been less than half a century since Robert W. Holley et al. used 140 kg of commercial baker's yeast to characterize the first noncoding RNA (ncRNA), alanine tRNA. Now, 48 years later, advancements in genomic technologies have enabled scientists to study genomes, transcriptomes, and proteomes, on an unprecedented and high-throughput scale, and even at the single cell resolution. These discoveries have completely changed the classical view of the central dogma of molecular biology, as we now understand that protein coding genes account for less than 2 % of human genome, however, the vast majority of the genome is transcribed (Clark et al. 2011) (Lander et al. 2001). This means that the bulk of the genome encodes for ncRNA molecules, which can be further categorized into housekeeping and regulatory ncRNAs. The latter can be broadly classified based on their size as small ncRNAs (<200 bp) and long noncoding RNAs (lncRNAs) (>200 bp) (Nagano and Fraser 2011; Ponting et al. 2009). Many of the small ncRNAs have been identified and their mechanism of action has been heavily studied. However, the journey to study the lncRNAs has just begun (Gupta et al. 2010; Wilusz et al. 2009; Derrien et al. 2012).

Xist gene was one of the first lncRNA genes that were characterized using conventional molecular techniques such as RT-PCR, slot blot, and northern blot assays. The great interest on defining the underlying mechanism for dosage compensation and X chromosome inactivation led to a breakthrough in finding the regulatory roles for genes expressed from untranslated genomic regions in humans. Searching for X chromosome inactivation-associated genes, Huntington F. Willard and his team generated the first Xist cDNA probe, which was originally obtained from a placental cDNA library. This probe was further used for Xist transcript expression analysis in human and mouse. Expression profiling across in human

F. Jahaniani · V. Rao · S. Nevins · D. Spacek · N. Bharadwaj · J. Reuter · M. Snyder (✉)
Department of Genetics, Stanford University, Stanford, CA, USA
e-mail: mpsnyder@stanford.edu

A. M. Khalil and J. Coller (eds.), *Molecular Biology of Long Non-coding RNAs*,
DOI: 10.1007/978-1-4614-8621-3_7, 

male and female lymphoblast cell lines and somatic cell hybrids showed Xist is specific to the inactive X chromosome. They could also show that this transcript has several alternatively spliced isoforms. Due to high number of stop codons in the reading frame, lack of a potential ORF longer than >300 bp, and also low degree of sequence similarity to other known coding exons, they concluded that Xist is found in an untranslated genomic region and doesn't have protein coding potential (Brown et al. 1991).

Over the past few decades, many new methods have been developed for genome-wide transcriptome analyses. The development of techniques such as DNA microarray and tiling array was a milestone for comprehensive and precise mapping of human RNA coding region and verification of predicted genes. Assessing the RNA coding region on human chromosomes 21 and 22 resulted in identification of many novel transcripts, revealing much higher RNA coding capacity for human genome than was predicted before (Kampa et al. 2004). Moreover, the development of genome-wide high-resolution tiling arrays brought about the idea that ncRNA made up a significant portion of human transcripts and might have regulatory function (Kampa et al. 2004; Cheng et al. 2005).

To better understand lncRNAs and their function, studying their genomic organization, modifications, cellular locations, and tissue expression profiles has been the focus of many academic and industrial research laboratories. This concerted effort has resulted in the development of advanced biochemical and molecular assays and computational tools to bring this unknown part of genome to the light. There are many examples of lncRNAs being essential to distinct cellular mechanisms including regulation of gene expression (Rinn et al. 2007), dosage compensation (Bernstein and Allis 2005; Plath et al. 2003), genomic imprinting (Kretz et al. 2013), nuclear organization and compartmentalization (Batista and Chang 2013; Clemson et al. 2009), and nuclear-cytoplasmic trafficking in a number of organisms including humans (Willingham et al. 2005). These studies suggest the existence of elaborate networks of regulatory interactions between lncRNAs and their protein-coding counterparts, which together can have a large impact on human health. Recently, a number of reports have shown that many lncRNAs are dysregulated in a variety of human diseases (Gupta et al. 2010; Batista and Chang 2013). Many studies utilize unbiased genome-wide assays for the identification of single nucleotide polymorphisms (SNPs) and copy number polymorphisms nearby lncRNA loci that are associated with certain diseases (Jiang et al. 2012). Together, these discoveries have provided ample evidence for the association of lncRNAs to human health and diseases. Further study into lncRNA is needed to provide insight into the mechanism underlying lncRNAs-associated diseases and to help find biomarkers for early detection as well as the development of lncRNAs-based drug targets.

Our current understanding of lncRNAs has greatly benefitted from existing biomolecular tools (Fig. 1). These tools have allowed the discoveries of lncRNAs as a key component in cell fate during development, organization of protein complexes for proper activation/deactivation, and the onset of pathological conditions. Microarray assays allow for high-throughput analysis with medium

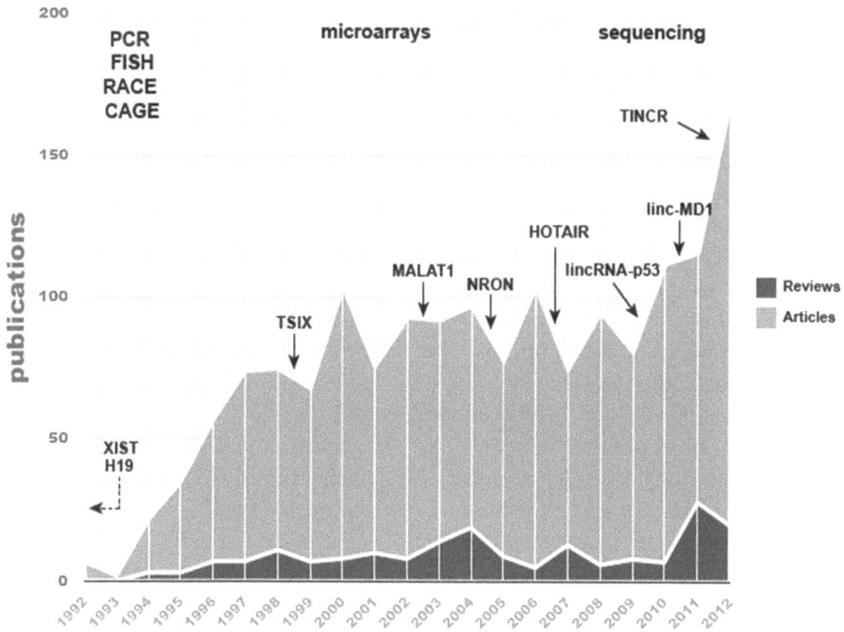


**Fig. 1** Timeline of lncRNA discovery with advancing technologies. An advanced PubMed search was performed for article published before 2013 and containing one of the following MESH terms: "lncRNAs", "lincRNA", "long noncoding RNA", "long noncoding RNA". Dominant technology shifts and the discovery of specific lncRNA are highlighted

sensitivity and specificity (Tang et al. 2007). Array-based approaches results are often validated using qRT-PCR to further quantify samples, which cannot be done well by microarray (Benes and Castoldi 2010). Since its advent, RNAseq (or Whole Transcriptome Shotgun Sequencing; WTSS) has been a tool to study lncRNA with high sensitivity and specificity in a genome-wide manner (Nagalakshmi et al. 2008; Mortazavi et al. 2008; Wang et al.2009; Cabili et al. 2011). Studies and discoveries of lncRNAs using these methods have presented lncRNAs as a fascinating topic of investigation. (Since then, there) There has been more lncRNA research focused on expanding the technologies that exist to discover previously unknown lncRNAs, often using different pulldown strategies to enrich for the interactions among lncRNAs, DNA, proteins, or other RNAs in their native physiological and pathological condition (Rinn et al. 2007). RNAi based knockdown techniques have provided novel platforms to elucidate lncRNA functions. Furthermore, researchers have been developing new genomics and bioinformatics tools that build upon these standard tools and expand the ability to discover and quantify new lncRNAs (Derrien et al. 2012).

In this chapter, we will briefly discuss the novel technologies and ongoing improvements in the existing genome-wide biochemical and computational strategies. These innovations will enable better insights to the complicated network of regulatory lncRNAs in physiological and pathological conditions.

## 2 RNA *In situ* Hybridization

A large body of genetic and biochemical work has shown that the majority of lncRNAs are expressed in a spatiotemporally controlled manner, often at very low levels (Rinn et al. 2007; Mercer et al. 2008). This precise tissue-specific expression pattern might be indicative of their biological importance and could hold some clue to their functional significance. However, the low level expression of lncRNAs makes the detection of their subcellular localization difficult. Methods such as microarray, qRT-PCR, and RNA *in situ* hybridization have helped overcome this problem and been applied for lncRNAs expression profiling.

*In situ* hybridization was originally developed as a powerful tool for localization and visualization of DNA and RNA molecules in their original location in the cell by Joseph G. Gall in 1969 (Gall and Pardue 1969). In this method cultured cells or sections of tissue (Rinn et al. 2007) are first fixed and then hybridized to a complementary DNA or RNA probe. This technique also can be applied on formalin-fixed paraffin-embedded tissues (Chisholm et al. 2012). These radioactive or fluorescent-tagged single-stranded nucleic acid probes are (the latter refers to fluorescence *in situ* hybridization (FISH)) hybridized to the targeted DNA or RNA sequence, and the location of the gene on a chromosome or a transcript in the cell can be visualized using a confocal fluorescence microscope. Chromogenic *in situ* hybridization (CISH) is an alternative form of FISH, which can also be used for visualization of the subcellular localization of lncRNAs in a wide variety of biological samples including cells, formalin-fixed, paraffin-embedded (FFPE) tissues, blood or bone marrow smears (Rapicavoli et al. 2011).

Using a high-throughput colorimetric RNA *in situ* hybridization, a group of scientists including Paul Allen and David Anderson created the Allen Brain Atlas, to map gene expression of more than 20,000 transcripts for entire mouse brain. The catalog has provided researchers with a broader view of the differential gene expression pattern across the entire nervous system. Mercer et al, have utilized the Allen Brain Atlas and identified 849 lncRNAs with specific expression patterns within adult mouse brain (Mercer et al. 2008). They also found that these lncRNAs are derived from a variety of genomic loci including intergenic, intronic, and imprinted, with some overlapping the protein-coding genes in converging or diverging direction. These tissue and cell specific expression patterns underscore the biological significance of lncRNAs and their role in increasing the complexity of the higher eukaryotes including human (Taft et al. 2007). The results of this study are available in a searchable database (http://jsm-research.imb.uq.edu.au/abancrna). The authors, however, raised this point that they only have been focusing on 4 % of the known non-coding transcriptome, and future in-depth studies are needed, using the recently published lncRNAs catalog (Derrien et al. 2012), to better understand the complex nature of lncRNAs and protein-coding RNAs interaction in the brain biology (Mehler and Mattick 2007).

Despite the power of RNA *in situ* hybridization (RISH) to identify the subcellular localization of lncRNAs, the conventional *in situ* hybridization is not sensitive enough to be used as a quantitative approach for gene expression profiling. Concerted efforts from industrial research laboratories have sought to resolve this issue and increase the quantitative strength of RNA in situ. Affymetrix has introduced Quanti Gene View RNA Assays, as a high sensitive RNA *in situ* hybridization assay, suitable for quantitative visualization of single-molecule RNA in low- or high-throughput experiments. Using the branched DNA (bDNA) signal amplification technology, an original signal can be amplified up to 8000-fold, which allows for the detection of low levels of lncRNA expression (Collins et al. 1997). This technique enables researchers to detect as few as two and four lncRNA molecules per cell (http://www.panomics.com/products/rna-insitu-analysis/view-rna-overview).

More recently, Biosearch Technologies, Inc also introduced Stellaris FISH, known as Single Molecule RNA FISH, which is useful for the accurate detection and quantification of long RNA molecules in a thin layer of tissue sample, in singleplex or multiplex assay. The amplification of the fluorescent signal in this technique is based on the multiple fluorescent-labeled oligos that are designed to specifically target a single RNA molecule. This helps to reduce the background noise and maximize the efficiency of detection of the targeted-RNA molecule in a wide-field fluorescent microscopy image. (https://www.biosearchtech.com/display.aspx?catid=227&pageid=215)

## 3 Microarray

Microarray experiments rely on the similar biological principles as situ hybridization. DNA microarrays consist of more than thousands of fluorescent-, silver-, or chemiluminescent-labeled complementary DNA probes that are covalently attached to a solid surface. Under high-stringency conditions, these probes hybridize to the targeted DNA or cDNA sample and the specific interaction can be visualized in a semi-quantitative manner allowing for the measurement of the expression of many genes or genomic regions. Recent lncRNAs research has benefitted greatly from microarray technology. However, microarrays can only detect known lncRNA transcripts, and the discovery of novel lncRNAs demands other techniques.

Biotech companies have helped the research community to overcome some of the limitations of using microarray in lncRNAs studies to investigate the spatio-temporal expression pattern of lncRNAs in physiological and pathological condition with a number of new tools. Life Technologies NCode™ Non-coding RNA Arrays and GeneChip® Human Gene ST Arrays are two examples of such tools which allow whole-transcript analysis. NCode™ Noncoding RNA Arrays was designed to simultaneously profile the expression of coding and lncRNAs transcripts. It measures the expression of over 17,000 human lncRNAs or over 10,000 mouse lncRNAs. The design of GeneChip® Human Gene ST Array has also enabled whole-transcriptome analysis, with a particular focus on long intergenic ncRNA transcripts. It covers more than 30,000 coding transcripts and 11,000 long intergenic noncoding transcripts. GeneChip® Human Gene ST Array also contains probes to measure alternative splicing events/transcript variants.

The growing number of annotated human and mouse lncRNAs (Jia et al. 2010 ; Cabili et al. 2011; Guttman et al. 2009; Khalil et al. 2009) has resulted in the development of custom arrays, which contains probes specific to a limited number of lncRNAs relevant to the study of interest. Loewer et al. performed a customized microarray-based lncRNA genes expression profiling to identify lncRNAs associated with pluripotency (Loewer et al. 2010). Their custom microarray contained probes that were able to detect 900 human long intergenic noncoding RNAs (lincRNAs). Using this custom array and total RNA from four different fibroblast lines, their derivative iPSCs, and ESCs, Loewer and his colleagues performed whole transcriptomc analysis and identified a subset of 28 "iPSC-enriched" lincRNAs. With subsequent loss and gain of function experiments, they could further show that a 2.6 kb lincRNA, RoR (Regulator of Reprogramming), is iPSC-specific and is crucial for reprogramming.

Version 7 of GENCODE released a very comprehensive, high-quality catalog of human lncRNAs including 14,880 manually curated lncRNA transcripts (Harrow et al. 2012) using their microarray based-expression profiling across human body (Derrien et al. 2012). Their results have shown that a common pathway is involved in the generation of both protein-coding and long noncoding genes is transcripts. However, the latter is enriched for two-exon transcripts and mostly localized in the chromatin and nucleus. In groundbreaking work done by Derrien et al., as part of Encode consortium, human lncRNA expression was mapped across a wide range of human tissues and cell types, including nine brain regions, 17 other tissues from the adult body, and five common cell lines. The research team was then able to develop a custom microarray, containing multiple non-redundant probes to detect 9747 GENCODE version 3c-annotated lncRNA transcripts (Derrien et al. 2012). Furthermore this study was able to find positive correlation between the expression of lncRNAs and antisense coding genes (Derrien et al. 2012). This valuable catalog will help researchers further their investigations on lncRNAs distribution and functions.

The inherent incapability of microarrays to identify novel lncRNAs and also to examine the reliability of the microarray platform was significantly resolved by the invention of RNA-seq technique. However, due to higher cost of RNA-seq and the complexity of analyzing the generated data, microarray is still a very popular method in the lncRNA field.

## 4 Tiling Array

Like the standard microarray, tiling arrays hybridize target RNA or DNA to probes fixed on a solid surface. Tiling arrays are different, however, in the type of probes that are used. Rather than probing for known sequences across the entire genome as in traditional microarrays, tiling arrays probe for specific sequences within a contiguous region (Rinn and Chang 2012). Due to this difference, tiling arrays are often used to blanket (or tile) regions of the genome that have been sequenced, but whose functions are largely unknown. Although traditional microarrays are less useful for quantification, tiling arrays provide improved quantification of transcription products by adjusting the sequence overlap between probes. Tiling arrays are used to find expressed genes and to map the transcriptome (Bertone et al. 2005).

Tiling microarrays have been powerful in the discovery of lncRNA, as two independent studies have reported initial estimates that there may be many lncRNA genes as protein-coding genes (Kapranov et al. 2002; Rinn et al. 2003). In a paper published in February 2012, Chang et al. used RNA-Seq and tiling arrays to study the possible role for lncRNAs in the suppression of progenitor differentiation (Kretz et al. 2012). Chang et al. combined these complementary technologies to identify previously uncharacterized lncRNAs. They discovered a previously uncharacterized lncRNA, ANCR (antidifferentiation ncRNA). Depleting ANCR in progenitor cell populations induced differentiation. ANCR was significantly suppressed during differentiation of somatic tissue progenitor cells (Kretz et al. 2012), suggesting a novel role for lncRNA in maintaining homeostasis in human somatic cells.

Where traditional methods for gene prediction fall short, tiling arrays can detect small and rare RNA molecules with high resolution and sensitivity. With overlapping probes, tiling arrays can detect non-polyadenylated RNA and overall can create a more accurate picture of gene structure than can be produced with traditional microarrays.

## 5 RNA-Seq

RNA-Seq is a powerful tool based on the principles of next-generation deep-sequencing that can be applied to the detection and quantification of lncRNAs (Wang et al. 2009). The Snyder group while at Yale developed RNA-Seq, and subsequently, work by several groups led to the development of high-throughput methods involving RNA-Seq in the study of the transcriptome structure and dynamics (Nagalakshmi et al. 2008; Wang et al. 2009). Along with protein-coding RNA, lncRNA discovery and primary structure analyses have benefited from RNA-Seq development.

Genome-wide RNA-Seq has several advantages over classic microarray-based techniques including high resolution and detection of novel sequences (Wang et al. 2009). RNA-Seq has been combined with multiple techniques discussed throughout the chapter in determining lncRNA expression profiles, transcript boundaries, and regulatory function. LncRNA specific analyses also involve the typical steps of RNA-Seq (Lee and Kikyo 2012). First poly adenylated or total RNA depleted of rRNA (ribosomal RNA) and cDNA libraries are generated. Following sequencing, the reads obtained are aligned to started reference genomes using aligning tools such as Burrows-Wheeler Aligner and TopHat (Trapnell et al. 2009). Using a slew of bioinformatics tools, the reads are assembled and annotated. Novel lncRNAs are identified and annotated using databases such as ENCODE and FANTOM or lncRNA specific databases (Table 1 lists the currently available lncRNA databases). False positives are then identified using rigorous computational analyses. For example, protein-coding potential and evolutionary conservation of newly identified RNA are measured (Guttman et al. 2009). Subsequent experimental analyses are then performed for validation of RNA-Seq results.

In a large-scale study, RNA-Seq data from multiple human tissues as a part of the Illumina Human Body Map project and GENCODE lncRNAs were quantified (Derrien et al. 2012). This report showed that, lncRNAs demonstrated lower expression albeit with higher expression variability. Corroborating the findings of other studies, this study also reports that lncRNA transcript expression is more tissue-specific than protein-coding transcripts (Cabili et al. 2011). Similar RNA-Seq-based approach has been used to identify novel lncRNA and investigate their role in vital biological processes such as embryogenesis in model systems such as zebrafish and drosophila (Pauli et al. 2012; Young et al. 2012).

## 6 Deep RNA-Seq of Sub-cellular Fractions

Deep sequencing, based on the principles of RNA-Sequencing, allows for greatly increased sensitivity and accuracy by sequencing fragments multiple times in a short period of time. Subcellular fractionation, which requires homogenization (ex. needle/syringe or hypotonic shock) and fractionation of the homogenate, allows separation of the organelles based on their physical or biological properties (de Araujo and Huber 2007). Bhatt et al. studied transcript dynamics of the subcellular fractions using RNA-Seq (Bhatt et al. 2012). This allowed them much greater insights into transcription and RNA localizations, giving them a high-resolution view of promoter and chromatin properties, as well as regulation of coexpressed genes.

In a recent study by Tilgner et al. from Snyder group, splicing dynamics were shown to differ dramatically between protein-coding and non-proteincoding exons (Tilgner et al. 2012). They used an exon-based measure of splicing completion, called the completed splicing index (coSI). Using deep RNA-Seq of nuclear and cytoplasmic fractions, they found that lncRNAs as a class, including the well-known lncRNAs XIST, H19, and U50HG_SNHG5, had significantly lower coSI values. Lower coSI values for lncRNAs indicate that lncRNAs remain completely unspliced or have a larger proportion of primary transcripts that are not spliced. Two lncRNAs involved in imprinting, AIRN and KCNQ1OT1, remain predominantly unspliced as they remain in the nucleus. (Sleutels et al. 2002; Mancini-Dinardo et al. 2006). This novel, emerging method holds promise for lncRNA discovery.

**Table 1** Compiled database resources

| DB Name | Description | Website | Citation |
|---|---|---|---|
| Gencode v7 | Most complete human lncRNA annotation to date, comprising 9277 manually annotated genes producing 14,880 transcripts | http://big.crg.cat/ | |
| | bioinformatics_and_genomics/lncrna_data | (Derrien et al. 2012) | |
| ncRNA Database Resource | Categorizes 102 databases into 4 families RNA family, information source, information content and available search mechanisms | http://www.ime.usp.br/~durham/ncrnadatabases/index.php | (Paschoal et al. 2012) |
| lncRNA DB | Comprehensive list of lncRNAs that have been shown to have, or to be associated with, biological functions in eukaryotes, as well as messenger RNAs that have regulatory roles. Was known as RNAdb | www.lncrnadb.org | (Amaral et al. 2011) |
| ncRNA.org | A collection of databases and bioinformatics tools specialized for functional RNA | http://www.ncrna.org/About | (Mituyama et al. 2009) |
| Noncode | Includes the first integrated collection of expression and functional lncRNA data obtained from reannotated microarray studies in a single database | http://noncode.org/NONCODERv3/guide.htm | (Bu et al. 2011) |
| rFam | Collection of ncRNA families with conserved RNA secondary structure. Each family is represented by a multiple sequence alignment, predicted secondary structure, and covariance model | http://rfam.sanger.ac.uk/ | (Burge et al. 2012) |
| nRed | The noncoding RNA Expression database provides gene expression information for thousands of long ncRNAs in human and mouse. This database contains both microarray and *in situ* hybridization data | http://jsm-research.imb.uq.edu.au/nred/cgi-bin/ncrnadb.pl | (Dinger et al. 2009) |
| NPInter | NPInter extensively covers functional interactions between noncoding RNAs and protein-related biomacromolecules in six model organisms (E.coli, yeast, worm, fly, mouse, and human) | http://www.bioinfo.org.cn/NPInter/ | (Wu et al. 2006) |
| ChipBase | Platform for decoding transcription factors, binding maps, expression profiles and transcriptional regulation of lncRNA, miRNAs, other ncRNAs(snoRNAs, tRNAs, snRNAS, etc.) and protien-coding genes from CHIP-Seq Data | http://deepbase.sysu.edu.cn/chipbase/ | (Yang et al. 2012) |

(continued)

**Table 1** (continued)

| DB Name | Description | Website | Citation |
|---|---|---|---|
| H-InvDB | Comprehensive annotation resource of human genes and transcripts, and consists of two main views and six subdatabases | http://www.h-invitational.jp/ | (Yamasaki et al. 2010) |
| Noncoding RNA Database | Collection of currently available sequence data on RNAs, which have no protein-coding capacity and have been implicated in regulation of cellular processes | http://biobases.ibch.poznan.pl/ncRNA/ | (Szymański et al. 2003) |
| Functional RNA DB | fRNAdb is a comprehensive compilation of noncoding RNA sequences including known ncRNAs, acquired from other sequence databases | http://www.ncrna.org/frnadb/ | (Mituyama et al. 2009) |
| Functional lncRNA Database | Repository of mammalian lncRNA | http://www.valadkhanlab.org/database.php/ | (Niazi and Valadkhan 2012) |
| LNCipedia | Integrated database of 32,000+ human annotated transcripts | http://www.lncipedia.org/ | (Volders et al. 2013) |
| lncRNA and Disease Database | Curates experimentally supported lncRNA-disease association data, and integrated tools for predicting novel lncRNA-disase associations | http://cmbi.bjmu.edu.cn/lncrnadisease | (Chen et al. 2013) |
| Diana-lncBase | Predicted and experimentally verified miRNA–lncRNA interactions. | http://62.217.127.8/DianaTools/index.php?r=lncBase/index | (Paraskevopoulou et al. 2013) |
| Genecards | Database extracts and integrates a selected subset of gene related transcriptomic, genetic, proteomic, functional, and disease information | http://www.genecards.org/ | (Belinky et al. 2013) |
| PLncDB | a comprehensive genomic view of Arabidopsis lncRNAs for the plant research community | http://chualab.rockefeller.edu/gbrowse2/homepage.html | (Jin et al. 1068) |
| iseeRNA | High-throughput screening of lincRNAs from transcriptome sequencing data | http://sunlab.lihs.cuhk.edu.hk/iSeeRNA/ | (Sun et al. 2013) |
| Noncoder | Web interface for exon array-based detection of lncRNA | http://noncoder.mpi-bn.mpg.de | (Gellert et al. 2013) |

(continued)

**Table 1** (continued)

| DB Name | Description | Website | Citation |
|---|---|---|---|
| ncPRO-seq | Tool for annotation and profiling of ncRNAs using deep-sequencing data | https://ncpro.curie.fr/ | (Chen et al. 2012) |
| regRNA 2.0 | Integrated web server for identifying functional RNA motifs in an input RNA sequence | http://regrna2.mbc.nctu.edu.tw/ | (Chang et al. 2013) |
| Hugo Gene Nomenclature Committee | Standardized nomenclature for all human genes | http://www.genenames.org/rna/LNCRNA | (Wright and Bruford, 2011) |
| lncRScan | Pipeline consists of five steps for detecting novel long noncoding RNAs from a set of candidate transcripts annotated by Cuffcompare | https://code.google.com/p/lncrscan/ | (Sun et al. 2012) |
| miRcode | Human microRNA target predictions based on the comprehensive GENCODE gene annotation, including >10,000 long noncoding RNA genes | http://www.mircode.org/mircode/ | (Jeggari et al. 2012) |

## 7 Boundary Determination of lncRNAs

Demarcating transcript boundaries is a crucial step in lncRNA studies. Due to the length and structural complexities of lncRNA, identifying the transcriptional start sites and ends are challenging. Recently developed technologies, however, have enabled discovery of transcription start site (TSS) and poly-A tails that have lead to efficient full-length lncRNA cloning for use in functional studies and identification of lncRNA isoforms.

Rapid amplification of cDNA ends (RACE) is a procedure used to acquire a full cDNA sequence when the sequence is only partially known. The protocol begins with a whole or partial cDNA template between a known internal site and unknown sequences at the 5′ or 3′ end. This methodology has been termed “one-sided” PCR or “anchored” PCR because of its single-sided specificity. Specific sequences of cDNA are directly amplified by PCR using gene-specific primers that anneal to known exon sequences and adapter primers that target the 3′ poly-A tail for 5′ RACE or an appended homopolymer tail for 3′ RACE. This allows for rapid determination of transcript boundary sequences and exon information (Kapranov et al 2005). By generating unique gene-specific or isoform primers, multiple lncRNAs and their isoforms may be analyzed simultaneously (Broadbent et al. 2011).

More recently in a technique called RNA ligation mediated-RACE (RLM-RACE), RNA oligonucleotides are added to the 5′ end of the transcript that ensures full-length cDNA amplification and use of universal primers in 5′ amplification (Scotto-Lavino et al. 2006). When combined with tiling arrays, qRT-PCR or sequencing, RACE can be used to characterize transcripts in the human transcriptome. Strand of origin, start and termination positions, lengths and genomic positions of exons, and maximal lengths of the transcript and the extent of the genome covered by RACE-associated exons are all valuable information that can be gained using the RACE/Array approach. Moreover, RACE-based techniques can also be used to elucidate the mechanism and machinery involved in lncRNA processing (Broadbent et al. 2011).

Cap analysis gene expression (CAGE) is another tool used in studying RNA transcript boundaries. CAGE is used to determine TSS on a full-length mRNA transcript by sequencing short sequence tags beginning at the 5′ end (Shiraki et al. 2003). These short sequence tags of about 20 nucleotides are sequenced to detect transcription start sites on a genome-wide scale (Wilusz et al. 2009). CAGE determines the location of transcription events in addition to expression levels. When combined with high throughput sequencing technologies (also called DeepCAGE or CAGE-Seq), this approach can be used as a powerful tool for studying genome-wide lncRNA transcriptional regulation by promoters, around specific transcription start sites in multiple tissues (Valen et al. 2009). RNA paired end ditags (PET) approaches are more recent sequence-based techniques used in the identification of lncRNA transcript boundaries (Derrien et al. 2012).

The GENCODE consortium utilized CAGE-based approaches for validation and found CAGE support for lncRNA TSS, albeit at a lower level (<15 %) than protein-coding genes (Derrien et al. 2012). Additionally, Fejes-Toth et al. studied the post-transcriptional processing of lncRNA using DeepCAGE. A significant number of CAGE tags were found in exonic regions and in splice junctions, alluding to the possibility that lncRNAs are processed into small RNAs. These small RNAs then gain a 5′ cap structure following post-transcriptional cleavage and are thereby detected by DeepCAGE (Djebali et al. 2012). This insight from using DeepCAGE led to the observation that cleavage of lncRNAs into many smaller RNAs allows a single lncRNA to have multiple distinct functions and locations in the cell.

Given that several lncRNAs are precursors to microRNAs and small RNAs, techniques such as Parallel analysis of RNA end sequencing (PARE-Seq) can be used in the study of lncRNA degradome and to identify functional end products of lncRNA processing (German et al. 2009).

Majorities of lncRNAs are polyadenylated at their 3′ ends (Cheng et al. 2005). Therefore, the methods used in identification of mRNA polyadenylation sites can be applied to lncRNA studies. Polyadenylation Site Sequencing (PAS-Seq) is a recently developed deep-sequencing-based method that focuses on 3′ end identification of mRNA. While the current technique has several limitations including inability to process low transcript input as in case of lncRNAs, ongoing improvements in library preparation and sequencing techniques and more robust alignment softwares have ensured more efficient lncRNA poly(A) site or junction determination.

## 8 FragSeq: Transcriptome-Wide RNA Structure Probing Using High-Throughput Sequencing

Classical approaches to determine lncRNA structures rely on probing one RNA molecule at a time with enzymes, chemicals, and electrophoresis to identify structurally important positions. Chemicals or nucleases react with RNA bases depending on the structural context of these bases to help distinguish between that bases participate in base pairing and other interactions from those that do not (Knapp 1989). Recent advances in probing by selective 2′-hydroxyl acylation analyzed by primer extension (SHAPE) (Low and Weeks 2010) enables faster, higher-quality probing focused on one RNA sequence per experiment.

Conversely, computational structure prediction methods allow rapid, large-scale analyses of many RNA sequences. These methods, rooted in comparative sequence analysis, require several RNA sequences with conserved structures. There also exist methods that can predict structure from a single sequence. These methods are useful for RNAs for which structural homologs are not known or ones that undergo lineage specific structure changes. They can determine theoretical folds for RNA sequences using thermodynamic models (Machado-Lima et al. 2008). While powerful, these methods suffer from ambiguity because often several distinct structures can be predicted from different sequences.

To combine the best of both worlds, speed of computational methods and quality of RNA probing experiments, Underwood et al. (2010) developed fragmentation sequencing (FragSeq), a high-throughput RNA structure probing method that uses high-throughput RNA sequencing of fragments digested with nuclease P1, which targets and cleaves single stranded nucleic acids. Bioinformatics' analysis is then used to deduce cut sites (phosphate backbone scissions) and assign cut scores. By modifying the well-established RNA structure (Reuter and Mathews 2010) to use FragSeq's assigned cutting scores, allows researchers to easily guide computational structure prediction. This level of analysis provides what they call an 'RNA Accessibility Profile', similar to DNase hypersensitivity assays on chromatin (Crawford et al. 2006). Applying their method to mice' naked RNAs they were able to deduce structural data for both known and new ncRNAs. This technology will allow lncRNA researchers to push transcriptome analysis beyond sequencing and reveal lncRNA structural features and help provide clues to their underlying biology.

## 9 Copy Number Variation and Single Nucleotide Polymorphism

SNPs constitute one of the most common forms of genetic variation in human genome (Reich et al. 2003). SNPs occurring in functional regions of genome are more likely prone to cause phenotypic changes or have a role in susceptibility to disease. Thereby, one might hypothesize that some of SNPs targeting lncRNAs might also be linked to some of the known human diseases. Indeed, results from a number of studies support this hypothesis, showing that SNPs at lncRNAs are related to human phenotypic differences and also complex diseases such as cancer and coronary artery disease (Pasmant et al 2007). However, more effort is needed to map lncRNAs- associated SNPs that contribute to disease states.

Copy number variation (CNV), another form of structural variation is the product of genomic DNA alterations. CNVs result in deletion or duplication of certain chromosomes, which subsequently causes loss or gain of function of the dosage-sensitive genes. The extent to which CNV impact on human health and contributes to human disease have been under investigation by researchers involved in projects such as The CNV Project, Global CNV assessment (Barnes et al. 2008) and High-resolution CNV discovery (Sebat et al. 2007). These studies and many more have shown that while most of the known CNV have no observable consequence, some CNVs may result in human phenotypic and behavioral variation, also disease susceptibility (Sebat et al. 2007; Mefford et al 2010; Swaminathan et al. 2012; Hirsch et al. 2003).

Techniques including fluorescent
*in situ* hybridization (Duan et al. 2013), comparative genomic hybridization (Friedman et.al. 2006), array comparative genomic hybridization (Mefford et al 2010), and next-generation sequencing (Yoon et al. 2009) have been used for the detection of CNVs (Duan et al. 2013). As technologies for CNVs detection have been improving, the impact of lncRNAs CNV on gene expression and recent human evolution has began to unravel.

In an effort to find breast cancer-associated lncRNAs, Xiaowei Chen team at Fox Chase Cancer Center in Philadelphia, for the first time investigated the lncRNA associated genomic aberrations (Jiang et al. 2012). To perform their study, they took advantage of a high-density SNP array, the Illumina Human Omni5 Beadchip that consists of about 4.3 million SNPs and provides a comprehensive view of the intergenic portion of the genome. This high level of resolution enabled them to identify 122 lincRNA-associated somatic CNVs that were specific to the 7 breast cancer tumors they have included in their study. Interestingly, lincRNA-associated CNVs are mostly enriched for copy number losses and concentered to the ends of chromosomes. This is in contrast to the protein-coding CNVs that are scattered through each chromosome. The authors also proposed that the observed low level of lincRNAs loci-genomic abnormality might be due to the role of lincRNAs in cell growth and survival, which is important for the development of both normal and cancerous tissues. They finally validated some of their finding by searching for the expression pattern of the identified and affected lincRNAs in the published expression dataset (Cabili et al. 2011) and also performing qPCR.

This novel approach can be applied to identify lincRNA-associated CNV that might be the underlying mechanism for other common human diseases that can further be used for translational research and therapeutically approaches.

## 10 RNP Analysis of lncRNA

With the discovery of many novel lncRNAs that play crucial roles in different regulatory networks, the mechanisms of lncRNA function have garnered more attention. Studies have demonstrated that several lncRNAs exist as ribonucleoprotein (RNP) complexes or conditionally interact with proteins and mediate their trans-regulatory function (Moran et al. 2012). For example, lncRNA-p21 interacts with the chromatin modifying complex polycomb repressive complex 2 (PRC2) and regulates DNA methylation at specific chromatin regions (Huarte et al.2010). In other cases, for example, the lncRNA *HOTAIR* interacts with both PRC2 and other complexes and acts as a structural or scaffolding component in RNP complexes (Guttman et al. 2009). Other classes of lncRNAs such as *Gas5* modulate transcriptional regulation by directly interacting with DNA-binding proteins (Kino et al. 2010).

To date several *in vitro* and *in vivo* methods have been developed to study RNA–protein interactions (Niranjanakumari et al. 2002). Despite the great value of in vitro techniques to map the RNA–protein interaction sites, they may fail to capture some of the physiological interactions, (Niranjanakumari et al. 2002; Brooks and Rigby 2000). To overcome this limitation, researchers have developed in vivo assays such as RNA Immunoprecipitation (RIP) to study RNA–proteins interactions in their native physiological and pathological conditions (Brooks and Rigby 2000).

Delineation of these lncRNA–protein interactions have greatly benefitted from co-immunoprecipitation-based techniques such as RIP, CLIP, RNA-ChIP, and ChIRP, where complexes containing lncRNAs have been isolated by using antibodies against interacting proteins. Typically, following immunoprecipitation-based pull-down, the interacting RNA–proteins are cross-linked and the RNA molecules are subsequently isolated (Moran et al. 2012). Structural and functional interactions of lncRNAs can be precisely determined by augmenting these procedures with existent sequencing and microarray technologies (Fig. 2). The following sections discuss the currently available techniques that utilize immunoprecipitation-based approaches in lncRNA studies.

# 11 RNA Immunoprecipitation

RIP was originally developed based on the chromatin immunoprecipitation (ChIP) to pull out all the RNA species that are specifically bound to a RNA binding protein of interest, assuming that their biological in vivo interaction wouldn't be affected during the capturing procedure (Mili and Steitz 2004).

In this method, either the whole cell lysate or the nuclear pellet is suspended in RIP buffer. Antibody binding is carried out by overnight incubation of the cell lysate with a specific antibody against the protein of interest. The target RNA binding protein is then captured using protein A/G beads along with the bound RNA molecules. After stringent washes, the RNA can be isolated from the complex with trizol or any other commercially available RNA isolation kits (such as QIAGENRNeasy Mini Kit) (Valen et al. 2009; Brooks and Rigby 2000) The pool of target RNAs can further be applied to downstream processes such as RTPCR for the assessment of panel of genes or microarray analysis/RNA-Seq for genome-wide mapping of RNA–protein interaction (Rinn et al. 2007). With the RIP assay, researchers can identify the subset of RNAs that are interacting with a particular protein, and are possibly are co-regulated or performing similar functions. Digestion with RNase H (digests RNA in RNA-DNA hybrids) and DNase I can be used to exclude RNAs with indirect interactions.

One of the proposed mechanisms of function for some of the lncRNAs found using RIP is that they are acting as scaffold between proteins (Gupta et al. 2010; Khalil et al. 2009; Collins 2008). A work done by Rinn et al. showed that PRC2 consists of transcription factors or other effector molecules along with lncRNA

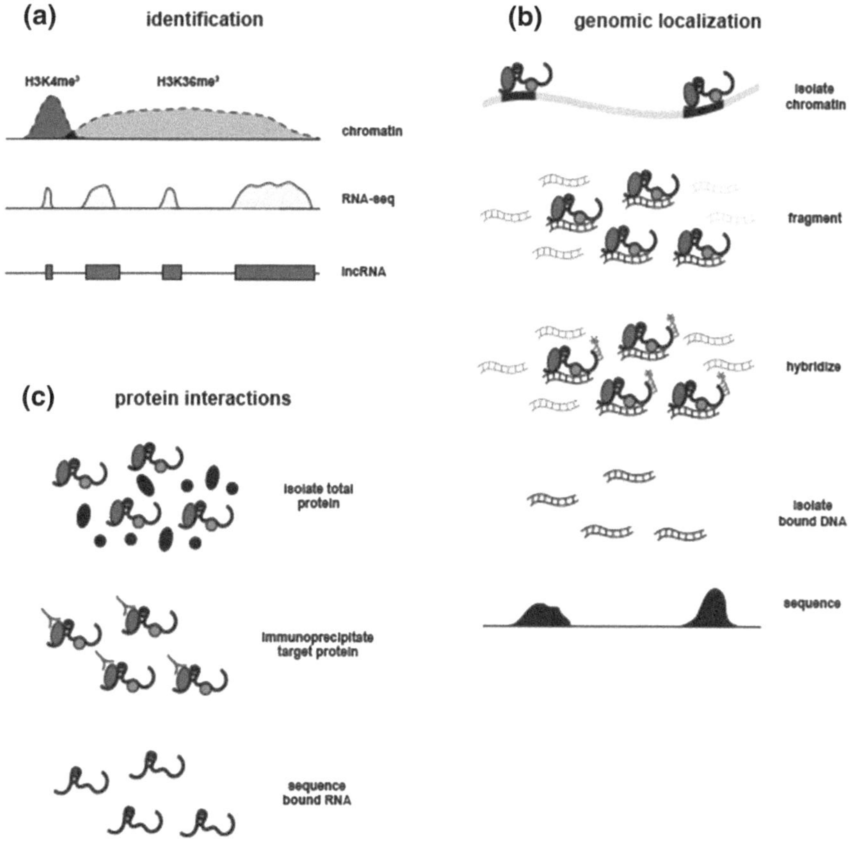


**Fig. 2** Role of sequencing in the study of lncRNAs. **a** H3K4me$^3$ and H3K36me$^3$ chromatin signature and RNA-Sequencing is used for the systematic discovery of lncRNAs in various cells, tissues, and organisms. **b** Chromatin isolation by RNA purification (ChIRP) can be used to determine global localization of the particular lncRNA. Briefly, chromatin is isolated, fragmented, and hybridized with biotinylated DNA probes corresponding to lncRNA of interest. Finally, sequencing of the isolated DNA is performed to genomically localize the lncRNA. **c** In RNA- immunoprecipitation followed by sequencing (RIP-Seq), target proteins are immunoprecipitated from total proteins extracts, and the bound RNA is identified using sequencing

species (Rinn et al. 2007). These long noncoding transcripts may contain multiple protein or DNA binding motifs, which enable them specifically to interact with different incorporated components of an unassembled RNP complex, bring them closer to each other to facilitate the assembly of these RNPs. They also can guide chromatin modifiers to particular location on the chromatin to activate or repress gene expression.

Many research teams to identify known or novel lncRNAs in different RNP complexes have used the RIP approach, combined with microarray analyses or RNA-Seq. Using a specific antibody against LSD1, Tsai et al. could show that the lncRNA HOTAIR binds to LSD1 protein, through its 3′ 700 nucleotides (Gupta et al. 2010). LSD1 is a part of REST/Co-REST complex. It has been previously shown that HOTAIR also binds the polycomb complex PRC2 (Rinn et al. 2007). The authors further could show that the 5′ end of HOTAIR is responsible for its interaction with the PRC2 (Gupta et al. 2010). Both PRC2 and REST complexes are involved in gene silencing. This data supports the proposed mechanism of action of lncRNAs-mediated RNP assembly. In this case, two distinct transcriptional repressor complexes were tethered by the lncRNA HOTAIR, leading to their co-binding on target genes for further transcriptional inactivation.

## 12 Cross-Linked Immunoprecipitation

Despite the power of RIP technique for in vivo study of RNA–protein interaction, this approach has several potential limitations including high rate of detecting nonspecific interactions. Also, the conventional RIP can only isolate lncRNA complexes from the soluble fraction of a whole cell lysate and identifying the RNA sequence responsible for protein binding can be difficult to do (Heyne et al. 2012).

Cross-linking of cultured cells with UV or formaldehyde prior to cell lysis can capture RNA–protein interaction in their physiological context. It also inhibits their dissociation during whole cell extract preparation (Niranjanakumari et al. 2002; Selth et al. 2009). Inspired by RIP-Seq, CLIP-Seq (cross-linking immunoprecipitation sequencing) is a powerful technique, which was originally designed for accurate genome-wide mapping of NOVA1 and NOVA2 -RNA interaction (Licatalosi et al 2008).

In standard CLIP experiments, UV-crosslinked cell extract is subjected to mild RNase digestion to retain only the fraction of RNA regions that are interacting with protein of interest. Partially digested crosslinked RNA fragments are then immunoprecipitated, 5′ radiolabeled, and tagged with a 3′ linker. The RNA fragments–protein complex is further purified by SDS gel electrophoresis and transferred to a nitrocellulose membrane for autoradiography and band excision. Ultimately, Proteinase K treatment releases the RNA fragments bound to protein (Ule et al. 2005). The purified RNA pool can be used for high-throughput sequencing application for transcriptome-wide identification of RNA binding site analysis (Licatalosi et al. 2008). CLIP-Seq has been used extensively as a powerful tool to study lnRNA–protein interaction in normal and disease situation (M. Huarte et al 2010).

Sònia Guil et al. used a EZH2-specific monoclonal antibody in CLIP-Seq analyses to map EZH2–RNA interactions in human cancer cells, identified a number of intronic lncRNAs directly interacting with PRC2 complex (Guil et al. 2012). EZH2 is the core component of PRC2 with histone methyltransferase activity and catalyze H3K27 trimethylation. Marking the chromatin with H3K27me3, PRC2 mediates transcriptional silencing for genomic regions that contain Polycomb response elements. This is a crucial step to maintain proper cell identity during development and differentiation (Rinn et al. 2007; Sparmann and van Lohuizen2006). Aberrant expression of EZH2 has been observed in some human cancers and its overexpression has been linked to cancer progression and metastasis (Chase and Cross 2011; Bu et al. 2012).

PAR-CLIP (Photoactivatable-ribonucleoside-enhanced CLIP) and iCLIP (individual-nucleotide resolution CLIP) are the new versions of CLIP to identify the precise binding sites of RBPs (RNA binding proteins) (Konig et al. 2011). In PAR-CLIP, modified nucleotides are added to cell culture to increase the efficiency of UV crosslinking and precise identification of the crosslinked nucleotide (looking for a Uracil to Cytosine conversion) (Hafner et al. 2010). Very similar to CLIP assay, iCLIP technique, however, takes advantage of one extra intramolecular circularization step that allows binding sites mapping at single nucleotide resolution (Konig et al. 2010).

## 13 RNA-ChIP

The advent of high-throughput deep-sequencing techniques has led to multiple trans- and cis- acting transcription factors binding sites identification. More recently, however, with the discovery of lncRNAs, several protein–lncRNA interactions on the chromatin have shown to be crucial for epigenetic regulation. Modifications to existing ChIP-based techniques and coupled with sequencing have led to the discovery of novel roles and protein–RNA and chromatin–lncRNA interactions. Routine RNA-ChIP procedures involve formaldehyde fixing of RNA–protein and chromatin complexes, followed by DNaseI treatment and RNA sonication and immunoprecipitation. Unlike ChIP or ChiP-Seq, the chromatin is completely degraded by DNaseI treatment and the remaining RNA is analyzed by RT-PCR or microarray analysis. Using this technique the roles of several lncRNAs such as ANRIL and Mistral have been shown to play a role in chromatin modifying complexes and regulating downstream target gene expression (Yap et al. 2010; Bertani et al. 2011). These technologies may be modified and combined with extant high-throughput techniques such as microarray and RNA-Seq and used in the discovery of novel lncRNAs and in understand the role of lncRNA and their variants in diseases and development.

## 14 Chromatin Isolation by RNA Purification

As discussed above, researchers have shown that lncRNA can recruit chromatin modifiers to specific site in the genome to regulate chromatin status and gene regulation. Chromatin Isolation by RNA Purification (ChIRP) is designed by Howard Change lab to capture and identify RNA-interacting proteins, DNA or

RNAs in their cellular context. It also maps RNA-chromatin occupancy sites throughout the genome with high sensitivity and low background (Chu et al. 2011). In apposite to RIP, Clip and RNA-chIP approaches, the RNA–chromatin–protein complex is not immunoprecipitated using protein specific antibodies. ChIRP characterizes the whole interactome of a given RNA molecule using tiling antisense oligonucleotides targeting the entire length of the RNA molecule of interest within the RNP complex. Specific assays are then performed to identify and quantify the associated molecules, DNA by ChIRP-seq, RNA by ChIRP-RNA-seq, and protein by ChIRP-protein-mass spectrometry (Chu et al. 2011). It is also the first method that has been developed to identify RNA–RNA interactions (Guttman and Rinn 2012).

Chu et al. applied ChIRP-Seq to drosophila dosage compensation system, a well-characterized RNP complex that consists of two lncRNAs, roX1 and roX2 and the male-specific lethal (MSL) proteins and is involved in overexpression of genes located on male single X chromosome (Lucchesi et al. 2005). They could identify 308 rox2-X chromosome specific binding sites at higher resolution, confirming almost 90 % of the previous known rox2 occupancy sites on X-chromosome (Alekseyenko et al. 2008). Their results also supported the idea that the drosophila dosage compensation regulates male x-chromosome gene expression via enhancing transcriptional elongation. Based on their data, ChIRP-Seq can be potentially used to map in vivo genomic occupancy of any lncRNA with known primary sequences in an unbiased way (Chu et al. 2012), therefore increasing the current understanding of the functional significance of lncRNAs in mammalian genome.

## 15 High-Throughput Loss of Function by RNA Interference

One of the main aspects of lncRNA studies is in understanding its function. Are lncRNAs just a by-product of transcription or precursor molecules, do they interact in cis or in trans, do they regulate chromatin remodeling complexes or transcription factor activity? These are pertinent questions that arise while studying lncRNAs. One method to address these questions is through RNAi knockdown of lncRNA targets. However, very few large-scale loss-of-function studies have been performed to date. Only recently, Guttmann et al. have shown that lncRNAs play an important role in the pluripotency regulatory circuit using an lncRNA knock-down approach (Guttman et al. 2011).

Using lentivirus-based short-hairpin RNA (shRNA) against the known lncRNAs in mouse embryonic stem cells ($\sim$226), loss-of-function effects were assayed. By comparing these knockdowns to shRNAs against protein-coding regions as positive control, the researchers were able to find changes in transcriptional regulation. Similar lentiviral-based shRNA systems that target known lncRNAs are currently being developed (McCarthy 2012). Following lncRNA knockdown, using relevant markers or reporter systems, loss-of-function effects on specific biological processes or networks may be elucidated. For example, in ES cells, Nanog expression serves as an indicator of pluripotency and shRNA-based lncRNA-ROR knockdown alters Nanog expression and pluripotency maintenance (Guttman et al. 2011). The lncRNA knockdown expression pattern significantly correlated with the expression patterns seen in ES cells undergoing induced differentiation. Attributing function to lncRNA also relies on incorporating existing gene expression database, high-throughput microarray analyses, and downstream RNA-Seq analyses subsequent to RNAi-based knockdown.

Similar studies using RNAi-based lncRNA knockdown in other regulatory pathways such as the p53 DNA repair response have been reported and lncRNA expression and functional databases have been generated based on these findings (Huarte et al. 2010). Moreover, lncRNA specific siRNA databases such as those provided by the RNAi Consortium provide a platform to design robust, high-throughput and hypothesis-driven lncRNA functional studies. Lincode is an Ingenuity powered, commercially available siRNA database, that can be used to generate pathway-specific cocktails of siRNA. Customized lncRNA-specific RNAi screening libraries that can improve efficiency of loss-of-function studies are currently available (http://www.thermoscientificbio.com/rnai-and-custom-rna-synthesis/sirna/lincode-sirna/).

More recently, an endo-ribonuclease-based siRNA technique called combined knockdown and localization of noncoding RNA (c-KLAN) has been developed to efficiently study lncRNA knockdown and localization. The lnc-esiRNA (lncRNA-endo-ribonuclease-based siRNA) were generated using an algorithm called design and quality control of (e)siRNAs' (DEQOR) that has previously been used to design robust esi-RNA against protein-coding transcripts. This versatile technology not only ensures minimal off-target effects in knockdown studies, the lnc-esiRNAs can also be used as templates to generate RISH (probes to study lncRNA localization (Chakraborty et al. 2012).

# 16 Bioinformatics Approaches

## *16.1 Bioinformatic Tools for lncRNA Discovery and Annotation*

The advent of massive parallel sequencing techniques that generate millions of reads has significantly improved the speed and efficiency of lncRNA discovery and functional studies. However, these large datasets have necessitated the development of powerful computational tools to analyze them. Moreover, large-scale collaborative projects including ENCODE and GENCODE and 1000 genomes project, involving several research groups worldwide have generated massive publicly available datasets (Djebali et al. 2012; Derrien et al. 2012). These massive genomic and transcriptomic studies have greatly benefitted from newly developed computational tools. Using a slew of bioinformatic and annotation softwares, these studies have successfully identified many novel lncRNAs alongside snoRNAs and other types of ncRNAs. Moreover, these consortia also include studies that focus on cell-type specific and conditional expression of lncRNAs and their target genes, therefore, computational tools and systems biology-based approaches that combine expression data, facilitate understanding of lncRNA roles in biologically relevant regulatory networks are needed.

Several comprehensive tools including Scripture and ncFANs have been used in ab initio transcription reconstruction (Guttman et al. 2010; Liao et al. 2011). These methods enable gene structure elucidation of previously identified lncRNA and novel lncRNA discovery. Bioinformatic tools are crucial in understanding phylogenetic conservation of specific genomic regions and extant computational tools such as phastCons and statistical methods such composite of multiples test (CMS) have been used in assigning evolutionary pathways for lncRNA (Grossman et al. 2013; Siepel et al. 2005). With the exponential increase in lncRNA-based studies and novel findings in lncRNA targets, functions, and regulation, the role of computational approaches cannot be overstated. Consequently, the existing bioinformatics and annotation tools are constantly updated to generate newer versions to keep abreast of the advances in the experimental tools in lncRNA studies. The table below summarizes the annotation tools and other softwares that are the currently used in lncRNA discovery and analyses (Table 2).

## *16.2 lncRNAs :Public Database*

In the age of high-throughput next generation sequencing, it is becoming increasingly complex and important to sort through the vast amount of data produced. This breadth of data poses several challenges ranging from simple annotations to more complex functional annotations. Various databases have emerged, each with their own rationalization of information and specific nomenclature. One of the first databases developed for ncRNA, Noncoding RNA Database, was developed in 2003 by Barciszewski's group (Szymanski et al. 2003). This database was the first repository containing nucleotide sequences, in FASTA format, short descriptions of the activities of particular ncRNAs, GenBank accession numbers, and literature references. When this database was created in 2003, the number of unique mammalian ncRNAs was less than 40. As of 2011, the database holds over 30,000 unique sequences. As you can see, the field of lncRNA is growing exponentially, and the research data produced is growing with it. NGS integrated with bioinformatics will lead to a detailed description of lncRNAs and further understanding of this unexplored world (Da Sacco et al. 2012).

**Table 2** Annotation tools and other softwares

| Tool name | Description | Reference |
|---|---|---|
| EricScript | Discovering chimeric transcripts in paired-end RNA-Seq data | (Benelli et al. 2012) |
| GA TExplorer | Integrated gene loci explorer and probe mapping tool that provides lncRNA probe visualization in the genomic context | (Risueno et al. 2010) |
| Scripture | A comprehensive method for ab initio transcription reconstruction from sequencing reads | (Guttman et al. 2010) |
| ncRNAscout | An ab initio ncRNA finder utilizing both sequence motifs and structural parameters | (Bao et al. 2012) |
| smyRNA | An ab initio ncRNA gene finder, based on identifying sequence motifs that have potential structural roles | (Salari et al. 2009) |
| CPAT | An alignment free tool to differentiate between protein-coding and noncoding genes | (Wang et al. 2013) |
| ncFANs | A web interface that provides tools for functional annotation and functional enrichment of lncRNA | (Liao et al. 2011) |
| phyloCSF | A comparative genomics method to distinguish protein-coding and noncoding regions | (Lin et al. 2011) |
| GraphClust | An alignment-free structural clustering of local RNA secondary structures | (Heyne et al. 2012) |
| MXSCARNA | Pairwise alignment tool that separately aligns 5' parts and 3' parts of the stem candidates | (Tabei et al. 2008) |
| Profile-csHMMS | Markov model-based method for RNA secondary structure searches | http://ieeexplore.ieee.org/xpl/articleDetails.jsp?arnumber=4537407 |
| phastCons | Markov model-based method that estimates the probability that each nucleotide belongs to a conserved element, based on the multiple alignment. | (Siepel et al. 2005) |
| CNVFinder | An algorithm designed to detect copy number variants (CNVs) in the human population from large-insert clone DNA microarray | (McCallum and Wang 2013; Fiegler et al. 2006) |
| CNVTools | A collection of packages useful in the analysis of copy number variants (CNV) | (Barnes et al. 2008; Glessner et al. 2013) |

## 17 Conclusion

The accumulating body of work on lncRNA biology makes it seem surprising that until recently, lncRNAs were largely considered to be transcriptomic "junk". Nevertheless, conquering new frontiers in lncRNA biology relies heavily on emerging molecular and genomic technology (Table 3). This chapter provides a snapshot of existing and constantly evolving experimental and computational approaches that play a crucial role in improving our understanding of lncRNA structure and function.

With the ever-increasing interest in lncRNA biology, academic research consortia and biotechnology companies alike have focused on developing reagents, kits, and bioinformatics tools to enable robust lncRNA discovery, validation, and functional characterization and to generate disease and pathway-specific lncRNA databases. Table 3 represents a selected list of emerging technologies to detect and quantify lncRNAs, building on previous methods. Starting from technologies for improved lncRNA isolation from a variety of challenging samples while maintaining sequence and RNA structural integrity, to improved bioinformatics tools that aid in powerful systematic analyses, these state-of-the-art approaches help conduct lncRNA studies rapidly at economically feasible rates. Incorporating the technological innovations in qPCR, array-based technologies and sequencing strategies from general transcriptomic studies can generate lncRNA-regulated conditional expression profiles with unprecedented efficiency.

Functional studies focusing on lncRNA interaction with cellular components and intracellular localization are benefitting from novel discoveries in the field of microscopy, including development of new detection reagents, dyes, and high precision instruments. Owing to the development of improved antibodies, oligonucleotides, and other reagents, pull-down techniques have reduced the amount of ambiguous, false positive discoveries. Multiplexed lncRNA knockdown approaches with increased specificity ensure minimal 'off-target' effects. When coupled with high throughput deep sequencing techniques including DeepCAGE, CLIP-Seq, and subcellular fraction deep-sequencing, and the associated annotation pipelines, these procedures can be performed rapidly to provide specific insights to lncRNA interactions and functions. More importantly, a majority of these procedures including sample preparation and library generation can now be automated, thus reducing the time and cost involved in lncRNA research (Meldrum 2000a, b; Ramskold et al. 2012).

Additionally, the ongoing improvements to extant the existing bioinformatic software and development of new computational tools have reduced time and computational memory constraints in data analyses. Experimental procedures and existing lncRNA databases are constantly updated to incorporate the novel findings without compromising quality of the data obtained. Given the advances in the biomolecular techniques, single-cell analyses have emerged as novel avenues in genomic and transcriptomic research. This will address the inherent problem caused by heterogeneity within samples (Ramskold et al. 2012; Mustafi et al.

**Table 3** Future methods to detect and quantify noncoding RNAs

| | | |
|---|---|---|
| 3'-end Sequencing for Expression Quantification (3SEQ) | A sequencing technique used for precise quantification of expression levels across the whole genome. 3SEQ captures 3' polyadenylated ends of RNA fragments. Studies have begun to use 3SEQ to do transcriptional profiling of lncRNAs | (Beck et al. 2010; Brunner et al. 2012) |
| Parallel Analysis of RNA Structure (PARS) | Increases the throughput of collecting experimental RNA structure data independent of RNA size. PARS identifies double and single stranded regions of RNAs in solution using traditional and next generation RNA sequencing techniques. This method allows researchers to collect genome-wide RNA structural data at the single nucleotide level. This dramatically increases the number of RNA secondary structures that can be probed | (Kertesz et al. 2010) |
| RNA SHAPE | A recently developed technique, 2'-hydroxyl acylation analyzed by primer extension (SHAPE), allows for the probing of RNA structure in living cells (in vivo) and assessing changes in RNA structure across different cell states. This method can help researchers study the effect of lncRNA on gene regulation and cell function | (Spitale et al. 2013) |
| Nanopore-Based RNA Detection | Applied voltage is used to pull ncRNAs through either biological or solid state nanopores. As the molecules translocate, they partially block the ion flow through the pore, detected as a drop in the measured current. Solid State pores are being tested for lncRNA detection | (Wang et al. 2011; Wanunu et al. 2010) |
| Fluorescence Quenching on Graphene Oxide | Graphene Oxide bound ssDNA and fluorescent dyes are hybridized with ncRNAs, which are protected from adsorption to the graphene. Sensitivity can be enhanced by isothermal strand-displacement amplification | (Dong et al. 2012) |
| Single Molecule Detection Based on Dual Fluorescent Labeling | Homogeneous assay using two specific dye colored probes. When both probes bind to ncRNA, the detector will acquire two peaks, whereas unbound probes will appear as single peaks | (Neely et al. 2006) |
| Bioluminescence | Competitive hybridization assays with bioluminescent labeled DNA probes | (Cissell et al. 2008) |

(continued)

**Table 3** (continued)

| | | |
|---|---|---|
| Silicon Nanowires (SiNWs) | Label-free and direct hybridization assay for ultrasensitive detection using silicon nanowires with peptide nucleic acids serving as the complementary receptors. Change in resistance is directly correlated to the concentrations of hybridized ncRNA | (Zhang et al. 2009) |
| Nanoresonator Chip | ncRNA hybridized with metallic or silica nanowire bound cDNA are bound to mass amplifying gold nanoparticles. This large change in mass induces a large, optically measurable shift in resonance frequency of the nanoresonator | (Sioss et al. 2012) |
| SERS | RAMAN scattering spectroscopy used to detect the absorption of ncRNA to silver or gold nanoparticles | (Driskell et al. 2009) |
| Surface Plasmon Resonance | Sensitive optical detection method for detecting small changes of refractive index induced by the hybridization of ncRNA to immobilized capture DNA | (Fang et al. 2008; Nasheri et al. 2011; Sipova et al. 2010; Wark et al. 2008) |
| Electrochemical Detection | Electrocatalytic nanoparticle labeled ncRNAs are hybridized with complementary DNA immobilized on a chip. The hybridized nanoparticle-ncRNAs are able to catalyze an oxidation reaction, leading to a measurable increase of current | (Gao and Yu 2007; Gao and Yang 2006; Peng and Gao 2011; Cissell and Deo 2009) |
| Scanometric miRNA array | Enzymatically ligated miRNA to a universal linker followed by hybridization to a microarray. To detect captured targets, SNA-functionalized gold nanoparticles are subsequently hybridized. Gold with gold-enhancing solution can be used for signal amplification | (Alhasan et al. 2012) |
| Base Stacking Hybridization coupling with time-resolved fluorescence technology | With an immobilized DNA capture probe, ncRNA and fluorescent DNA tags can hybridize together. If no ncRNA is present, the fluorescent tag and probe will be disrupted when washing, as the binding energetically unfavorable | (Jiang et al. 2012) |
| TIRFM | cDNA probes hybridized to ncRNAs are labeled with fluorescent YOYO-1. The hybrids are imaged and quantified by an Electron-multiplying charge-coupled device-coupled TIRFM and single-molecule counting | (Chan et al. 2010) |

2013). Soon the dynamics of lncRNA expression and function may be traced at single-cell level or with single-molecule precision. Based on the wide expanse of emerging technologies summarized here it is easy to envision that, in the subsequent years, our understanding of lncRNA function and their role in diseases will improve exponentially. This may potentially lead to a transition from traditional trial-and-error practice of medicine to an effective personalized medicine.

## References


Clark, M. B., et al. (2011). The reality of pervasive transcription. *PLoS Biology, 9*, e1000625; discussion e1001102.

Friedman J.M., et al. (2006). Oligonucleotide microarray analysis of genomic imbalance in children with mental retardation. *Am J Hum Genet. 79*, 500.

Nagano, T., & Fraser, P. (2011). No-nonsense functions for long noncoding RNAs. *Cell, 145*, 178.

Ponting, C. P., Oliver, P. L., & Reik, W. (2009). Evolution and functions of long noncoding RNAs. *Cell, 136*, 629.

Gupta, R. A., et al. (2010). Long non-coding RNA HOTAIR reprograms chromatin state to promote cancer metastasis. *Nature, 464*, 1071.

Wilusz, J. E., Sunwoo, H., & Spector, D. L., (2009). Long noncoding RNAs: functional surprises from the RNA world. *Genes & Development 23*, 1494.

Derrien, T., et al. (2012). The GENCODE v7 catalog of human long noncoding RNAs: analysis of their gene structure, evolution, and expression. *Genome Research, 22*, 1775.

Brown, C. J., et al. (1991). A gene from the region of the human X inactivation centre is expressed exclusively from the inactive X chromosome. *Nature, 349*, 38.

Kampa, D., et al. (2004) Novel RNAs identified from an in-depth analysis of the transcriptome of human chromosomes 21 and 22. *Genome Research, 14*, 331.

Cheng, J., et al. (2005). Transcriptional maps of 10 human chromosomes at 5-nucleotide resolution. *Science, 308*, 1149.

Rinn, J. L., et al. (2007). Functional demarcation of active and silent chromatin domains in human HOX loci by noncoding RNAs. *Cell, 129*, 1311.

Bernstein, E., & Allis, C. D. (2005). RNA meets chromatin. *Genes & Development, 19*, 1635.

Plath, K., et al. (2003). Role of histone H3 lysine 27 methylation in X inactivation. *Science, 300*, 131.

Kretz, M., et al. (2013). Control of somatic tissue differentiation by the long non-coding RNA TINCR. *Nature, 493*, 231.

Batista, P. J., & Chang, H. Y. (2013). Long noncoding RNAs: cellular address codes in development and disease. *Cell, 152*, 1298.

Clemson, C. M., et al. (2009). An architectural role for a nuclear noncoding RNA: NEAT1 RNA is essential for the structure of paraspeckles. *Molecular Cell, 33*, 717.

Willingham, A. T., et al. (2005) A strategy for probing the function of noncoding RNAs finds a repressor of NFAT. *Science, 309*, 1570.

Jiang, L., Duan, D., Shen, Y., & Li, J. (2012). Direct microRNA detection with universal tagged probe and time-resolved fluorescence technology. *Biosensors & Bioelectronics, 34*, 291.

Tang, X., et al. (2007). A simple array platform for microRNA analysis and its application in mouse tissues. *Rna, 13*, 1803.

Benes, V., & Castoldi, M. (2010). Expression profiling of microRNA using real-time quantitative PCR, how to use it and what is available. *Methods, 50*, 244.

Nagalakshmi, U., et al. (2008). The transcriptional landscape of the yeast genome defined by RNA sequencing. *Science, 320*, 1344.

Mortazavi, A., Williams, B. A., McCue, K., Schaeffer, L., & Wold, B. (2008). Mapping and quantifying mammalian transcriptomes by RNA-Seq. *Nature Methods, 5*, 621.

Wang, Z., Gerstein, M., & Snyder, M. (2009). RNA-Seq: a revolutionary tool for transcriptomics. *Nature Reviews. Genetics, 10*, 57.

Cabili, M. N., et al. (2011). Integrative annotation of human large intergenic noncoding RNAs reveals global properties and specific subclasses. *Genes & Development, 25*, 1915.

Mercer, T. R., Dinger, M. E., Sunkin, S. M., Mehler, M. F., & Mattick, J. S. (2008). Specific expression of long noncoding RNAs in the mouse brain. *Proceedings of the National Academy of Sciences of the United States of America, 105*, 716.

Gall, J. G., & Pardue, M. L. (1969). Formation and detection of RNA-DNA hybrid molecules in cytological preparations. *Proceedings of the National Academy of Sciences of the United States of America, 63*, 378.

Chisholm, K. M., et al. (2012). Detection of long non-coding RNA in archival tissue: correlation with polycomb protein expression in primary and metastatic breast carcinoma. *PLoS ONE, 7*, e47998.

Rapicavoli, N. A., Poth, E. M., Zhu, H., & Blackshaw, S. (2011). The long noncoding RNA Six3OS acts in trans to regulate retinal development by modulating Six3 activity. *Neural development, 6*, 32.

Taft, R. J., Pheasant, M., & Mattick, J. S. (2007). The relationship between non-protein-coding DNA and eukaryotic complexity. *BioEssays : News and Reviews in Molecular, Cellular and Developmental Biology, 29*, 288.

Mehler, M. F., & Mattick, J. S. (2007). Noncoding RNAs and RNA editing in brain development, functional diversification, and neurological disease. *Physiological Reviews, 87*, 799.

Collins, M. L., et al. (1997). A branched DNA signal amplification assay for quantification of nucleic acid targets below 100 molecules/ml. *Nucleic Acids Research, 25*, 2979.

Jia, H., et al. (2010). Genome-wide computational identification and manual annotation of human long noncoding RNA genes. *Rna, 16*, 1478.

Guttman, M., et al. (2009). Chromatin signature reveals over a thousand highly conserved large non-coding RNAs in mammals. *Nature, 458*, 223.

Khalil, A. M., et al. (2009). Many human large intergenic noncoding RNAs associate with chromatin-modifying complexes and affect gene expression. *Proceedings of the National Academy of Sciences of the United States of America, 106*, 11667.

Loewer, S., et al. (2010). Large intergenic non-coding RNA-RoR modulates reprogramming of human induced pluripotent stem cells. *Nature Genetics, 42*, 1113.

Harrow, J., et al. (2012). GENCODE: the reference human genome annotation for The ENCODE Project. *Genome Research, 22*, 1760.

Rinn, J. L., & Chang, H. Y. (2012). Genome regulation by long noncoding RNAs. *Annual Review of Biochemistry, 81*, 145.

Bertone, P., Gerstein, M., & Snyder, M. (2005). Applications of DNA tiling arrays to experimental genome annotation and regulatory pathway discovery. *Chromosome Research : An International Journal on the Molecular, Supramolecular and Evolutionary Aspects of Chromosome Biology, 13*, 259.

Kapranov, P., et al. (2002) Large-scale transcriptional activity in chromosomes 21 and 22. *Science, 296*, 916.

Rinn, J. L., et al. (2003), The transcriptional activity of human Chromosome 22. *Genes & Development, 17*, 529.

Kretz, M., et al. (2012). Suppression of progenitor differentiation requires the long noncoding RNA ANCR. *Genes & Development, 26*, 338.

Lander E.S., et al. (2001). Initial sequencing and analysis of the human genome. *Nature, 409*, 860. http://www.ncbi.nlm.nih.gov/pubmed/11237011.

Lee, C., & Kikyo, N. (2012). Strategies to identify long noncoding RNAs involved in gene regulation. *Cell & Bioscience, 2*, 37.

Trapnell, C., Pachter, L., & Salzberg, S. L. (2009). TopHat: discovering splice junctions with RNA-Seq. *Bioinformatics, 25*, 1105.

Pauli, A., et al. (2012). Systematic identification of long noncoding RNAs expressed during zebrafish embryogenesis. *Genome Research, 22*, 577.
Young, R. S., et al. (2012). Identification and properties of 1,119 candidate lincRNA loci in the Drosophila melanogaster genome. *Genome Biology and Evolution, 4*, 427.
de Araujo, M. E., & Huber, L. A. (2007). Subcellular fractionation. *Methods in Molecular Biology, 357*, 73.
Bhatt, D. M., et al. (2012). Transcript dynamics of proinflammatory genes revealed by sequence analysis of subcellular RNA fractions. *Cell, 150*, 279.
Tilgner, H., et al. (2012). Deep sequencing of subcellular RNA fractions shows splicing to be predominantly co-transcriptional in the human genome but inefficient for lncRNAs. *Genome Research, 22*, 1616.
Sleutels, F., Zwart, R., & Barlow, D. P. (2002). The non-coding Air RNA is required for silencing autosomal imprinted genes. *Nature, 415*, 810.
Mancini-Dinardo, D., Steele, S. J., Levorse, J. M., Ingram, R. S., & Tilghman, S. M. (2006). Elongation of the Kcnq1ot1 transcript is required for genomic imprinting of neighboring genes. *Genes & Development, 20*, 1268.
Kapranov, P., et al. (2005). Examples of the complex architecture of the human transcriptome revealed by RACE and high-density tiling arrays. *Genome Research, 15*, 987.
Broadbent, K. M., et al. (2011). A global transcriptional analysis of Plasmodium falciparum malaria reveals a novel family of telomere-associated lncRNAs. *Genome Biology, 12*, R56.
Scotto-Lavino, E., Du, G., & Frohman, M. A. (2006). Amplification of 5' end cDNA with 'new RACE'. *Nature Protocols, 1*, 3056.
Shiraki, T., et al. (2003). Cap analysis gene expression for high-throughput analysis of transcriptional starting point and identification of promoter usage. *Proceedings of the National Academy of Sciences of the United States of America, 100*, 15776.
Valen, E., et al. (2009). Genome-wide detection and analysis of hippocampus core promoters using DeepCAGE. *Genome Research, 19*, 255.
Djebali, S., et al. (2012). Landscape of transcription in human cells. *Nature, 489*, 101.
German, M. A., Luo, S., Schroth, G., Meyers, B. C., & Green, P. J. (2009). Construction of Parallel Analysis of RNA Ends (PARE) libraries for the study of cleaved miRNA targets and the RNA degradome. *Nature Protocols, 4*, 356.
Knapp, G. (1989). Enzymatic approaches to probing of RNA secondary and tertiary structure. *Methods in Enzymology, 180*, 192.
Low, J. T., & Weeks, K. M. (2010). SHAPE-directed RNA secondary structure prediction. *Methods, 52*, 150.
Machado-Lima, A., del Portillo, H. A., & Durham, A. M. (2008). Computational methods in noncoding RNA research. *Journal of Mathematical Biology, 56*, 15.
Underwood, J. G., et al. (2010). FragSeq: transcriptome-wide RNA structure probing using high-throughput sequencing. *Nature Methods, 7*, 995.
Reuter, J., & Mathews, D. (2010). RNAstructure: software for RNA secondary structure prediction and analysis. *BMC Bioinformatics, 11*, 129.
Crawford, G. E., et al. (2006). Genome-wide mapping of DNase hypersensitive sites using massively parallel signature sequencing (MPSS). *Genome Research, 16*, 123.
Reich, D. E., Gabriel, S. B., & Altshuler, D. (2003). Quality and completeness of SNP databases. *Nature genetics, 33*, 457.
Pasmant, E., et al. (2007). Characterization of a germ-line deletion, including the entire INK4/ARF locus, in a melanoma-neural system tumor family: identification of ANRIL, an antisense noncoding RNA whose expression coclusters with ARF. *Cancer Research, 67*, 3963.
Barnes, C., et al. (2008). A robust statistical method for case-control association testing with copy number variation. *Nature Genetics, 40*, 1245.
Sebat, J., et al. (2007). Strong association of de novo copy number mutations with autism. *Science, 316*, 445.
Mefford, H. C., et al. (2010). Genome-wide copy number variation in epilepsy: novel susceptibility loci in idiopathic generalized and focal epilepsies. *PLoS Genetics, 6*, e1000962.

Swaminathan, S., et al. (2012). Analysis of copy number variation in Alzheimer's disease in a cohort of clinically characterized and neuropathologically verified individuals. *PLoS ONE, 7*, e50640.

Hirsch, F. R., et al. (2003). Epidermal growth factor receptor in non-small-cell lung carcinomas: correlation between gene copy number and protein expression and impact on prognosis. *Journal of Clinical Oncology : Official Journal of the American Society of Clinical Oncology, 21*, 3798.

Duan, J., Zhang, J. G., Deng, H. W., & Wang, Y. P. (2013). Comparative studies of copy number variation detection methods for next-generation sequencing technologies. *PLoS ONE, 8*, e59128.

Yoon, S., Xuan, Z., Makarov, V., Ye, K., & Sebat, J. (2009). Sensitive and accurate detection of copy number variants using read depth of coverage. *Genome Research, 19*, 1586.

Moran, V. A., Perera, R. J., & Khalil, A. M. (2012). Emerging functional and mechanistic paradigms of mammalian long non-coding RNAs. *Nucleic Acids Research, 40*, 6391.

Huarte, M., et al. (2010). A large intergenic noncoding RNA induced by p53 mediates global gene repression in the p53 response. *Cell, 142*, 409.

Kino, T., Hurt, D. E., Ichijo, T., Nader, N., & Chrousos, G. P. (2010). Noncoding RNA gas5 is a growth arrest- and starvation-associated repressor of the glucocorticoid receptor. *Science Signaling, 3*, ra8.

Niranjanakumari, S., Lasda, E., Brazas, R., & Garcia-Blanco, M. A. (2002). Reversible cross-linking combined with immunoprecipitation to study RNA-protein interactions in vivo. *Methods, 26*, 182.

Brooks, S. A., & Rigby, W. F. (2000). Characterization of the mRNA ligands bound by the RNA binding protein hnRNP A2 utilizing a novel in vivo technique. *Nucleic Acids Research, 28*, E49.

Mili, S., & Steitz, J. A. (2004). Evidence for reassociation of RNA-binding proteins after cell lysis: implications for the interpretation of immunoprecipitation analyses. *Rna, 10*, 1692.

Collins, K. (2008). Physiological assembly and activity of human telomerase complexes. *Mechanisms of Ageing and Development, 129*, 91.

Heyne, S., Costa, F., Rose, D., & Backofen, R. (2012). GraphClust: alignment-free structural clustering of local RNA secondary structures. *Bioinformatics, 28*, i224.

Selth, L. A., Gilbert, C., & Svejstrup, J. Q. (2009). RNA immunoprecipitation to determine RNA-protein associations in vivo. *Cold Spring Harbor protocols, 2009*, pdb prot5234.

Licatalosi, D. D., et al. (2008). HITS-CLIP yields genome-wide insights into brain alternative RNA processing. *Nature, 456*, 464.

Ule, J., Jensen, K., Mele, A., & Darnell, R. B. (2005). CLIP: a method for identifying protein-RNA interaction sites in living cells. *Methods, 37*, 376.

Guil, S., et al. (2012). Intronic RNAs mediate EZH2 regulation of epigenetic targets. *Nature Structural & Molecular Biology, 19*, 664.

Sparmann, A., & van Lohuizen, M. (2006). Polycomb silencers control cell fate, development and cancer. *Nature Reviews. Cancer, 6*, 846.

Chase, A., & Cross, N. C. (2011). Aberrations of EZH2 in cancer. *Clinical Cancer Research : an Official Journal of the American Association for Cancer Research, 17*, 2613.

Bu, D., et al. (2012). NONCODE v3.0: integrative annotation of long noncoding RNAs. *Nucleic Acids Research, 40*, D210.

Konig, J., Zarnack, K., Luscombe, N. M., & Ule, J. (2011). Protein-RNA interactions: new genomic technologies and perspectives. *Nature Reviews. Genetics, 13*, 77.

Hafner, M., et al. (2010). Transcriptome-wide identification of RNA-binding protein and microRNA target sites by PAR-CLIP. *Cell, 141*, 129.

Konig, J., et al. (2010). iCLIP reveals the function of hnRNP particles in splicing at individual nucleotide resolution. *Nature Structural & Molecular Biology, 17*, 909.

Yap, K. L., et al. (2010). Molecular interplay of the noncoding RNA ANRIL and methylated histone H3 lysine 27 by polycomb CBX7 in transcriptional silencing of INK4a. *Molecular Cell, 38*, 662.

Bertani, S., Sauer, S., Bolotin, E., & Sauer, F. (2011). The noncoding RNA Mistral activates Hoxa6 and Hoxa7 expression and stem cell differentiation by recruiting MLL1 to chromatin. *Molecular Cell, 43*, 1040.

Chu, C., Qu, K., Zhong, F. L., Artandi, S. E., & Chang, H. Y. (2011). Genomic maps of long noncoding RNA occupancy reveal principles of RNA-chromatin interactions. *Molecular Cell, 44*, 667.

Guttman, M., & Rinn, J. L. (2012). Modular regulatory principles of large non-coding RNAs. *Nature, 482*, 339.

Lucchesi, J. C., Kelly, W. G., & Panning, B. (2005). Chromatin remodeling in dosage compensation. *Annual Review of Genetics, 39*, 615.

Alekseyenko, A. A., et al. (2008) A sequence motif within chromatin entry sites directs MSL establishment on the Drosophila X chromosome. *Cell, 134*, 599.

Chu, C., Quinn, J., & Chang, H. Y. (2012). *Chromatin isolation by RNA purification (ChIRP).* Journal of Visualized Experiments: JoVE.

Guttman, M., et al. lincRNAs act in the circuitry controlling pluripotency and differentiation. *Nature, 477*, 295.

McCarthy, N. (2012). Epigenetics. Going places with BANCR. *Nature Reviews. Cancer, 12*, 451.

Chakraborty, D., et al. (2012). Combined RNAi and localization for functionally dissecting long noncoding RNAs. *Nature Methods, 9*, 360.

Guttman, M., et al. (2010). Ab initio reconstruction of cell type-specific transcriptomes in mouse reveals the conserved multi-exonic structure of lincRNAs. *Nature Biotechnology, 28*, 503.

Liao, Q., et al. (2011). ncFANs: a web server for functional annotation of long non-coding RNAs. *Nucleic Acids Research, 39*, W118.

Grossman, S. R., et al. (2013). Identifying recent adaptations in large-scale genomic data. *Cell, 152*, 703.

Siepel, A., et al. (2005). Evolutionarily conserved elements in vertebrate, insect, worm, and yeast genomes. *Genome Research, 15*, 1034.

Benelli, M., et al. (2012). Discovering chimeric transcripts in paired-end RNA-seq data by using EricScript. *Bioinformatics, 28*, 3232.

Risueno, A., Fontanillo, C., Dinger, M. E., & De Las Rivas, J. (2010). GATExplorer: genomic and transcriptomic explorer; mapping expression probes to gene loci, transcripts, exons and ncRNAs. *BMC Bioinformatics, 11*, 221.

Bao, M., Cervantes Cervantes, M., Zhong, L., & Wang, J. T. (2012). Searching for non-coding RNAs in genomic sequences using ncRNAscout. *Genomics, Proteomics & Bioinformatics, 10*, 114.

Salari, R., et al. (2009). smyRNA: a novel Ab initio ncRNA gene finder. *PLoS ONE, 4*, e5433.

Wang, L., et al. (2013). CPAT: Coding-Potential Assessment Tool using an alignment-free logistic regression model. *Nucleic Acids Research, 41*, e74.

Lin, M. F., Jungreis, I., & Kellis, M. (2011). PhyloCSF: a comparative genomics method to distinguish protein coding and non-coding regions. *Bioinformatics, 27*, i275.

Tabei, Y., Kiryu, H., Kin, T., & Asai, K. (2008). A fast structural multiple alignment method for long RNA sequences. *BMC Bioinformatics, 9*, 33.

McCallum, K. J., & Wang, J. P. (2013). Quantifying copy number variations using a hidden Markov model with inhomogeneous emission distributions. *Biostatistics, 14*, 600.

Fiegler, H., et al. (2006). Accurate and reliable high-throughput detection of copy number variation in the human genome. *Genome Research, 16*, 1566.

Glessner, J. T., Li, J., & Hakonarson, H. (2013). ParseCNV integrative copy number variation association software with quality tracking. *Nucleic Acids Research, 41*, e64.

Szymanski, M., Erdmann, V. A., & Barciszewski, J. (2003). Noncoding regulatory RNAs database. *Nucleic Acids Research, 31*, 429.

Da Sacco, L., Baldassarre, A., & Masotti, A. (2012). Bioinformatics Tools and Novel Challenges in Long Non-Coding RNAs (lncRNAs) Functional Analysis. *International Journal of Molecular Sciences, 13*, 97.

Paschoal, A. R., et al. (2012). Non-coding transcription characterization and annotation: A guide and web resource for non-coding RNA databases. *RNA Biology, 9*, 274.

Amaral, P. P., Clark, M. B., Gascoigne, D. K., Dinger, M. E., & Mattick, J. S. (2011). lncRNAdb: a reference database for long noncoding RNAs. *Nucleic Acids Research 39*, D146.

Mituyama, T., et al. (2009). The Functional RNA Database 3.0: databases to support mining and annotation of functional RNAs. *Nucleic Acids Research, 37*, D89.

Bu, D., et al. (2011). NONCODE v3.0: integrative annotation of long noncoding RNAs. *Nucleic Acids Research, 39,* d146–d151.

Burge, S. W., et al. (2012). Rfam 11.0: 10 years of RNA families. *Nucleic Acids Research, 41*, D226.

Dinger, M. E., et al. (2009). NRED: a database of long noncoding RNA expression. *Nucleic Acids Research, 37*, D122.

Wu, T., et al. (2006). NPInter: the noncoding RNAs and protein related biomacromolecules interaction database. *Nucleic Acids Research, 34*, D150.

Yang, J.-H., Li, J.-H., Jiang, S., Zhou, H., & Qu, L.-H. (2012). ChIPBase: a database for decoding the transcriptional regulation of long non-coding RNA and microRNA genes from ChIP-Seq data. *Nucleic Acids Research, 41,* D177.

Yamasaki, C., et al. (2009). H-InvDB in 2009: extended database and data mining resources for human genes and transcripts. *Nucleic Acids Research, 38*, D626.

Szymański, M., Erdmann, V. A., & Barciszewski, J. (2003). Noncoding regulatory RNAs database. *Nucleic Acids Research, 31*, 429.

Niazi, F., & Valadkhan, S. (2012). Computational analysis of functional long noncoding RNAs reveals lack of peptide-coding capacity and parallels with 3' UTRs. *Rna, 18*, 825.

Volders, P.-J., et al. LNCipedia: a database for annotated human lncRNA transcript sequences and structures. *Nucleic Acids Research, 41*, D246.

Chen, G., et al. (2013). LncRNADisease: a database for long-non-coding RNA-associated diseases. *Nucleic Acids Research, 41*, D983.

Paraskevopoulou, M. D., et al. (2013). DIANA-LncBase: experimentally verified and computationally predicted microRNA targets on long non-coding RNAs. *Nucleic Acids Research, 41*, D239.

Belinky, F., et al. (2013). Non-redundant compendium of human ncRNA genes in GeneCards. *Bioinformatics, 29*, 255.

Jin, J., Liu, J., Wang, H., Wong, L., & Chua, N.-H. (2013). PLncDB: plant long non-coding RNA database. *Bioinformatics, 29*, 1068.

Sun, K., et al. (2013). iSeeRNA: identification of long intergenic non-coding RNA transcripts from transcriptome sequencing data. *BMC Genomics, 14*, 1.

Gellert, P., Ponomareva, Y., Braun, T., & Uchida, S. (2013). Noncoder: a web interface for exon array-based detection of long non-coding RNAs. *Nucleic Acids Research, 41*, e20.

Chen, C.-J., et al. (2012). ncPRO-seq: a tool for annotation and profiling of ncRNAs in sRNA-seq data. *Bioinformatics, 28*, 3147.

Chang, T.-H., et al. (2013). An enhanced computational platform for investigating the roles of regulatory RNA and for identifying functional RNA motifs. *BMC Bioinformatics, 14*, 1.

Wright, M. W., & Bruford, E. A. (2011). Naming 'junk': human non-protein coding RNA (ncRNA) gene nomenclature. *Human Genomics, 5*, 90.

Sun, L., et al. (2012). Prediction of novel long non-coding RNAs based on RNA-Seq data of mouse Klf1 knockout study. *BMC Bioinformatics, 13*, 331.

Jeggari, A., Marks, D. S., & Larsson, E. (2012). miRcode: a map of putative microRNA target sites in the long non-coding transcriptome. *Bioinformatics*, *28,* 2062–2063.

Beck, A. H., et al. (2010). 3′-end sequencing for expression quantification (3SEQ) from archival tumor samples. *PLoS ONE, 5*, e8768.

Brunner, A. L., et al. (2012). Transcriptional profiling of long non-coding RNAs and novel transcribed regions across a diverse panel of archived human cancers. *Genome Biology, 13*, R75.

Kertesz, M., et al. (2010). Genome-wide measurement of RNA secondary structure in yeast. *Nature, 467*, 103.

Spitale, R. C., et al. (2010). RNA SHAPE analysis in living cells. *Nature Chemical Biology, 9*, 18.

Wang, Y., Zheng, D., Tan, Q., Wang, M. X., & Gu, L. Q. (2011). Nanopore-based detection of circulating microRNAs in lung cancer patients. *Nature Nanotechnology, 6*, 668.

Wanunu, M., et al. Rapid electronic detection of probe-specific microRNAs using thin nanopore sensors. *Nature Nanotechnology, 5*, 807.

Dong, H., et al. (2012). Highly sensitive multiple microRNA detection based on fluorescence quenching of graphene oxide and isothermal strand-displacement polymerase reaction. *Analytical Chemistry, 84*, 4587.

Neely, L. A., et al. (2006). A single-molecule method for the quantitation of microRNA gene expression. *Nature Methods, 3*, 41.

Cissell, K. A., Rahimi, Y., Shrestha, S., Hunt, E. A., & Deo, S. K. (2008). Bioluminescence-based detection of microRNA, miR21 in breast cancer cells. *Analytical Chemistry, 80*, 2319.

Zhang, G. J., Chua, J. H., Chee, R. E., Agarwal, A., & Wong, S. M. (2009). Label-free direct detection of MiRNAs with silicon nanowire biosensors. *Biosensors & Bioelectronics, 24*, 2504.

Sioss, J. A., et al. (2012). Nanoresonator chip-based RNA sensor strategy for detection of circulating tumor cells: response using PCA3 as a prostate cancer marker. *Nanomedicine : Nanotechnology, Biology, and Medicine, 8*, 1017.

Driskell, J. D., Primera-Pedrozo, O. M., Dluhy, R. A., Zhao, Y., & Tripp, R. A. (2009). Quantitative surface-enhanced Raman spectroscopy based analysis of microRNA mixtures. *Applied Spectroscopy, 63*, 1107.

Fang, S., Lee, H. J., Wark, A. W., & Corn, R. M. (2006). Attomole microarray detection of microRNAs by nanoparticle-amplified SPR imaging measurements of surface polyadenylation reactions. *Journal of the American Chemical Society, 128*, 14044.

Nasheri, N., et al. (2011). An enzyme-linked assay for the rapid quantification of microRNAs based on the viral suppressor of RNA silencing protein p19. *Analytical Biochemistry, 412*, 165.

Sipova, H., et al. (2010). Surface plasmon resonance biosensor for rapid label-free detection of microribonucleic acid at subfemtomole level. *Analytical Chemistry, 82*, 10110.

Wark, A. W., Lee, H. J., & Corn, R. M. (2008). Multiplexed detection methods for profiling microRNA expression in biological samples. *Angewandte Chemie, 47*, 644.

Gao, Z., & Yu, Y. H. (2007). Direct labeling microRNA with an electrocatalytic moiety and its application in ultrasensitive microRNA assays. *Biosensors & Bioelectronics, 22*, 933.

Gao, Z., & Yang, Z. (2006). Detection of MicroRNAs Using Electrocatalytic Nanoparticle Tags. *Analytical Chemistry, 78*, 1470.

Peng, Y., & Gao, Z. (2011). Amplified detection of microRNA based on ruthenium oxide nanoparticle-initiated deposition of an insulating film. *Analytical Chemistry, 83*, 820.

Cissell, K. A., & Deo, S. K. (2009). Trends in microRNA detection. *Analytical and Bioanalytical Chemistry, 394*, 1109.

Alhasan, A. H., et al. (2012). Scanometric microRNA array profiling of prostate cancer markers using spherical nucleic acid-gold nanoparticle conjugates. *Analytical Chemistry, 84*, 4153.

Chan, H. M., Chan, L. S., Wong, R. N., & Li, H. W. (2010). Direct quantification of single-molecules of microRNA by total internal reflection fluorescence microscopy. *Analytical Chemistry 82*, 6911.

Meldrum, D. (2000a). Automation for genomics, part one: preparation for sequencing. *Genome Research 10*, 1081.

Meldrum, D. (2000b). Automation for genomics, part two: sequencers, microarrays, and future trends. *Genome Research, 10*, 1288.

Ramskold, D., et al. (2012). Full-length mRNA-Seq from single-cell levels of RNA and individual circulating tumor cells. *Nature Biotechnology, 30*, 777.

Mustafi, D., et al. (2013). Evolutionarily conserved long intergenic noncoding RNAs in the eye. *Human Molecular Genetics*, 22, 2992.

# Long Non-coding RNAs and Nuclear Body Formation and Function

**Ellen Fortini, Ruohan Li and Archa H. Fox**

## 1 General Introduction

In the past decade we have made a quantum leap in our understanding of the genetics of complex organisms, with the discovery that the nonprotein coding regions of our genomes are transcribed into tens of thousands of long noncoding RNA (lncRNA) molecules. However, while we know of their identity, deciphering the functions of these lncRNAs has been, and is continuing to be a challenge. In this chapter we focus on one of the well-characterized functions of specific lncRNAs: to form subnuclear structures and/or influence the function of subnuclear bodies. These findings have been important to the field of lncRNA biology, as the ability to place specific lncRNAs within the context of known nuclear architecture has given many clues as to the roles of these lncRNAs, and has also affirmed their functional relevance. So, what do lncRNAs do in subnuclear bodies? The mechanisms range from dynamic induction of nuclear bodies to sequester or modify nuclear proteins involved in splicing and transcription, to lncRNA enrichment in subnuclear bodies directing the recruitment of gene loci to influence their transcriptional environment. The formation and enrichment of lncRNAs in subnuclear bodies has thus become one more example of the myriad different ways that lncRNAs regulate gene expression.

Here we discuss lncRNAs with defined nuclear localizations, and separate them into two classes (Fig. 1). First, there are the lncRNAs whose role is to form subnuclear bodies as essential structural scaffolds, these include mammalian

E. Fortini · R. Li · A. H. Fox (✉)
Western Australian Institute for Medical Research/Centre for Medical Research, The University of Western Australia, Rear, 50 Murray St, Perth, WA 6000, Australia
e-mail: archa.fox@waimr.uwa.edu.au

E. Fortini
e-mail: ellen.fortini@waimr.uwa.edu.au

R. Li
e-mail: ruohan.li@waimr.uwa.edu.au

A. M. Khalil and J. Coller (eds.), *Molecular Biology of Long Non-coding RNAs*,
DOI: 10.1007/978-1-4614-8621-3_8, 

NEAT1 in paraspeckles, primate Satellite III (SatIII) transcripts in nuclear stress bodies (nSBs), *drosophila hsr-ω* RNA in omega speckles, and mammalian neuronal MIAT in gomafu speckles. The second class of nuclear lncRNAs have been observed to localize to particular subnuclear sites, but are not essential for the nucleation or formation of the subnuclear structures they associate with. For these lncRNAs, their enrichment within subnuclear bodies may reflect an aspect of their function that is associated with nuclear organization. Examples here include MALAT1 in nuclear speckles, as well as TUG1 and (potentially) HOTAIR in polycomb bodies. In this chapter we focus on each of these well-studied examples, and describe the history, structure, and functions of the subnuclear bodies and their associated lncRNAs, to build up a picture of the insights being gained in this important nexus between lncRNA biology and nuclear organization.

## 2 LncRNAs that Form Structural Scaffolds for Subnuclear Bodies

In recent years, it has emerged that several types of subnuclear bodies are built on a lncRNA scaffold or backbone. A common theme seems to be that these lncRNAs nucleate the assembly of these bodies, in most cases by 'seeding' the bodies: recruiting abundant nuclear RNA-binding proteins to the site of lncRNA

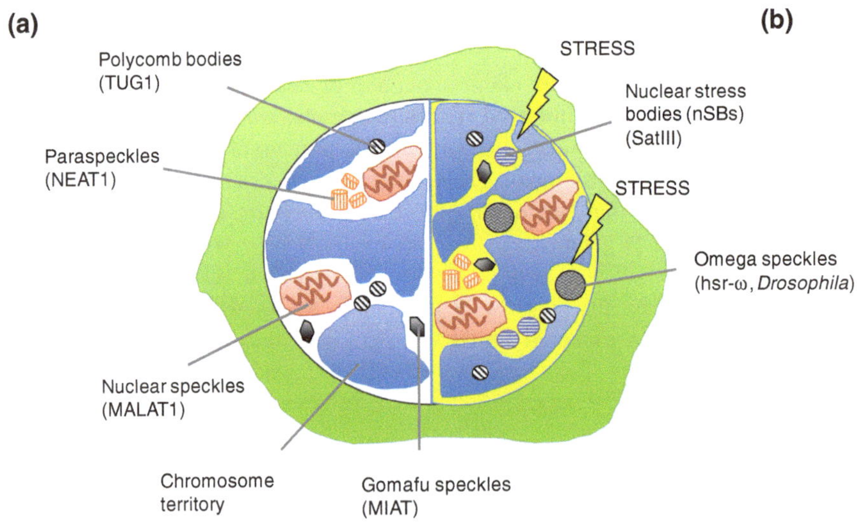


**Fig. 1** Nuclear bodies formed by, or associating with lncRNAs. **a** Under steady-state conditions, RNA FISH can be used to demonstrate that NEAT1 lncRNA is co-localised with paraspeckle markers, TUG1 resides within polycomb bodies, and MALAT1 is found in nuclear speckles. **b** Under stress, such as heat shock, specific lncRNAs are transcribed that nucleate additional subnuclear bodies. These include Satellite III lncRNA derived from pericentric heterochromatin in primates to form nuclear stress bodies, and *hsr-ω* RNA in *Drosophila* to form omega speckles

transcription to force a local high concentration of these molecules and start the process of body formation (Dundr and Misteli 2010). LncRNAs not only nucleate these bodies, but they appear to be an essential ongoing component for the maintenance of these structures. In terms of function, these subnuclear bodies are usually highly dynamic depending on the stress response of the cell, or the developmental stage of the tissue. There are usually many types of RNA or DNA binding proteins found associated with these subnuclear bodies, and studies have shown that in general the bodies are highly likely to be involved in transcriptional and post-transcriptional processes.

## 2.1 NEAT1 and Paraspeckles

Paraspeckles are mammalian subnuclear bodies that form around the NEAT1 (Nuclear Paraspeckle Assembly Transcript 1) lncRNA. Paraspeckles were first described as Interchromatin Granule Associated Zones, electron dense structures distinct from other nuclear bodies observed with the electron microscope in cultured cells (Visa et al. 1993). However, it was in 2002 that a clear marker protein, PSP1, or Paraspeckle protein 1, was found, and the term 'Paraspeckles' was coined to describe the subnuclear foci in which PSP1 was enriched (Fox et al. 2002). Additional paraspeckle proteins have since been identified, and these include the DBHS (Drosophila Behaviour Human Splicing) proteins related to PSP1–SFPQ and NONO, as well as a host of other RNA-binding proteins (Bond and Fox 2009; Fox et al. 2005; Naganuma et al. 2012; Prasanth et al. 2005). It is important to note that paraspeckle proteins, while enriched in paraspeckles, are also generally diffusely distributed in the nucleoplasm (Fox et al. 2002).

In the years following their identification, several early clues also suggested that RNA would likely be crucial to both paraspeckle structure and function: first, the paraspeckle proteins were all known RNA-binding proteins, and several only localized to paraspeckles via key RNA recognition Motifs (RRM), second, paraspeckles were sensitive to RNase treatment, third, they only formed in newly divided cells once RNA Polymerase II transcription was well established, and finally, they were disassembled by inhibition of RNA Polymerase II transcription (Dye and Patton 2001; Fox et al. 2002, 2005).

In 2009, three groups reported that paraspeckles were formed around the NEAT1 lncRNA, and that NEAT1 was an essential structural component of paraspeckles (Clemson et al. 2009; Sasaki et al. 2009; Sunwoo et al. 2009). *NEAT1* is a mammalian-specific gene located on human chromosome 11q13 and mouse 19qA (Hutchinson et al. 2007; Sasaki et al. 2009). The *NEAT1* gene promoter triggers transcription of two major isoforms of RNA that overlap completely at their 5′-end, yet have very different 3′-ends (Hutchinson et al. 2007; Sasaki et al. 2009; Sunwoo et al. 2009). The shorter canonically polyadenylated isoform (3,700 nt in human, 3,100 nt in mouse), is termed NEAT1_v1 or MENε. The longer isoform, 23,000 nt in human and 20,500 nt in mouse, is termed NEAT1_v2 or

MEN$\beta$, and contains an unusual tRNA-like structure at its 3′-end that is recognized and cleaved by RNase P, to produce a 3′-end with a short genomically encoded poly(A)-rich sequence (Sunwoo et al. 2009). NEAT1_v2 is estimated to be at least fivefold less abundant than NEAT1_v1, and in many tissues and cell types, present at an even lower proportion (Sasaki et al. 2009; Sunwoo et al. 2009).

Transcription of NEAT1 lncRNA is the seed that triggers paraspeckle formation. This has been elegantly demonstrated with two main pieces of evidence: first, paraspeckles form in close proximity to the NEAT1 gene (Clemson et al. 2009), clustering near there, (although, once formed, they are capable of moving further afield), and, second, as inducible NEAT1 expression is sufficient to nucleate the formation of paraspeckles (Mao et al. 2011). In another fascinating twist, Spector and colleagues showed that it is not enough to simply have NEAT1 in the nucleus for paraspeckles to form, instead, NEAT1 has to be *actively* transcribed (Mao et al. 2011). Interestingly, in a variety of cultured cell lines, both NEAT1 isoforms clearly display the characteristic punctate localization typical of paraspeckles, co-localizing and co-purifying with DBHS proteins (Clemson et al. 2009; Hutchinson et al. 2007; Mao et al. 2011; Sasaki et al. 2009; Sreenivasa Murthy and Rangarajan 2010; Sunwoo et al. 2009). However, while both isoforms are found in paraspeckles, it is now generally accepted that transcription of the lower abundance NEAT1_v2, instead of the more abundant NEAT1_v1, is the critical factor for the assembly and maintenance of paraspeckles. Three pieces of evidence support this: siRNA specific for NEAT1_v2 is sufficient to ablate paraspeckles (Sasaki et al. 2009; Sunwoo et al. 2009); cells expressing endogenous NEAT1_v1, but not NEAT1_v2, lack paraspeckles; and overexpressed NEAT1_v2, but not NEAT1_v1 restores paraspeckles in NEAT1–/– Murine Embryonic Fibroblasts (MEFs) (Naganuma et al. 2012; Sasaki et al. 2009).

In line with the importance of NEAT1_v2 in paraspeckle formation, we also know from electron microscopy analysis that NEAT1_v2 RNA extends throughout the core of a paraspeckle, whereas NEAT1_v1 is only found at the periphery (Souquere et al. 2010). In fact, our understanding of the spatial organization of NEAT1 within paraspeckles is unrivalled by any other lncRNA in nuclear organization (Fig. 2). There are also some additional observations that suggest NEAT1_v1 may play a greater role in paraspeckle formation when artificially tethered to the chromatin at high levels: when NEAT1_v1 is post-transcriptionally targeted to a specific genomic location this can also recruit paraspeckle proteins efficiently, presumably forming de novo paraspeckles (Shevtsov and Dundr 2011). Whilst it is not known if these de novo paraspeckles are functional, these data raise the possibility that the function of NEAT1_v2 is to provide a binding platform for NEAT1_v1, for it to reach a local high concentration in order to allow paraspeckle proteins to associate with the RNA and form stable RNA-protein complexes (Nakagawa and Hirose 2012; Shevtsov and Dundr 2011).

While NEAT1 is essential for paraspeckle formation, so to are a number of paraspeckle proteins. For example, siRNA against the DBHS proteins SFPQ and NONO results in paraspeckle disassembly and a reduced stability of NEAT1_v2 (Sasaki et al. 2009). However, it is important to note that paraspeckle proteins,

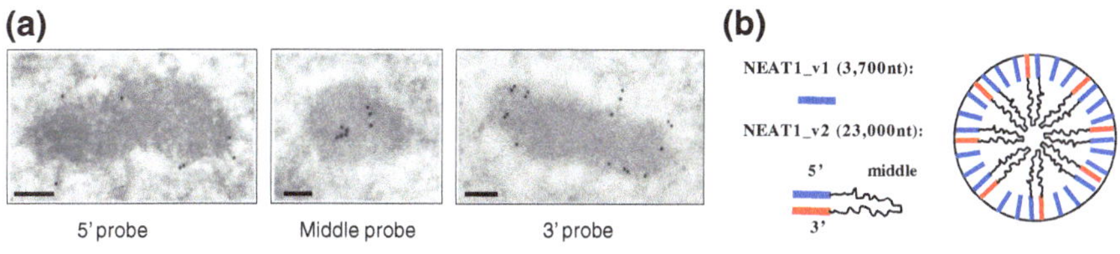


**Fig. 2** LncRNAs can have an ordered spatial arrangement within subnuclear bodies (eg. NEAT1 in paraspeckles) **a** Electron microscopy of HeLa nuclear sections, coupled with In situ hybridisation using probes to different regions within NEAT1. The probes are visualised with gold-conjugated antibody (small black dots). The 5' and 3' ends of NEAT1 are found at the periphery of paraspeckles, but the middle of the RNA is at the centre. Scale bars 100 nm. **b** A model of the arrangement of NEAT1_v1 and NEAT1_v2 isoforms in a cross section of a paraspeckle. Figure is courtesy of Gerard Pierron, CNRS, France

while an essential factor for making paraspeckles, do not have the capacity to nucleate paraspeckle formation: immobilizing DBHS proteins to chromatin could not effectively recruit NEAT1 to form de novo paraspeckles (Mao et al. 2011), suggesting a sequential assembly of different components that starts with NEAT1 transcription. At present there are approximately 40 proteins identified that are enriched in paraspeckles, mostly having RNA or DNA binding domains. Many of those proteins are indispensible for the formation of paraspeckles, or maintaining the stability for NEAT1 (Naganuma et al. 2012; Sasaki et al. 2009). One area that is still largely unknown is the molecular details of paraspeckle protein interactions with NEAT1. Structural studies on the essential DBHS paraspeckle proteins have revealed a novel dimer consisting of four RRM motifs held in a brace position by a coiled coil domain (Passon et al. 2012), however the RNA-binding modalities of these dimers are not yet known. In addition, the DBHS protein SFPQ has been shown to interact with several other lncRNAs, besides NEAT1 (Li et al. 2009; Takayama et al. 2013; Wu et al. 2013). Hirose and colleagues recently revealed that the function for some paraspeckle proteins is in establishing NEAT1_v2 production (Naganuma et al. 2012). They showed that the paraspeckle protein HNRNPK competes with paraspeckle-associated RNA cleavage proteins to determine if NEAT1 transcripts are cleaved and polyadenylated after 3.7 kb, or if this process is prevented to allow transcription to continue and generate NEAT1_v2 (Naganuma et al. 2012). This mechanism suggests a constant competition for the production, stabilization, and degradation of NEAT1_v2, which is in turn closely linked to paraspeckle formation.

While we know a considerable amount about the formation, components, and structure of paraspeckles, we have a poorer understanding of paraspeckle function. Mice lacking NEAT1, devoid of paraspeckles, have no gross phenotype, indicating that their function is unlikely to be crucial for development (Nakagawa et al. 2011). Nakagawa and colleagues (2011) have thus far produced the most comprehensive mapping of NEAT1 expression in tissues, using in situ hybridization against NEAT1 on mouse tissues, and have found that while most cells express NEAT1_v1, NEAT1_v2 is only found in a distinct subpopulation of cells. In silico,

RNA-seq datasets show widespread and abundant NEAT1 expression in most of the cell lines and tissues examined (Gibb et al. 2011), as well as indicating dynamic regulation of NEAT1 in various models of cellular differentiation (Sunwoo et al. 2009). However, there are exceptions to the rule, and NEAT1 is expressed at extremely low levels in embryonic stem cells (Chen and Carmichael 2009; Gibb et al. 2011; Nakagawa et al. 2011).

In terms of the molecular function of paraspeckles, the best evidence suggests that sequestering both RNA and protein components may be the route to influencing gene expression. In 2005, a specific nuclear retained mRNA was identified that partially co-localized in paraspeckles (Prasanth et al. 2005). This mRNA contains a long 3′-untranslated region (UTR), with Adenosine to Inosine (A-to-I) edited inverted Alu repeats that are a binding site for the paraspeckle proteins NONO and SFPQ (Prasanth et al. 2005; Zhang and Carmichael 2001). Specific stresses resulted in the edited RNA translocating to the cytoplasm, with a concomitant increase in translation (Prasanth et al. 2005). It has also been demonstrated that knockdown of NEAT1 alters the nuclear retention of these inverted Alu repeat RNAs (Chen and Carmichael 2009). Aspects of this nuclear retention mechanism could also be applied to other genes with inverted repeats in their 3′-UTRs, including Lin28, Nicn1, and Apobec3G (Chen and Carmichael 2009; Mao et al. 2011), however, it has also been found that some other genes with A-to-I edited inverted Alu repeats in their 3′-UTRs may undergo export to the cytoplasm where they are translationally repressed (Capshew et al. 2012; Fitzpatrick and Huang 2012). This repression appears to be mediated by cytoplasmic stress granules, which can form under heat shock stress (Capshew et al. 2012; Fitzpatrick and Huang 2012). Recently it has been postulated that an additional molecular function for paraspeckles could be the sequestration of paraspeckle proteins such as SFPQ (Nakagawa and Hirose 2012). This is interesting as the sequestration of nuclear proteins have been either hypothesized or well documented for other nuclear bodies that also rely on essential structural lncRNA component for their assembly.

## *2.2 Satellite III LncRNA and Nuclear Stress Bodies*

Nuclear stress bodies (nSBs) are formed around stress-induced lncRNAs transcribed from SatIII repetitive pericentromeric heterochromatin. NSBs were first identified when heat shock responsive transcription factor (HSF1) was observed to accumulate in large foci at pericentromeric heterochromatin after heat shock, chemical, and hypertonic stresses (Denegri et al. 2001; Jolly et al. 1997; Mähl et al. 1989; Sarge et al. 1993). These accumulation sites were formed primarily on the 9q12 loci of human chromosome 9, but also chromosome 12 and 15, which contain long tandem repeats of SatIII DNA (Denegri et al. 2002; Jolly et al. 2002). The nSBs were sensitive to RNase treatment, and also required ongoing RNA transcription for their maintenance, suggesting that RNA might play a structural role in their assembly (Chiodi et al. 2000; Weighardt et al. 1999). In 2002, Jolly and colleagues (2002) reported that HSF1 bound to the SatIII DNA and facilitated transcription of SatIII lncRNA. Indeed, under heat shock, these heterochromatic DNA regions shifted to euchromatin, marked by active histone modification marks, reinforcing the finding that the SatIII loci were becoming transcriptionally active following stress (Rizzi et al. 2004). Once transcribed, the SatIII ncRNA transcripts remain locally associated with the chromatin, and are bound by a number of pre-mRNA processing factors to form the nSBs, including SF2/ASF, SRp30c, and 9G8, and small nuclear ribonucleoproteins (snRNPs) (Denegri et al. 2001; Jolly et al. 2004; Metz et al. 2004). Interestingly, HSF1, the transcription factor responsible for up-regulating the RNAs, can also be found in nSBs (Shevtsov and Dundr 2011).

SatIII lncRNAs can have a variable length from either 2,000–5,000 nt, to no more than 10,000 nt (Biamonti and Caceres 2009; Jolly et al. 2004; Rizzi et al. 2004). This variable length of RNA likely results from the repetitive SatIII sequence, the multiple transcription start sites inside the array of tandem repeats, or the close contact with those bound splicing factors which have found to cause splicing of the lncRNA (Metz et al. 2004; Valgardsdottir et al. 2005). The SatIII RNA is absolutely required for nSB formation: knockdown of SatIII lncRNA abolishes the recruitment of the protein splicing factors to the nSBs. However, Sat III knockdown does not prevent the initial accumulation of HSF1 (Metz et al. 2004; Valgardsdottir et al. 2005). Recent studies have demonstrated that the immobilization of SatIII lncRNA transcript artificially onto chromatin can recruit HSF1, SAF-B, and SF2/ASF to form de novo nSBs (Shevtsov and Dundr 2011). Interestingly, heat shock resulting in the massive upregulation of SatIII lncRNA is accompanied by a global deacetylation of chromatin in the rest of the rest of the nucleus (Fritah et al. 2009).

The specific function of nSBs remains a matter for speculation. Whatever the function, it is possible it is highly complex and unique to primates, as SatIII elements appeared late in evolution, being primate specific (Denegri et al. 2002; Jarmuż et al. 2007). One possible function for nSBs is that they sequester RNA-binding proteins and RNAs to prevent them from circulating freely or performing their normal functions under heat shock conditions. This might be in line with the global suppression of transcription, altered splicing functions, and suppression of translation after heat shock, (with the exception of ongoing expression and translation of the heat shock responsive genes) (Lindquist 1986). Heat shock proteins rarely have introns in their genes, and they undergo a dramatic increase in expression and translation following heat shock stress, without great reliance on splicing factors (Lindquist 1986). It is therefore interesting to ponder if mobilizing active transcriptional power to the production of SatIII lncRNA, and then trapping particular splicing factors and tRNAs in the nSBs, might aid cells to prevent unnecessary or even harmful transcriptional, splicing, or translational events following heat shock (Biamonti and Vourc'h 2010; Metz et al. 2004). As with many other nuclear bodies, there remain many unanswered questions about the functions of these structures.

## 2.3 *hsr-ω* and Omega Speckles

In *Drosophila* there is a well-studied lncRNA-induced subnuclear structure termed 'omega speckles' that are nucleated by the Heat Shock RNA omega (*hsr-ω* or 93D). The *hsr-ω* gene locus is conserved among *Drosophila* species, but has not been found in other types of organisms. The *hsr-ω* gene contains two short exons (~475 and 700 bp in *D. melanogaster*) separated by a 700 bp intron, followed by a long stretch (5–15 kb) of short (280 bp in *D. melanogaster*) tandem repeats (Jolly and Lakhotia 2006). The overall gene may span 10–20 kb long, and produces two major transcripts and one precursor transcript. The major cytoplasmic transcript, termed *hsr-ω*-c, is less than 2,000 nt long, and contains the spliced exons with a polyadenylated 3′-end. The long nuclear transcript *hsr-ω*-n spans the entire length of the gene, including the intron, and is also polyadenylated (Bendena et al. 1991; Garbe et al. 1986; Ryseck et al. 1987). Therefore, *hsr-ω*-c could be considered a shorter spliced and overlapping version of *hsr-ω*-n. *Hsr-ω*-c appears to have a 23–27 amino acids open reading frame, but its sequence is not conserved and product is undetectable (Bendena et al. 1991; Garbe et al. 1986; Lakhotia and Sharma 1995; Ryseck et al. 1987). The *hsr-ω* gene is active in all cell types and at various developmental stages of *Drosophila*, and can be one of the most active loci under heat shock or amide stresses (Bendena et al. 1991; Mutsuddi and Lakhotia 1995; Prasanth et al. 2000; Tapadia and Lakhotia 1997).

The long *hsr-ω*-n transcript has been the most closely studied RNA of the *hsr-ω* group. *Hsr-ω*-n has a rapid turnover in the nucleus under normal conditions, but under stresses that might inhibit general nuclear transcription, it is rapidly up-regulated and accumulates with increased stability (Bendena et al. 1989; Hogan et al. 1995; Lakhotia and Sharma 1995). *Hsr-ω*-n was found colocalized with a variety of hnRNPs, forming a variable number of 'omega speckles' (Lakhotia et al. 1999; Prasanth et al. 2000). Without active transcription of *hsr-ω*, omega speckles cannot form (Prasanth et al. 2000). Similar to paraspeckles, omega speckles can be found both next to the locus of *hsr-ω*, or away from the locus (Lakhotia et al. 1999; Mao et al. 2011; Prasanth et al. 2000). It is particularly important to note that in normal conditions, most of the omega speckle proteins are present in both omega speckles, and at other nucleoplasmic locations that are usually transcriptionally active (Lakhotia et al. 1999; Prasanth et al. 2000). However, under stressful conditions, these minor sites rapidly disappear and the omega speckle proximal to the gene locus becomes enlarged. This stress-induced enlargement is accompanied by the translocation of omega speckle proteins, such as HRB87F (*Drosophila* orthologue of HNRNPA1) and HRB57A (*Drosophila* orthologue of HNRNPK), from their chromatin binding sites to the enlarged omega speckles, followed by a reduction of transcriptional activity at their previous binding sites (Buchenau et al. 1997; Dangli and Bautz 1983; Dangli et al. 1983; Hovemann et al. 1991; Lakhotia et al. 1999; Prasanth et al. 2000; Samuels et al. 1994; Zu et al. 1998). These data suggest a potential involvement of omega speckles in regulating the trafficking and availability of hnRNPs and other related RNA-binding proteins in the nucleus (Prasanth et al. 2000). This mechanism is similar to the sequestration hypothesis suggested for both paraspeckles and nSBs, where the transcription of the nucleating lncRNAs results in the accumulation of proteins in those bodies, thus altering their original localization and function. This sequestration might be a protection, or a temporary storage mechanism for those proteins, so that they can quickly resume normal function after the stress has been relieved (Jolly and Lakhotia 2006; Lakhotia et al. 1999; Prasanth et al. 2000).

A major focus of omega speckle research in the past decade has been determining the physiological significance of *hsr-ω*. Flies that are *hsr-ω* null are mostly embryonic lethal, with some flies hatching that are very weak and lacking omega speckles, suggesting that *hsr-ω* has a critical role in the development of *Drosophila* and assembly of omega speckles (Prasanth et al. 2000). The overall expression level of *hsr-ω* also seems to be critical, as its overexpression in whole flies results in polyglutamine (Poly-Q) induced neurodegeneration (Mallik and Lakhotia 2009; Sengupta and Lakhotia 2006), and its overexpression in the cyst cells of testis leads to male sterility (Rajendra et al. 2001). However, it is not clear yet if and how omega speckles are critically involved in causing the abnormal phenotypes resulted by the deletion or overexpression of *hsr-ω*.

The difference between nSBs, paraspeckles, and omega speckles lies in the different lncRNA identities, induced under different conditions, to nucleate different sets of proteins. For example, Serine/Arginine (SR) proteins, which are frequently found in nSBs, are not found in hnRNP containing omega speckles (Jolly and Lakhotia 2006). Intriguingly, SR proteins are generally considered as competitors of hnRNPs in pre-mRNA splicing, and yet both nSBs and omega speckles can be rapidly induced by heat shock stress (Jolly and Lakhotia 2006). Another interesting connection is that the *drosophila* homologs of two paraspeckle proteins, NONO and HNRNPK, were also shown to associate with *hsr-ω*-n, which might indicate conservation of functions shared by the two subnuclear bodies (Prasanth et al. 2000; Zimowska and Paddy 2002). Finally, there is a similarity in gene structure, such that, as with the *hsr-ω* transcripts, the paraspeckle nucleating lncRNA NEAT1_v1 and v2 also share their 5′- end, with NEAT1_v2 and *hsr-ω*-n being longer and containing repetitive sequences.

## 2.4 MIAT and Gomafu Speckles

MIAT (Myocardial infarction associated transcript) lncRNA is also known as GOMAFU, or retinal noncoding RNA 2, however, here we will use the official HGNC (Human Gene Nomenclature Committee) symbol MIAT when referring to this lncRNA. MIAT was originally identified as an lncRNA differentially expressed during the development of mouse retina cells (Blackshaw et al. 2004; Ishii et al. 2006). MIAT is also widely expressed throughout the nervous system in development and adulthood (Sone et al. 2007). MIAT contains multiple spliced exons, with a final transcript size of approximately 10,000 nt, and is polyadenylated, however, despite its mRNA like characteristics, it is not exported to the cytoplasm and instead concentrates in a number of 'gomafu' speckles in the cell nucleus (Sone et al. 2007) ('gomafu' means 'speckled' in Japanese). While it is yet to be demonstrated that gomafu speckles depend on MIAT for their formation and maintenance, we have placed them in the category of lncRNAs forming subnuclear bodies, since the best marker is indeed the MIAT lncRNA. There is one known protein component of gomafu speckles: the pre-mRNA splicing factor SRSF1, although, as with many other bodies of this type, SRSF1 is also found outside the speckles as well (Tsuiji et al. 2011). There is evidence that SRSF1 interacts directly with MIAT through tandem UACUAAC repeats in the RNA (Tsuiji et al. 2011). Interestingly, it appears that MIAT recruits SRSF1 to gomafu speckles through these repeats, however, this recruitment is not required for gomafu speckle formation, as overexpression of MIAT lacking the SRSF1 binding sites was still localized there (Tsuiji et al. 2011). Recent exciting work has shown nevertheless that the interaction with SRSF1 is key to a novel role for MIAT in schizophrenia (Barry et al. 2013). Mattick and colleagues found that in schizophrenia MIAT is down-regulated, resulting in altered alternative splicing mediated by SRSF1. The model put forward suggests that in normal neurons, MIAT recruits key splicing factors to gomafu speckles in a sequestration model reminiscent of the postulated function of paraspeckles, nSBs, and omega speckles; however, when MIAT is down-regulated these speckles disperse, resulting in altered splicing activities of the released proteins (Barry et al. 2013). It will be important for future studies to test this model by detailed examination of the nuclear organization of these splicing components in the relevant schizophrenic cell types.

## 3 LncRNAs that are Enriched Within Nuclear Bodies/Complexes, but are not an Essential Structural Component

Thus far we have considered examples of lncRNAs that are essential components of the subnuclear bodies they nucleate. In addition, over recent years biologists have utilized RNA Fluorescent in situ hybridization (FISH) technology to probe the subcellular localization of many different lncRNAs, and in several cases have observed distinct subnuclear patterns. In some cases, these patterns of localization have been subsequently identified as co-localizing with a known subnuclear structure (e.g., MALAT1 in nuclear speckles, TUG1 in polycomb bodies), whilst in other cases these patterns of localization are unique. A common theme in these cases is that the subnuclear structure appears to form irrespective of the lncRNA. However, there are indications that the presence of the lncRNAs inside the subnuclear bodies is nevertheless important for their function.

### 3.1 MALAT1 and Nuclear Speckles

Nuclear speckles (also known as splicing speckles) are distinct subnuclear domains that are defined by the co-localization of snRNPs and the pre-mRNA splicing factor SC-35 (Spector and Lamond 2011; Thiry 1995). There are 20–50 irregularly shaped nuclear speckles in a typical mammalian nucleus, located within the interchromatin space, and a large number of additional pre-mRNA splicing-related proteins are also enriched there (Mintz 1999).

A major function of nuclear speckles is acting as a reservoir for splicing proteins, rather than the site of actual splicing per se. This is supported by evidence that there is little active splicing occurring within the nuclear speckles (reviewed in Spector and Lamond 2011). Rather, it is thought that splicing happens in a co-transcriptional manner at transcription sites (Zhang et al. 1994). The key pre-mRNA splicing SR proteins are targeted in and out of nuclear speckles to transcription sites via their selective phosphorylation (Misteli 1998; Misteli et al. 1997). Another function of nuclear speckles relates to their frequent close proximity to highly expressed genes, suggesting that they are enhancing processing of the resulting transcripts.

In 2007 a specific nuclear speckle lncRNA, MALAT1 (Metastasis Associated Lung Adenocarcinoma Transcript 1, also known as NEAT2) was observed to co-localise with nuclear speckle marker proteins (Hutchinson et al. 2007). MALAT1 is an unspliced approximately 8,000nt lncRNA that exhibits broad tissue expression, and is associated with tumorigenesis and metastasis (Gutschner et al. 2013). Interestingly, the *MALAT1* gene is located in a syntenically conserved fashion in close proximity to the *NEAT1* gene (11q13 in human and 19qA in mouse). Although MALAT1 has clear co-localization with nuclear speckle markers, it is not essential for their formation. Nuclei of mice lacking MALAT1 still contain nuclear speckles (Eissmann et al. 2012; Nakagawa et al. 2012; Zhang et al. 2012), and siRNA against MALAT1 does not disrupt nuclear speckles in cultured cells (Clemson et al. 2009), although it can alter the recruitment of various nuclear speckle proteins to these domains by regulating the phosphorylation of SR proteins (Lin et al. 2011; Tripathi et al. 2010).

MALAT1 is targeted to nuclear speckles through interactions with various proteins: knockdown of RNPS1, SRm160, or IBP160, which are well-known mRNA processing factors, resulted in MALAT1 becoming diffusely distributed within the nucleoplasm (Miyagawa et al. 2012). There are contrasting reports indicating the importance of different regions of MALAT1 to nuclear speckle targeting: Tripathi et al. found that overexpression of any 2 kb segment of MALAT1 resulted in its targeting to nuclear speckles (Tripathi et al. 2010), whereas Miyagawa et al. expressed smaller 1 kb fragments of MALAT1 and observed a more significant role for a region of MALAT1 towards its 3′-end that is predicted to form a binding site for key splicing proteins (Miyagawa et al. 2012).

Besides influencing splicing proteins, how else might MALAT1 exert its function on gene expression? An interesting study has shown that MALAT1 can recruit particular gene loci to the surface of nuclear speckles, in competition with other lncRNA-enriched subnuclear structures (Yang et al. 2011). In this seminal study, the authors showed that in response to growth signals, MALAT1 participates in a gene activation program through binding unmethylated polycomb protein, to sequester polycomb-associated genes to the surface of nuclear speckles. In contrast, in a repressive environment, genes with an associated methylated polycomb protein are recruited to polycomb group (PcG) bodies through interaction with the TUG1 lncRNA (more of which below). This interplay between subnuclear localization sites and gene expression status gives an intriguing insight into the myriad ways that lncRNAs may be affecting gene expression through as yet undiscovered mechanisms.

Beyond these studies, other researchers have defined the mechanism that MALAT1 uses to enhance cellular proliferation, through its involvement in regulating the expression and/or pre-mRNA processing of oncogenic transcription factors, especially those that control mitotic progression (Tripathi et al. 2013). Given that an important role for MALAT1 in cell growth, proliferation, synaptogenesis and cancer is now well defined, it is fascinating that MALAT1 is not required for mouse development, as seen with the viability of the MALAT1 knockout mice with no gross phenotype (Bernard et al. 2010; Eissmann et al. 2012; Nakagawa et al. 2012; Tripathi et al. 2010; Zhang et al. 2012). It is interesting to speculate that there exist compensatory mechanisms in vivo to account for these effects. Indeed, recent work has indicated that either MALAT1, or SRSF1, can 'seed' nuclear speckles, suggesting they compensate for each other, and this may explain the intact nuclear speckles and unimpaired nuclear speckle function in MALAT1 knockout mice (Nakagawa et al. 2012).

## *3.2 TUG1 and Polycomb Bodies*

Polycomb (PCG) bodies are defined as subnuclear foci enriched in the chromatin-associated polycomb group proteins (Pirrotta and Li 2012). PcG bodies vary in size, shape, and number from cell type to cell type, likely reflecting the gene activity of polycomb-regulated genes. It is generally thought that PcG bodies form near to localized clusters of PcG-regulated genes, or as a result of interaction with insulator proteins at PcG-regulated genes (reviewed in Pirrotta and Li 2012).

It has been speculated that PcG bodies may have some dependency on lncRNA for their formation or function, largely due to the growing number of reports indicating that individual lncRNAs can associate with PcG proteins. In this context it is of interest that a recent report has identified TUG1 as a PcG localized lncRNA (Yang et al. 2011). TUG1 (Taurine upregulated gene 1) is a conserved mammalian lncRNA that was first found up-regulated in mouse post natal retinal cells following taurine treatment, with evidence that it promotes proliferation through chromatin regulation (Young et al. 2005). TUG1 was subsequently observed in clear defined speckles in the nucleus and cytoplasm of several human and mouse tissues (Khalil et al. 2009). In 2011, Yang and colleagues (2011) showed that TUG1 associates with a variety of proteins associated with transcriptional repression including the PcG proteins, and that TUG1 localized within PcG bodies. As indicated above, TUG1 is involved in directing the recruitment of gene loci to PcG bodies, via interactions with methylated PcG and its associated gene targets (Yang et al. 2011).

Another lncRNA with a potential involvement in PcG bodies is HOTAIR. The HOTAIR lncRNA acts as a scaffold to recruit chromatin-modifying complexes to their site of action (Wang and Chang 2011). HOTAIR is expressed from the HOXC locus and its mechanism of action includes recruiting the PcG protein PRC2 to multiple loci, playing crucial roles in development and cancer metastasis (Gupta et al. 2010; Kogo et al. 2011). In cancer cells, high HOTAIR expression is associated with increased metastasis, as it redirects chromatin-modifying complexes to suppress metastasis suppressor genes and pro-apoptotic factors (Tsai et al. 2010). RNA FISH against HOTAIR in human foreskin fibroblasts revealed a pattern of distinct foci found throughout the nucleus and cytoplasm, however, it is yet to be determined if these nuclear foci overlap PcG bodies, or represent distinct structures (Khalil et al. 2009). It is likely that these foci could be co-located with the gene loci regulated by HOTAIR, and the organization of HOTAIR into these bodies may enhance the efficiency of the regulation. It will be important in the future to determine the composition and role of these HOTAIR foci in the function of this important lncRNA.

## 4 Concluding Remarks

While there is only at present a handful of lncRNAs known to associate or form subnuclear bodies, these molecules have nevertheless provided a wealth of information about the mechanisms that lncRNAs can use when enriched in subnuclear bodies to alter gene expression (Fig. 3). It is also highly likely that the small number of lncRNAs described here may in fact represent the tip of the iceberg, in terms of the number of lncRNAs that will eventually emerge as associating or forming subnuclear structures. This is likely considering that most lncRNAs are found enriched in the nucleus and are tissue, developmental stage, or cell-type specific, and their localization, if indeed examined at all, are yet to be studied in the relevant cell type. Given this likelihood, it is with confidence that the efforts of researchers in the field of nuclear organization be redoubled to identify function for subnuclear structures, as this will continue to be important in increasing our understanding of lncRNAs that form them and localize to these bodies.

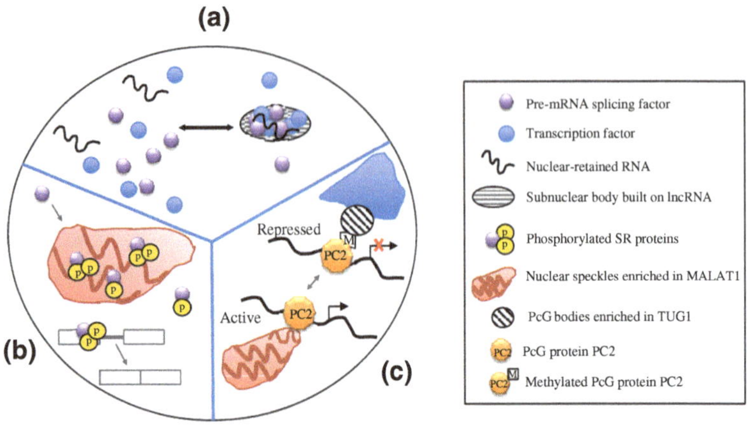


**Fig. 3** Functions of lncRNAs in subnuclear bodies. **a** Several subnuclear bodies (paraspeckles, omega speckles, nSBs, and gomafu speckles), formed by lncRNAs, act to sequester nuclear proteins, thereby reducing their availability within the nucleoplasm and affecting transcriptional and alternative splicing regulation by these factors. These bodies may also be involved in retaining specific RNAs within the cell nucleus. **b** MALAT1 presence in nuclear speckles has been demonstrated to influence the phosphorylation of pre-mRNA splicing factors, thereby affecting alternative splicing in the cell. **c** TUG1 in polycomb bodies and MALAT1 in nuclear speckles can both bind Polycomb group protein PRC2 (although TUG1 binds the methylated PRC2), resulting in the recruitment of gene loci to active (nuclear speckles) or repressed (polycomb bodies) environments

## References

Barry, G., Briggs, J., Vanichkina, D., Poth, E., Beveridge, N., Ratnu, V., et al. (2013). The long noncoding RNA Gomafu is acutely regulated in response to neuronal activation and involved in schizophrenia-associated alternative splicing. *Molecular Psychiatry,*. doi:10.1038/mp.2013.45.

Bendena, W. G., Garbe, J. C., Traverse, K. L., Lakhotia, S. C., & Pardue, M. L. (1989). Multiple inducers of the Drosophila heat shock locus 93D (hsr omega): Inducer-specific patterns of the three transcripts. *The Journal of Cell Biology, 108*, 2017–2028.

Bendena, W. G., Ayme-Southgate, A., Garbe, J. C., & Pardue, M. L. (1991). Expression of heat-shock locus hsr-omega in nonstressed cells during development in Drosophila melanogaster. *Developmental Biology, 144*, 65–77.

Bernard, D., Prasanth, K. V., Tripathi, V., Colasse, S., Nakamura, T., Xuan, Z. Y., et al. (2010). A long nuclear-retained noncoding RNA regulates synaptogenesis by modulating gene expression. *The EMBO Journal, 29*, 3082–3093.

Biamonti, G., & Caceres, J. F. (2009). Cellular stress and RNA splicing. *Trends in Biochemical Sciences, 34*, 146–153.

Biamonti, G., & Vourc'h, C. 2010. Nuclear stress bodies. *Cold Spring Harbor Perspectives in Biology*, *2*, a000695.

Blackshaw, S., Harpavat, S., Trimarchi, J., Cai, L., Huang, H. Y., Kuo, W. P., et al. (2004). Genomic analysis of mouse retinal development. *PLoS Biology, 2*, 1411–1431.

Bond, C. S., & Fox, A. H. (2009). Paraspeckles: Nuclear bodies built on long noncoding RNA. *The Journal of Cell Biology, 186*, 637–644.

Buchenau, P., Saumweber, H., & Arndt-Jovin, D. J. (1997). The dynamic nuclear redistribution of an hnRNP K-homologous protein during Drosophila embryo development and heat shock. Flexibility of transcription sites in vivo. *The Journal of Cell Biology, 137*, 291–303.

Capshew, C. R., Dusenbury, K. L., & Hundley, H. A. (2012). Inverted Alu dsRNA structures do not affect localization but can alter translation efficiency of human mRNAs independent of RNA editing. *Nucleic Acids Research, 40*, 8637–8645.

Chen, L.-L., & Carmichael, G. G. (2009). Altered nuclear retention of mRNAs containing inverted repeats in human embryonic stem cells: Functional role of a nuclear noncoding RNA. *Molecular Cell, 35*, 467–478.

Chiodi, I., Biggiogera, M., Denegri, M., Corioni, M., Weighardt, F., Cobianchi, F., et al. (2000). Structure and dynamics of hnRNP-labelled nuclear bodies induced by stress treatments. *Journal of Cell Science, 113*, 4043–4053.

Clemson, C. M., Hutchinson, J. N., Sara, S. A., Ensminger, A. W., Fox, A. H., Chess, A., et al. (2009). An architectural role for a nuclear noncoding RNA: NEAT1 RNA is essential for the structure of paraspeckles. *Molecular Cell, 33*, 717–726.

Dangli, A., & Bautz, E. F. (1983). Differential distribution of nonhistone proteins from polytene chromosomes of Drosophila melanogaster after heat shock. *Chromosoma, 88*, 201–207.

Dangli, A., Grond, C., Kloetzel, P., & Bautz, E. F. (1983). Heat-shock puff 93 D from Drosophila melanogaster: Accumulation of a RNP-specific antigen associated with giant particles of possible storage function. *The EMBO Journal, 2*, 1747–1751.

Denegri, M., Chiodi, I., Corioni, M., Cobianchi, F., Riva, S., & Biamonti, G. (2001). Stress-induced nuclear bodies are sites of accumulation of pre-mRNA processing factors. *Molecular Biology of the Cell, 12*, 3502–3514.

Denegri, M., Moralli, D., Rocchi, M., Biggiogera, M., Raimondi, E., Cobianchi, F., et al. (2002). Human chromosomes 9, 12, and 15 contain the nucleation sites of stress-induced nuclear bodies. *Molecular Biology of the Cell, 13*, 2069–2079.

Dundr, M., & Misteli, T. 2010. Biogenesis of nuclear bodies. *Cold Spring Harbor Perspectives in Biology, 2*, a000711.

Dye, B. T., & Patton, J. G. (2001). An RNA recognition motif (RRM) is required for the localization of PTB-associated splicing factor (PSF) to subnuclear speckles. *Experimental Cell Research, 263*, 131–144.

Eissmann, M., Gutschner, T., Hammerle, M., Gunther, S., Caudron-Herger, M., Gross, M., et al. (2012). Loss of the abundant nuclear noncoding RNA MALAT1 is compatible with life and development. *RNA Biology, 9*, 1076–1087.

Fitzpatrick, T., & Huang, S. (2012). 3′-UTR-located inverted Alu repeats facilitate mRNA translational repression and stress granule accumulation. *Nucleus, 3*, 359–369.

Fox, A. H., Lam, Y. W., Leung, A. K. L., Lyon, C. E., Andersen, J., Mann, M., et al. (2002). Paraspeckles: A novel nuclear domain. *Current Biology, 12*, 13–25.

Fox, A. H., Bond, C. S., & Lamond, A. I. (2005). P54nrb forms a heterodimer with PSP1 that localizes to paraspeckles in an RNA-dependent manner. *Molecular Biology of the Cell, 16*, 5304–5315.

Fritah, S., Col, E., Boyault, C., Govin, J., Sadoul, K., Chiocca, S., et al. (2009). Heat-shock factor 1 controls genome-wide acetylation in heat-shocked cells. *Molecular Biology of the Cell, 20*, 4976–4984.

Garbe, J. C., Bendena, W. G., Alfano, M., & Pardue, M. L. (1986). A Drosophila heat shock locus with a rapidly diverging sequence but a conserved structure. *Journal of Biological Chemistry, 261*, 16889–16894.

Gibb, E. A., Vucic, E. A., Enfield, K. S. S., Stewart, G. L., Lonergan, K. M., Kennett, J. Y., et al. (2011). Human cancer long noncoding RNA transcriptomes. *PLoS ONE, 6*, e25915.

Gupta, R. A., Shah, N., Wang, K. C., Kim, J., Horlings, H. M., Wong, D. J., et al. (2010). Long noncoding RNA HOTAIR reprograms chromatin state to promote cancer metastasis. *Nature, 464*, 1071–1076.

Gutschner, T., Hammerle, M., Eissmann, M., Hsu, J., Kim, Y., Hung, G., et al. (2013). The noncoding RNA MALAT1 is a critical regulator of the metastasis phenotype of lung cancer cells. *Cancer Research, 73*, 1180–1189.

Hogan, N. C., Slot, F., Traverse, K. L., Garbe, J. C., Bendena, W. G., & Pardue, M. L. (1995). Stability of tandem repeats in the Drosophila melanogaster Hsr-omega nuclear RNA. *Genetics, 139*, 1611–1621.

Hovemann, B. T., Dessen, E., Mechler, H., & Mack, E. (1991). Drosophila snRNP associated protein P11 which specifically binds to heat shock puff 93D reveals strong homology with hnRNP core protein A1. *Nucleic Acids Research, 19*, 4909–4914.

Hutchinson, J., Ensminger, A., Clemson, C., Lynch, C., Lawrence, J., & Chess, A. (2007). A screen for nuclear transcripts identifies two linked noncoding RNAs associated with SC35 splicing domains. *BMC Genomics, 8*, 39.

Ishii, N., Ozaki, K., Sato, H., Mizuno, H., Saito, S., Takahashi, A., et al. (2006). Identification of a novel noncoding RNA, MIAT, that confers risk of myocardial infarction. *Journal of Human Genetics, 51*, 1087–1099.

Jarmuż, M., Glotzbach, C. D., Bailey, K. A., Bandyopadhyay, R., & Shaffer, L. G. (2007). The evolution of satellite III DNA subfamilies among primates. *The American Journal of Human Genetics, 80*, 495–501.

Jolly, C., Morimoto, R., Robert-Nicoud, M., & Vourc'h, C. (1997). HSF1 transcription factor concentrates in nuclear foci during heat shock: Relationship with transcription sites. *Journal of Cell Science, 110*, 2935–2941.

Jolly, C., Konecny, L., Grady, D. L., Kutskova, Y. A., Cotto, J. J., Morimoto, R. I., et al. (2002). In vivo binding of active heat shock transcription factor 1 to human chromosome 9 heterochromatin during stress. *The Journal of Cell Biology, 156*, 775–781.

Jolly, C., Metz, A., Govin, J., Vigneron, M., Turner, B. M., Khochbin, S., et al. (2004). Stress-induced transcription of satellite III repeats. *The Journal of Cell Biology, 164*, 25–33.

Jolly, C., & Lakhotia, S. C. (2006). Human sat III and Drosophila hsr$\omega$ transcripts: A common paradigm for regulation of nuclear RNA processing in stressed cells. *Nucleic Acids Research, 34*, 5508–5514.

Khalil, A. M., Guttman, M., Huarte, M., Garber, M., Raj, A., Rivea Morales, D., et al. (2009). Many human large intergenic noncoding RNAs associate with chromatin-modifying complexes and affect gene expression. *Proceedings of the National Academy of Sciences of the United States of America, 106*, 11667–11672.

Kogo, R., Shimamura, T., Mimori, K., Kawahara, K., Imoto, S., Sudo, T., et al. (2011). Long noncoding RNA HOTAIR regulates polycomb-dependent chromatin modification and is associated with poor prognosis in colorectal cancers. *Cancer Research, 71*, 6320–6326.

Lakhotia, S. C., & Sharma, A. (1995). RNA metabolismin situ at the 93D heat shock locus in polytene nuclei of Drosophila melanogaster after various treatments. *Chromosome Research, 3*, 151–161.

Lakhotia, S. C., Ray, P., & Rajendra, T. K. (1999). The noncoding transcripts of hsr-omega gene in Drosophila: Do they regulate trafficking and availability of nuclear RNA-processing factors? *Current Science, 77*, 553–563.

Li, L., Feng, T. T., Lian, Y. Y., Zhang, G. F., Garen, A., & Song, X. (2009). Role of human noncoding RNAs in the control of tumorigenesis. *Proceedings of the National Academy of Sciences of the United States of America, 106*, 12956–12961.

Lin, R., Roychowdhury-Saha, M., Black, C., Watt, A. T., Marcusson, E. G., Freier, S. M., et al. (2011). Control of RNA processing by a large noncoding RNA over-expressed in carcinomas. *FEBS Letters, 585*, 671–676.

Lindquist, S. (1986). The heat-shock response. *Annual Review of Biochemistry, 55*, 1151–1191.

Mähl, P., Lutz, Y., Puvion, E., & Fuchs, J. P. (1989). Rapid effect of heat shock on two heterogeneous nuclear ribonucleoprotein-associated antigens in HeLa cells. *The Journal of Cell Biology, 109*, 1921–1935.

Mallik, M., & Lakhotia, S. C. (2009). RNAi for the large noncoding hsr$\omega$ transcripts suppresses polyglutamine pathogenesis in *Drosophila* models. *RNA Biology, 6*, 464–478.

Mao, Y. S., Sunwoo, H., Zhang, B., & Spector, D. L. (2011). Direct visualization of the co-transcriptional assembly of a nuclear body by noncoding RNAs. *Nature Cell Biology, 13*, 95–101.

Metz, A., Soret, J., Vourc'h, C., Tazi, J., & Jolly, C. (2004). A key role for stress-induced satellite III transcripts in the relocalization of splicing factors into nuclear stress granules. *Journal of Cell Science, 117*, 4551–4558.

Mintz, P. J. (1999). Purification and biochemical characterization of interchromatin granule clusters. *The EMBO Journal, 18*, 4308–4320.

Misteli, T., Cáceres, J. F., & Spector, D. L. (1997). The dynamics of a pre-mRNA splicing factor in living cells. *Nature, 387*, 523–527.

Misteli, T. (1998). Serine phosphorylation of SR proteins is required for their recruitment to sites of transcription in vivo. *The Journal of cell biology, 143*, 297–307.

Miyagawa, R., Tano, K., Mizuno, R., Nakamura, Y., Ijiri, K., Rakwal, R., et al. (2012). Identification of cis- and trans-acting factors involved in the localization of MALAT-1 noncoding RNA to nuclear speckles. *RNA, 18*, 738–751.

Mutsuddi, M., & Lakhotia, S. C. (1995). Spatial expression of the hsr-omega (93D) gene in different tissues of Drosophila melanogaster and identification of promoter elements controlling its developmental expression. *Developmental Genetics, 17*, 303–311.

Naganuma, T., Nakagawa, S., Tanigawa, A., Sasaki, Y. F., Goshima, N., & Hirose, T. (2012). Alternative 3′-end processing of long noncoding RNA initiates construction of nuclear paraspeckles. *The EMBO Journal, 31*, 4020–4034.

Nakagawa, S., Naganuma, T., Shioi, G., & Hirose, T. (2011). Paraspeckles are subpopulation-specific nuclear bodies that are not essential in mice. *The Journal of Cell Biology, 193*, 31–39.

Nakagawa, S., & Hirose, T. (2012). Paraspeckle nuclear bodies—useful uselessness? *Cellular and Molecular Life Sciences, 69*, 3027–3036.

Nakagawa, S., Ip, J. Y., Shioi, G., Tripathi, V., Zong, X., & Bosserhoff, A. K. (2012). Malat1 is not an essential component of nuclear speckles in mice. *RNA, 18*, 1487–1499.

Passon, D. M., Lee, M., Rackham, O., Stanley, W. A., Sadowska, A., Filipovska, A., et al. (2012). Structure of the heterodimer of human NONO and paraspeckle protein component 1 and analysis of its role in subnuclear body formation. *Proceedings of the National Academy of Sciences of the United States of America, 109*, 4846–4850.

Pirrotta, V., & Li, H.-B. (2012). A view of nuclear Polycomb bodies. *Current Opinion in Genetics & Development, 22*, 101–109.

Prasanth, K. V., Rajendra, T. K., Lal, A. K., & Lakhotia, S. C. (2000). Omega speckles—a novel class of nuclear speckles containing hnRNPs associated with noncoding hsr-omega RNA in Drosophila. *Journal of Cell Science, 113*, 3485–3497.

Prasanth, K. V., Prasanth, S. G., Xuan, Z., Hearn, S., Freier, S. M., Bennett, C. F., et al. (2005). Regulating gene expression through RNA nuclear retention. *Cell, 123*, 249–263.

Rajendra, T. K., Prasanth, K. V., & Lakhotia, S. C. (2001). Male sterility associated with overexpression of the noncoding hsr$\omega$ gene in cyst cells of testis of Drosophila melanogaster. *Journal of Genetics, 80*, 97–110.

Rizzi, N., Denegri, M., Chiodi, I., Corioni, M., Valgardsdottir, R., Cobianchi, F., et al. (2004). Transcriptional activation of a constitutive heterochromatic domain of the human genome in response to heat shock. *Molecular Biology of the Cell, 15*, 543–551.

Ryseck, R.-P., Walldorf, U., Hoffmann, T., & Hovemann, B. (1987). Heat shock loci 93D of Drosophila melanogaster and 48B of Drosophila hydei exhibit a common structural and transcriptional pattern. *Nucleic Acids Research, 15*, 3317–3333.

Samuels, M. E., Bopp, D., Colvin, R. A., Roscigno, R. F., Garcia-Blanco, M. A., & Schedl, P. (1994). RNA binding by Sxl proteins in vitro and in vivo. *Molecular and Cellular Biology, 14*, 4975–4990.

Sarge, K. D., Murphy, S. P., & Morimoto, R. I. (1993). Activation of heat shock gene transcription by heat shock factor 1 involves oligomerization, acquisition of DNA-binding activity, and nuclear localization and can occur in the absence of stress. *Molecular and Cellular Biology, 13*, 1392–1407.

Sasaki, Y. T. F., Ideue, T., Sano, M., Mituyama, T., & Hirose, T. (2009). MENε/β noncoding RNAs are essential for structural integrity of nuclear paraspeckles. *Proceedings of the National Academy of Sciences of the United States of America, 106*, 2525–2530.

Sengupta, S., & Lakhotia, S. C. (2006). Altered Expression of the noncoding *hsrω* Gene Enhances poly-Q Induced Neurotoxicity in *Drosophila. RNA Biology, 3*, 28–35.

Shevtsov, S. P., & Dundr, M. (2011). Nucleation of nuclear bodies by RNA. *Nature Cell Biology, 13*, 167–173.

Sone, M., Hayashi, T., Tarui, H., Agata, K., Takeichi, M., & Nakagawa, S. (2007). The mRNA-like noncoding RNA Gomafu constitutes a novel nuclear domain in a subset of neurons. *Journal of Cell Science, 120*, 2498–2506.

Souquere, S., Beauclair, G., Harper, F., Fox, A., & Pierron, G. (2010). Highly ordered spatial organization of the structural long noncoding NEAT1 RNAs within paraspeckle nuclear bodies. *Molecular Biology of the Cell, 21*, 4020–4027.

Spector, D. L., & Lamond, A. I. 2011. Nuclear speckles. *Cold Spring Harbor Perspectives in Biology, 3*, a000646.

Sreenivasa Murthy, U. M., & Rangarajan, P. N. (2010). Identification of protein interaction regions of VINC/NEAT1/Men epsilon RNA. *FEBS Letters, 584*, 1531–1535.

Sunwoo, H., Dinger, M. E., Wilusz, J. E., Amaral, P. P., Mattick, J. S., & Spector, D. L. (2009). MEN ε/β nuclear-retained noncoding RNAs are up-regulated upon muscle differentiation and are essential components of paraspeckles. *Genome Research, 19*, 347–359.

Takayama, K., Horie-Inoue, K., Katayama, S., Suzuki, T., Tsutsumi, S., Ikeda, K., et al. (2013). Androgen-responsive long noncoding RNA CTBP1-AS promotes prostate cancer. *The EMBO Journal, 32*, 1665–1680.

Tapadia, M. G., & Lakhotia, S. C. (1997). Specific induction of the hsrω locus of Drosophila melanogaster by amides. *Chromosome Research, 5*, 359–362.

Thiry, M. (1995). Nucleic acid compartmentalization within the cell nucleus by in situ transferase-immunogold techniques. *Microscopy Research and Technique, 31*, 4–21.

Tripathi, V., Ellis, J. D., Shen, Z., Song, D. Y., Pan, Q., Watt, A. T., et al. (2010). The nuclear-retained noncoding RNA MALAT1 regulates alternative splicing by modulating SR splicing factor phosphorylation. *Molecular Cell, 39*, 925–938.

Tripathi, V., Shen, Z., Chakraborty, A., Giri, S., Freier, S. M., Wu, X. L., et al. 2013. Long noncoding RNA MALAT1 controls cell cycle progression by regulating the expression of oncogenic transcription factor B-MYB. *PLoS Genetics, 9*, e1003368.

Tsai, M. C., Manor, O., Wan, Y., Mosammaparast, N., Wang, J. K., Lan, F., et al. (2010). Long noncoding RNA as modular scaffold of histone modification complexes. *Science, 329*, 689–693.

Tsuiji, H., Yoshimoto, R., Hasegawa, Y., Furuno, M., Yoshida, M., & Nakagawa, S. (2011). Competition between a noncoding exon and introns: Gomafu contains tandem UACUAAC repeats and associates with splicing factor-1. *Genes to Cells, 16*, 479–490.

Valgardsdottir, R., Chiodi, I., Giordano, M., Cobianchi, F., Riva, S., & Biamonti, G. (2005). Structural and functional characterization of noncoding repetitive RNAs transcribed in stressed human cells. *Molecular Biology of the Cell, 16*, 2597–2604.

Visa, N., Puvion-Dutilleul, F., Bachellerie, J. P., & Puvion, E. 1993. Intranuclear distribution of U1 and U2 snRNAs visualized by high resolution in situ hybridization: Revelation of a novel compartment containing U1 but not U2 snRNA in HeLa cells. The *European Journal of Cell Biology, 60*, 308–321.

Wang, K. C., & Chang, H. Y. (2011). Molecular mechanisms of long noncoding RNAs. *Molecular Cell, 43*, 904–914.

Weighardt, F., Cobianchi, F., Cartegni, L., Chiodi, I., Villa, A., Riva, S., et al. (1999). A novel hnRNP protein (HAP/SAF-B) enters a subset of hnRNP complexes and relocates in nuclear granules in response to heat shock. *Journal of Cell Science, 112*, 1465–1476.

Wu, C.-F., Tan, G.-H., Ma, C.-C., & Li, L. (2013). The noncoding RNA llme23 drives the malignant property of human melanoma cells. *Journal of Genetics and Genomics, 40*, 179–188.

Yang, L., Lin, C., Liu, W., Zhang, J., Ohgi, K. A., Grinstein, J. D., et al. (2011). ncRNA- and Pc2 methylation-dependent gene relocation between nuclear structures mediates gene activation programs. *Cell, 147*, 773–788.

Young, T. L., Matsuda, T., & Cepko, C. L. (2005). The noncoding RNA taurine upregulated gene 1 is required for differentiation of the murine retina. *Current Biology, 15*, 501–512.

Zhang, G., Taneja, K. L., & Singer, R. H. (1994). Localization of pre-mRNA splicing in mammalian nuclei. *Nature, 372*, 809–812.

Zhang, Z., & Carmichael, G. G. (2001). The fate of dsRNA in the nucleus: A p54nrb-containing complex mediates the nuclear retention of promiscuously A-to-I edited RNAs. *Cell, 106*, 465–476.

Zhang, B., Arun, G., Mao, Y. S., Lazar, Z., Hung, G. N., Bhattacharjee, G., et al. (2012). The lncRNA Malat1 is dispensable for mouse development but its transcription plays a cis-regulatory role in the adult. *Cell Reports, 2*, 111–123.

Zimowska, G., & Paddy, M. R. (2002). Structures and dynamics of Drosophila Tpr inconsistent with a static, filamentous structure. *Experimental Cell Research, 276*, 223–232.

Zu, K., Sikes, M. L., & Beyer, A. L. (1998). Separable roles in vivo for the two RNA binding domains of Drosophila A1-hnRNP homolog. *RNA, 4*, 1585–1598.

# Index

*Note:* Page numbers followed by "f" and "t" indicate figures and tables respectively



A. M. Khalil and J. Coller (eds.), *Molecular Biology of Long Non-coding RNAs*,
DOI: 10.1007/978-1-4614-8621-3, © Springer Science+Business Media New York 2013

FSC
www.fsc.org
MIX
Papier | Fördert gute Waldnutzung
FSC® C083411

Kolibri 360 GmbH
Ferdinand-Jühlke-Straße 7
99095 Erfurt, Deutschland
produktsicherheit@kolibri360.de